Planning Support Methods

Urban and Regional Analysis and Projection

Richard E. Klosterman
University of Akron

Kerry Brooks
Eastern Washington University

Joshua Drucker
University of Illinois at Chicago

Edward Feser
Oregon State University

Henry Renski
University of Massachusetts Amherst

ROWMAN & LITTLEFIELD

Lanham • Boulder • New York • London

Executive Editor: Susan McEachern
Editorial Assistant: Katelyn Turner
Senior Marketing Manager: Kim Lyons
Interior Designer: Ilze Lemesis

Credits and acknowledgments for material borrowed from other sources, and reproduced with permission, appear on the appropriate page within the text.

Rowman & Littlefield
An imprint of The Rowman & Littlefield Publishing Group, Inc.
4501 Forbes Boulevard, Suite 200, Lanham, MD 20706
www.rowman.com

Unit A, Whitacre Mews, 26-34 Stannary Street, London SE11 4AB, United Kingdom

British Library Cataloguing in Publication Information Available

Library of Congress Cataloging-in-Publication Data
Names: Klosterman, Richard E., author. | Brooks, Kerry, author. | Drucker, Joshua, author.
Title: Planning support methods : urban and regional analysis and projection / Richard E. Klosterman, University of Akron, Kerry Brooks, Eastern Washington University, Joshua Drucker, University of Illinois at Chicago, Edward Feser, Oregon State University, Henry Renski, University of Massachusetts Amherst.
Description: Lanham : Rowman & Littlefield Publishing Group, Inc., [2019] | Includes bibliographical references and index. viewed.
Identifiers: LCCN 2018004652 (print) | LCCN 2018020506 (ebook) | ISBN 9781442220300 (ebook) | ISBN 9781442220287 (hardcover : alk. paper) | ISBN 9781442220294 (pbk. : alk. paper)
Subjects: LCSH: City planning—Mathematical models. | Land use—Planning—Mathematical models. | Information storage and retrieval systems—Land use. | Regional planning—Data processing.
Classification: LCC HT166 (ebook) | LCC HT166 .K5846 2018 (print) | DDC 307.1/2160112—dc23
LC record available at https://lccn.loc.gov/2018004652

Brief Contents

Contents

9 Using Planning Support Methods   237

Figures, Maps, and Tables

Figures

Maps

Tables

Preface

> Planning is looking backward at what we can learn from experience, looking around at what we can learn from observation, . . . looking forward to see where we are going, . . . choosing where we want to go and figuring out how we can get there.[1]

Quantitative methods have played a central role in planning education and practice since the middle of the twentieth century when planners' vision of planning as an intuitive process of design was replaced by a new concept of planning as an applied science. Several texts have been published over the intervening years that introduced planning students and practitioners to an array of quantitative methods. The first widely used methods text, Walter Isard's 760-page *Methods of Regional Analysis* (Isard et al. 1960), was published in 1960. Several planning methods texts were published in the 1970s but *Urban Planning Analysis* (Krueckeberg and Silvers 1974) dominated planning education in the United States. Only two quantitative methods texts for planning students and practitioners have been published in North America in the last thirty years: *Community Analysis and Planning Techniques* (Klosterman 1990) and *Research Methods in Urban and Regional Planning* (Wang and Hofe 2007).

This book describes and applies the most important and widely used urban and regional analysis and projection methods. It updates and extends *Community Analysis and Planning Techniques* by including the trend, share, cohort-component, population, and employment analysis and projection methods covered in the earlier text. Recognizing the critical role that geographic information systems (GIS) play in planning education and practice, it reviews spatial analysis methods and describes land suitability methods. The book's appendices describe data sources for applying the methods in the book, US census geography, and the US Bureau of the Census's American Community Survey. More important, unlike recently published books such as Silva et al. (2015) and Lipovska and Stepenka (2016), the book applies the methods to an American city and county, demonstrating in detail how quantitative methods can be used to support community-based planning.

Planning Support Methods contains nine chapters and three appendices that describe and apply the quantitative planning methods at the core of planning education and practice. The book describes the methods' underlying assumptions, advantages, and limitations; it also provides guidelines for using the methods and criteria for selecting the most acceptable projection (or projections) and extensive lists of references for further reading. The text can be used in graduate and undergraduate methods courses and by planning practitioners and others interested in understanding and using planning methods.

Chapter 1 lays out the intellectual foundations for using the methods described in the remainder of the book. It argues that planning is a forward-thinking profession that should promote citizen-centered planning with the public, which involves as many people as possible, as completely as possible, in the decisions that affect their lives. It also suggests that planners should use simple, understandable, and easy-to-use methods like the ones described in this book to support these efforts.

Chapter 2 describes the City of Decatur, Georgia, and DeKalb County, in which it is located. The chapter provides background information on Decatur and describes the city's population and housing, employment and occupations, land uses, and block groups. The chapter uses population pyramids, Lorenz curves, and Gini coefficients to analyze and describe the city's population and economy.

Chapter 3 introduces concepts and terms that will be used throughout the book. It then describes four trend curves—linear, geometric, parabolic, and Gompertz—that identify past trends in an area's growth or decline and continue those trends to project the future. The chapter uses the trend curves to project Decatur and DeKalb County's 2040 populations and describes procedures for using trend curves and identifying the most appropriate trend projections.

Chapter 4 describes share projection methods that project an area's future population or employment, given projections for a larger area in which it is located. The constant-share, shift-share, share-of-change, and share-trend methods are used to project Decatur's 2040 population. The constant-share, shift-share, share-of-change, and adjusted-share-of-change methods are used to project the 2040 population of Decatur's thirteen block groups. The chapter concludes by evaluating the share projections for Decatur and its block groups.

Chapter 5 describes cohort-component projection methods that divide a population into age, sex, and racial or ethnic groups or cohorts and deal separately with the three components of population change—mortality, fertility, and migration. It describes past mortality, fertility, and migration trends for the United States. It projects Decatur's 2040 population for ten-year age cohorts and DeKalb County's 2040 population for five-year age cohorts, under different survival, fertility, and migration assumptions, and evaluates the projections.

Chapter 6 describes basic concepts and terms for understanding an area's economy and provides economic analysis guidelines. It uses location quotients and shift-share analysis to understand Decatur's aggregate economy, industrial composition, and industrial specialization. The chapter uses the share projection methods described in chapter 4 to project the 2040 employment in Decatur's major employment sectors. The chapter concludes with guidelines for using economic analysis and projection methods.

Chapter 7 provides an overview of the spatial analysis methods that underlie geographic information systems (GIS) applications. It describes: (1) fundamental spatial analysis concepts and terminology, (2) methods for locating spatial features, (3) the raster and vector data models, (4) methods for analyzing spatial and attribute data and representing spatial relationships, and (5) widely used spatial analysis methods.

Chapter 8 describes land suitability analysis methods that determine the suitability of one or more locations for one or more land uses. It introduces four widely used land suitability analysis methods—map overlay, binary selection, ordinal combination, and weighted combination. It uses the binary selection method to identify land parcels and buildings in Decatur's flood zones. It uses the ordinal combination and weighted combination methods to identify vacant parcels in DeKalb County that are suitable for residential and commercial development, given a range of suitability assumptions. It concludes by describing procedures for computing standardized suitability scores and opportunities for public involvement.

The final chapter provides guidelines for using the planning support methods described in the preceding chapters. Seven guidelines are described and applied to the analyses and projections for Decatur and DeKalb County: (1) use graphs and charts, (2) use maps, (3) document assumptions, (4) compare to projections by other organizations, (5) combine different information, (6) support other applications, and (7) use scenarios.

Appendix A describes the US Bureau of the Census's American Community Survey that replaced the US decennial population and housing census in 2010. Appendix B describes the geographic units and codes the US Bureau of the Census uses to collect and report its data. Appendix C describes free, online sources for data that can be used to apply the methods described in the book.

The book is a collaborative effort of the coauthors and other people who provided extremely helpful assistance and advice. Henry Renski coauthored chapter 5, Joshua Drucker and Edward Feser coauthored chapter 6, and Kerry Brooks coauthored chapter 8. Richard Klosterman coauthored these chapters and wrote the remainder of the book. Kerry Brooks and Richard Klosterman prepared the maps. Lewis D. Hopkins contributed greatly to the discussion of scenario planning in the final chapter and the remainder of the book. Terry Moore and Hyeon-Shic Shin made extremely helpful suggestions for improving large portions of the book. Any errors are the sole responsibility of the first author.

Terms in the glossary are printed in bold fonts when they first appear in the text. Additional resources for using this text and the methods it describes are available at PlanningSupport.org.

Note

[1] Isserman (2007, 195).

Foundations 1

Planning and predicting the future are essential human activities.[1] Individuals and families consult weather forecasts to plan their weekend activities and rely on financial advisors to plan their financial futures. Public and private organizations hire consultants to help identify actions and investments that will benefit them in the future. Local, regional, state, and national governments hire planners and other public policy experts to help them deal with an uncertain and politically contentious future. Rapid urbanization has brought half the world's population into cities and an array of public policy issues—climate change, housing affordability, and food security, to name a few—demand public attention. The increased interdependence, complexity, and uncertainty of a rapidly urbanizing world requires planning that helps public officials satisfy their diverse and vocal constituencies and assists private-sector actors to make and protect their investments.

Planners know that public sector planning is required to promote the collective interests of the community, consider the external effects of individual and group action, improve the information base for public and private action, and protect society's most needy members (Klosterman 1985). They also realize that current actions affect future outcomes and that taking no action is an action in itself. Planning assumes that individuals, organizations, and communities can control their fates and act to achieve desired futures and avoid undesirable ones. Without planning, communities can only stumble blindly into a future they make no effort to shape. With planning, the future is the result of underlying trends and the public's efforts to modify them.

However, planners also recognize the obstacles to public sector planning. Annual budgets and the short terms of elected officials focus public attention on issues immediately at hand. Property owners are understandably wary of public actions that may affect the value of their most precious assets and businesses, large and small, and place their particular interests above those of society at large. In the realm of political discourse, the rhetoric of protecting property rights and promoting individual freedom disparages government efforts to address the long-term collective interests of the community and protect the poor and powerless.

It is important to remember that people *like* planning. They prepare plans for their personal lives, for the organizations to which they belong, and for the companies that employ them. They care deeply about their neighborhoods and communities and are concerned about the future they are leaving their children and grandchildren. Given an opportunity, people turn out night after night to express their concerns and hopes for a better future. Arbitrary-seeming regulations and the lack of a meaningful role in shaping their world are what people understandably resist.

Planning is inherently about the future: it is a process of identifying and implementing actions and policies in the present that will make the future better. Unfortunately, planners have all too often abandoned their traditional role of helping communities consider what the future may—and should—be. Instead, many planners focus their efforts largely on pragmatic problem-solving and more immediate concerns of community development, code enforcement, and public/private development.

Planners' neglect of the future has several causes. The press of daily job requirements and the desire for job security understandably focus planners' attention on short-term concerns with clearly identified results. Today's political environment often makes idealism, visioning, and inspiration seem naively unrealistic. And the widespread adoption of social science methods has greatly enhanced planners' ability to understand the past and present but provides little assistance in helping communities engage in their collective futures.

Planning and the Future

> You spend your life walking backwards because you see the past not the future; that's why we trip.[2]

All predictions and projections are fictions about a future that has not occurred. Individuals and organizations cannot accurately predict their long-term futures, and there is no reason to assume that planners can foresee an unknowable future better than the public they serve. No model or method can determine what a community's future will be. At best, they can only provide explicit procedures for determining the implications that can be drawn from limited information about the past and the present, and plausible assumptions about the future. To ask for more is to require the impossible.

Planners' attempts to deal with the future are particularly difficult because they are often required to prepare long-term projections for small areas to satisfy the requirements of other agencies and support the needs of local clients. Short-term projections for small areas are relatively easy to prepare because demographic and economic changes over a two- or three-year period are generally small. Long-term projections for large areas are often not difficult to prepare because growth or decline in some subareas is often offset by countervailing changes in others. However, it is extremely difficult to project long-term changes for small areas where new residential developments and employment centers or changes in local land-use policies can lead to substantial population and employment changes over relatively short periods. Predicting the future is also more challenging in smaller communities where budgets and staff are limited, and the consequences of the projections are clearer to interested parties.

Planners' projections are also more difficult because they are generally prepared with information that is out of date, estimated, or collected at the wrong level of spatial and sectoral aggregation. Predictions in the natural sciences are based on well-developed bodies of **theory** and carefully controlled experiments that are impossible in the social sphere. Well-funded business analysts use large quantities of reliable data and sophisticated methods to predict short-term inventory and sales changes. As a result, it is not at all surprising that planners' projections for the long-term future of cities and neighborhoods inevitably prove to be wrong.

The difficulty of accurately projecting the future does not mean that projections are not useful or important. The question isn't whether projections turn out to be accurate but whether they provide information that helps inform the policy-making process, facilitate community understanding, and help the public deal with an unknown future. If the projected future is undesirable, planning is required to prevent the "most likely" future from occurring. If the future is uncertain, communities should recognize this fact and plan for dealing with an unknown future.

While the information available for considering the future is often bad, that is, incomplete and inaccurate, it is also the best available data. Planners' models and methods must accommodate the data available in a particular time and place and use these data as best they can. They can update their projections and improve their plans if more current and accurate information becomes available later.

The Role of Assumptions

Any projection model or method is loaded with assumptions.[3] Projecting the future by continuing past trends assumes that these trends will continue or will vary in a known way. But finding relationships that fit the past does not guarantee that they will apply in the future. Simple extrapolations of past trends into the future ignore many factors that may change in the future, including planners' efforts to identify and implement policies that support desired futures and avoid undesirable ones.

Implementing planning methods in a computer model does not eliminate the need to make assumptions and choices. Representing a complex reality in a computer model inevitably requires choices in specifying causal relationships, selecting input variables, and adjusting model outputs so that they are "reasonable" to modelers and their clients. All projections are built on hypotheses and assumptions that become increasingly unlikely the farther out the projection goes. Computer-based models and methods do not reduce the need to make assumptions; they only hide them in computer algorithms that are inaccessible outsiders.

The core assumptions about the future cannot be derived from a method or model. A model is only a vehicle for tracing the implications of the core assumptions chosen independently from, and prior to, the model or method that implements them. When the core assumptions are valid, the choice of methodology is either secondary or obvious. And when they are wrong, no level of model sophistication can generate good predictions from bad assumptions, an example of "garbage in, garbage out."

Projections prepared in a wide range of areas have routinely proven to be wrong, often embarrassingly so. Some of the most dramatic projection errors are the result of "assumption drag"—the assumption that current conditions and past trends will continue, despite recent changes. It is often difficult to know in advance whether changes in past trends represent fundamental (structural) changes or short-lived "quirks" and cyclical changes that will soon disappear. As a result, the most important—and difficult—question planners must ask is how will change itself change: What current trends are likely to change? How will they change? And when will they change?

The Politics of Projecting the Future

Planners, elected officials, and the public often assume—or at least claim—that planning models and methods are unbiased and politically neutral tools used by technical experts to objectively identify and evaluate public policies and action.[4] The appearance of authoritativeness is reinforced by the value- and politically neutral language of technical objectivity used to prepare, use, and evaluate planning information, analyses, and proposals. The choices planners must inevitably make to operationalize their models and apply their methods are similarly portrayed as purely technical judgments that do not promote the interests of some groups over others.

Portraying projections as the objective outcomes of value-neutral analysis allows analysts to maintain their professional identities as politically neutral experts. Advocates of particular policy positions gain support for their positions by relying on supposedly unbiased technical analyses that support their positions. Politicians who choose between conflicting policy proposals can accept projections that correspond to their preconceptions and ignore those that do not. All three sets of actors—analysts, advocates, and politicians—gain by pretending that projections are objective scientific statements and not political arguments for a particular position.

In fact, value-neutral technical expertise in the preparation and use of projections is more apparent than real. Projections are rarely produced in a vacuum. Instead, they are prepared for a client—a government agency, an organization, a private firm, or an interest group—who often has a vested interest in the projected numbers. Elected officials and developers generally prefer projections that show continued population and economic growth that will attract new residents, businesses, and public investments. Environmental groups often view rapid growth as a problem that threatens natural systems and community livability. At the local level, landowners and businesses prefer projections that direct growth toward—or away from—particular locations. In all these cases, the clients for whom the projections are prepared will often request—and even demand—projections that support policies and actions that benefit them.

Planners' models and methods are political in a bureaucratic sense because control of information is a powerful organizational resource that increases the power of technical experts and technically sophisticated groups. They are also useful in organizational settings for postponing difficult political choices, justifying decisions that have been made on other grounds, and adding the sheen of technological sophistication to public policy making. Consultants are hired not only for their expertise but also to provide someone to blame if things go wrong. Elected officials use planners' analyses and projections like drunks use a light post: for support, not enlightenment.

Planning analyses and projections are based not on universally accepted bodies of theory and uncontested facts, but rather on incomplete data, partial hypotheses, and assumptions about what the future will—and should—be. As a result, the preparation and use of planning models and methods necessarily involves numerous choices about selecting data, applying computational procedures, and analyzing, presenting, and distributing the analysis results. These choices are inherently political because they can help shape the analysis results, the perception of problems, and the definition of potential solutions.

Projections are inherently political because they can be either **self-fulfilling** or **self-defeating**. They can be self-fulfilling because the fact that an area is predicted to grow may stimulate growth by attracting people and firms seeking new opportunities and by providing the public infrastructure that allows growth to occur. Conversely, areas where decline is assumed will often do so because people and firms are reluctant to migrate to or invest in areas where their prospects appear to be poor. Projections can also be self-defeating by stimulating action that prevents an undesired future from occurring.

The **accuracy** of a projection can only be determined when the forecast date arrives. This makes them useful for political purposes because they can serve their intended purpose even though they may ultimately prove to be wrong. Planners can justify a major facility investment by exaggerating the demand for it and underestimating its costs. If the investment is made and the projections prove to be false, the political purpose has been achieved and planners can avoid responsibility by citing factors that were outside their control when the projections were prepared.

Planners should not only recognize that their analysis, projections, and plans are inherently political but should also see them as reflecting underlying assumptions that may support the competing perspectives and interests of contending parties. Viewed in this way, the future is not a single grand vision or the inevitable consequence of past trends but rather an object of public deliberation. Planning models and methods can play an essential role in this deliberation by testing a range of projections and policy proposals derived from the preferences and assumptions of different segments of the population.

The Promise of Scenario Planning

Recognizing the difficulty of predicting an unknowable future, a growing number of scholars have suggested that planners refocus their efforts from predicting what the future *will* be to describing what the future *may* be.[5] The possible futures are called **scenarios**. Scenarios are much more than projections that extend past trends to identify a single future. They describe *possible* futures, not necessarily *predicted* futures. They are plausible and divergent stories describing how the future may unfold and how it may be different from the past. Scenarios are not visions because not all the futures may be desirable; they are not projections because they describe a process of change, not a single point in time. Scenarios recognize the uncertainty, complexity, and unpredictability of forces shaping the future and provide a way to think carefully about the future without trying to predict what it will be.

Scenarios are used in the private sector to help organizations survive and prosper in the face of future contingencies and competition. In the public sector, they can involve a diverse group of people who work together to create a range of scenarios that may challenge the assumptions individuals may have about the future and allow them to jointly consider their shared future. A scenario describes one possible future. A collection of scenarios describes a range of structurally different, plausible, and challenging ways in which the future may unfold. Done well, scenario planning allows community members to learn about their community, identify what the community can control, recognize what it cannot control, and consider the implications of choices they make today. Scenario planning also utilizes not only tables and graphs but also images, videos, and personal narratives to portray alternative futures clearly and convincingly to the public.

In contrast to scenario planning, current practice often selects a single preferred future from a small set of alternative futures that generally include high/medium/low options—a desired future with two bookends—or two options, one of which is a baseline "business as usual" option that serves a foil for the preferred option. Assuming they have complete control over the future, planners then discard the less desirable options and attempt to make the desired future happen. Under fully realized scenario planning, there are no "good" and "bad" scenarios but rather a collection of plausible alternatives, some of which may include new, sometimes even uncomfortable, possibilities. By keeping alternative futures in mind, scenario planning encourages planners to act strategically in the face of uncertainty, just as wise investors employ a diverse portfolio of investments to prepare for a range of widely different financial futures.

Recognizing that a projection is only one of many possible futures invites community members to consider what the future can—and should—be. A successful scenario planning process creates an understanding of a place and an interest in its future. It encourages the public to think creatively about the future and cowrite compelling, evidence-based stories about the past, present, and future. Done well, scenario planning creates a strong image of a place that inspires subsequent public action.

Planning and the Public

> I participate; you participate; he participates; we participate; you participate. . . . *They profit.*[6]

Planners have long believed that the public should have a meaningful role in making the decisions that shape their lives.[7] Public participation is embodied in the American

Institute of Certified Planners' Code of Ethics and is routinely required in local, state, and federal planning mandates. It is widely seen as an essential tool for promoting citizens' right to be informed, consulted, and participate meaningfully in the decisions that affect them and for representing disadvantaged groups and issues that lack the support of organized interest groups.

Unfortunately, public participation efforts are often viewed by the public and planners alike as perfunctory, "going through the motions" rituals conducted primarily to satisfy legal requirements and obtain public support for agency proposals. The plans and proposals that shape policy discussion are often prepared by planning staff and consultants with little input from citizens or stakeholders. Public comments may be solicited, attitude surveys may be mailed, and volunteer working committees may be appointed. However, the public's formal role in the planning process is all too often limited to a public hearing held at the end of the process to hear citizens' comments, just before the action or policy is approved.

Planning for the Public

The public's limited role in contemporary planning practice is due, in part, to the professional role that has defined the profession since its inception. Reflecting its roots in architecture and landscape architecture, the early planning profession adopted the perspective of "planning as design," which assumed that planning a city is fundamentally the same as designing a building or a landscape. Guided by a naïve form of environmental determinism, the early planners assumed they could reform society by improving the physical environment and saw themselves as protecting the public interest from the self-interested and uninformed actions of politicians and private individuals. In the mid-twentieth century, the profession adopted a new ideal of "planning as applied science" that attempted to replace the intuitive designs of the planner-architects by the "scientific" methods and findings of the emerging fields of regional science, urban economics, and operations research.

Both approaches were based on an implicit model of planning *for* the public, which assumes that public-sector planning is like planning for a private firm or organization. In these settings, there is often a single client with well-defined objectives, long-established means for achieving those objectives, and centralized control over the resources needed to achieve them. Together, these factors allow designers and engineers to prepare blueprints providing detailed guidance for constructing the structure or landscape that will best serve their client's needs.

Planning for the public replaced the designer's client with the "the public" and their client's desires with their perceptions of the public interest. It replaced the client's image of a desired future by planners' projections for the city's future population, employment, land uses, and related infrastructure demands. The designer's empirically derived design principles and standards were replaced by generally accepted principles of "good professional practice." And the designer's blueprints were replaced by comprehensive plans that were assumed to guide public and private actions toward a shared image of the desired future.

Experience has revealed that public sector planning lacks all the conditions that characterize private sector planning. The long-term concerns of public sector planning are fundamentally different from the short-term focus of managing a business. A large and diverse public has conflicting goals and interests. The poorly defined guidelines of professional practice cannot address the increasingly complex issues facing planning. And the urban fabric is shaped by the actions of a diverse range of groups and organizations, largely outside the control of planners.

Planning with the Public

The ideal of professional-directed planning for the public has been augmented by an alternative ideal of citizen-based planning with the public. This ideal has several labels: consensus building, visioning, and anticipatory governance. These approaches assume that planning should not be the special providence of technical experts called "planners." Instead, they attempt to replace traditional top-down planning with efforts to involve more people in more decentralized ways, in more of the decisions that affect them.

Planning with the public requires that, whenever possible, planners involve as many people, as completely as possible, in the decisions that affect their lives. Unlike representative forms of governance, it encourages the direct participation of the representatives for all groups involved in an issue. Unlike traditional public hearings conducted at the end of the planning process, citizens are involved from the beginning: reviewing past and present conditions, identifying problems and opportunities, evaluating alternatives, and attempting to make decisions they can all support.

The people who live and work in a community have an intimate understanding of its past and present and are vitally concerned about its future. As a result, any attempt to plan for the community's future must actively seek input from community residents and stakeholders. This not only directly involves the people for whom the plans are being prepared; it encourages planners to learn from the people who are most familiar with local conditions and the realities in which their plans will be implemented. Involving the public in all stages of the planning process helps planners understand the concerns of stakeholders, identify potential collaborators and latent opposition, and consider issues that have traditionally been neglected. It also makes planners more politically effective by giving stakeholders a sense of ownership over the analyses that inform public decision making and creating social capital that helps get public policies adopted.

It is important to recognize, however, that the perceptions of individuals and groups are inevitably colored by their personal experiences and their limited understanding of the past and present. The residents of prosperous and growing areas generally assume that the future will be like present. In the face of overwhelming problems and crippling uncertainty, people living in declining communities often retreat to an idealized past rather than looking realistically at their current conditions and the most likely future. Community-based analysis and planning can be extremely helpful in both situations by providing elected officials and the public with authoritative information about the past and present. The goal is to augment the partial, poorly defined and often faulty tacit knowledge stored in individuals' mental models with the explicit knowledge of public information. Analysis and planning can also help reduce the influence of expedient viewpoints and counteract overly wishful or pessimistic thinking by clearly identifying the assumptions that underlie alternative futures and the implications of alternative public actions.

Taken seriously, planning with the public will be more challenging for planners and the communities they serve than the practices of the past. It will require much more than the popular "visioning" and "scenario planning" exercises in which a small number of self-selected citizens gather around tables and devote a few hours to placing chips or Lego blocks representing planner-generated land use demands to locations on maps of their community. The citizen-generated maps are then used to prepare one or more "Business as Usual" or "Trend" scenarios and an alternative "Smart Growth" scenario that is routinely selected as representing the public's image of a desired future.

Participants find these exercises engaging, empowering, and fun; they generate favorable media coverage and often increase public awareness of planning issues. However, it is unreasonable to assume that a diverse group of citizens can produce meaningful scenarios in a few hours without information on the feasibility and implications of alternative policy alternatives. In addition, these efforts rarely incorporate well-established principles for collective decision making, involve the continued participation of citizens, or provide any assurance that the exercise results will influence policy making. As a result, they all too often serve largely as exercises in unproductive wishful thinking or as mechanisms for promoting planners' preferred options for the future.

Planning Support Systems

The profession-centered model of planning for the public has been reflected in decades-long efforts to develop models of the city that mimicked architects' scale models and engineers' prototypes. Building on sophisticated theories of spatial and market interaction, planners developed complex models that attempted to capture the city in a computer. These efforts are reflected in a voluminous and esoteric literature and several large, complex, and expensive large-scale urban models that have been implemented in metropolitan areas around the world.[8] These efforts have been augmented in recent years by the development of GIS-based **planning support systems (PSS)** that abandon the attempt to model a complex reality for much simpler approaches.[9]

Planning support systems do not include general purpose tools for word processing, presentation, communication, and the like that planners use just like other professionals. Instead, they are specialized tools that support their efforts *as planners* to help communities deal with an uncertain future, promote the collective interests of a community, consider the external effects of individual and group action, improve the information base for public and private decision making, and consider the distributional effects of public and private action (Klosterman 1985).

Planning support systems reflect an ongoing evolution of planners' role and the role that information technologies can play in professional practice. The image of planning as an intuitive process of designing the physical city was replaced in the 1960s by the ideology—if not the reality—of planning as a value-neutral applied science. This paradigm shift was reflected in planning education with the addition of new students, faculties, and departments outside the traditional schools of design and the replacement of the ubiquitous design studios by equally prevalent courses in research design, statistics, and quantitative methods. The applied science "rational planning" model defined rationality in instrumental terms as finding the best means for achieving desired ends and planning as an iterative process of defining problems, identifying goals, generating alternatives, and evaluating those alternatives with respect to designated goals. Computers were assumed to play an important role in this process by collecting and storing the required data, providing urban models that described the present and projected the future, and helped identify the best plan from a range of alternatives.

The ideal of value-neutral "rational planning" was replaced in the 1970s by a new image of planning as politics. Planners realized that planning was inherently and inevitably "political"—helping determine who gets what, when, where, and how. They also acknowledged that computer-based information, models, and methods can help increase the power of administrative and technically sophisticated experts and hide inherently political choices in the selection of data analyses, the presentation, and distribution of results within supposedly technical analyses unfathomable to outsiders.

The realization that planning was inherently political was enriched in the 1980s by the new images of planning as communication, which recognized that planning involves more than the collection and distribution of information that can (presumably) improve public policy making. Planners not only analyze data and prepare plans, but they also negotiate, explain, and argue about planning rules, changes, and permissions, and administer rules and regulations. Quantitative information and analyses of course play an important role in these activities but so do advice-giving, story-telling, and the other rhetorical devices planners used to communicate with others. On this perspective, the *way* in which planners transmit information may be more important than what they say. For example, quantitative analyses, no matter how well meaning and technically correct, expressed in technical jargon and bureaucratic terms all too often separate planners from the planned for, increase the public's dependence on technical experts, and minimize planners' ability to learn from the public. The new conception of communicative rationality suggests that planning models and methods should support interactive, open, and ongoing processes of community debate and collective action.

Planning support systems are not closed black-box computer models that ingest raw data and automatically project the future, generate plans and proposals, and identify optimal policies and actions. Instead, they provide information infrastructures that use the best available data and simple, understandable, and easy-to-use methods, models, and procedures to support citizen-led processes of analysis, debate, planning, and action.

Simple and Complex Models and Methods

KISS (Keep It Simple, Stupid)[10]

The evaluation of simple and complex models and methods must begin with Box's Law: "All models are wrong; some models are useful" (Box and Draper 1987, 424).[11] The first part of the "law" is true by definition: a model is a simplification of reality that selectively focuses attention on some aspects of a complex reality and ignores others. A representation of reality that did not omit some details wouldn't be a model; it would be reality itself. As a result, all models are inevitably "wrong" in the sense that they are incomplete and leave out some aspects of reality.

However, Box's Law also suggests that the question isn't whether a model is correct in some absolute sense but, rather, whether it is useful for a particular purpose. A model can serve many purposes: to develop or test theory, to provide a sheen of technical sophistication to public policy making, to enhance the reputation and income of the modeler, or to support professional practice. While these objectives are all valid, models' primary value for planning practice lies in their ability to help planners and the communities they serve understand the past and the present and engage the future.

Complex models are large and expensive systems that take huge expenditures of time, manpower, and money to develop and refine.[12] They are uniquely capable of providing detailed answers to a wide range of policy questions and explicitly expressing causal relationships that are implicitly retained in the minds of analysts using less complex models. They are also useful for representing interconnected relationships, feedback loops, and other intricate relationships that are ignored by simpler models. These interactive relationships often have surprising and counterintuitive properties that are not apparent to analysts who do not use an explicit model to track their interactions. As a result, complex models are required for highly technical activities such as projecting transportation flows or estimating environmental impacts.

Overly simplistic models may introduce large specification errors by failing to represent adequately the processes that comprise the system being modeled. More complex models introduce more variables and longer causal chains in an attempt to better represent a complex reality. However, complex models require additional data that may not be measured correctly and are often not available, requiring the use of proxy variables. As more variables are added, these measurement errors may compound rapidly, overwhelming the errors resulting from not properly modeling underlying causal processes.

Analysts who rely on judgment and a simple model keep the richness of nuance and detail in their heads; complex model users rely instead on relationships that are expressed in formal mathematical relationships they may not understand. Experience suggests that the ability to incorporate local knowledge and up-to-date contextual assumptions is often more important than methodological sophistication in projecting the future. As a result, the advantages of formal explicitness of complex models are often offset by their inability to incorporate contextual knowledge about a particular time and place that cannot be expressed in clear-cut mathematical relationships.

An extensive body of research demonstrates convincingly that complex projection models and methods are no more accurate than simpler ones. It also suggests that forecast accuracy is not the only—and often not the most important—criterion for evaluating projections. Instead, it recognizes that simpler models and methods are often preferred because they are easier to use and understand, less expensive, more flexible, make better use of available data, and provide more timely information.

By focusing attention on the assumptions that underlie modeling outputs, simple models also promote more open and democratic policy making by making explicit the factual and political assumptions that complex models hide from view. By involving the public more directly in the modeling process, they not only limit the discretion of professionals but also reduce the knowledge differential between professionals and laymen, generating information that is more familiar to—and more likely to be accepted by—policy makers and the public.

Planning models will better support meaningful citizen participation if policy makers and the public understand and trust them. As a result, when possible, planners should develop and use models that are as simple as possible, but not simpler. While the model's detailed computational procedures will inevitably be too complex for nonexperts to understand, their underlying logic should be understandable and clear. Simple, easy-to-use planning models should allow analysts, policy makers, and, ideally, the public to consider a range of policy options and their likely impacts. This makes it less likely that projections will be accepted as inevitable and stimulates the consideration of alternative policy options.

Beyond Planning Support Systems

Planners should not only use simple and understandable models and methods that involve members of the public more directly in the planning process.[13] They should also pay more attention to the informal elements of projecting the future: the knowledge, judgment, and wisdom regarding the assumptions that shape the model results. Projecting well requires learning about a community, knowing its resources and potential, and understanding how it relates to other areas. Most importantly, planning models and methods should serve as *prostheses for the mind*, allowing planners and the communities they serve to understand the past and present and work together to prepare for an uncertain future.

Done well, projections can help communities understand where they are, how they got there, and where they would like to go in the future. Rather than preparing plans and proposals that they believe best serve the needs of their clients, planners should attempt to facilitate and inform a citizen-led policy process. Instead of blindly extending past trends, they should become authorities on the area for which they are planning: learning about its past and present, understanding the potential and limitations of its natural and organizational resources, and identifying the key factors that will impact its future development.

Citizen-led attempts to engage the future can provide an opportunity for planners and the communities they serve to cowrite compelling stories about the past and present and develop strategies for creating a better future. Done well, an analysis of the community's past and present will allow people to see themselves and their families, friends, and neighbors in the numbers. Compelling, evidence-based stories about the past, present, and desired futures embedded in scenarios can also set the stage for planning by creating a sense of place that inspires planning action to achieve a desired future.

Notes

[1] This chapter draws heavily on Klosterman (2013), which provides citations for many of the points made here.

[2] Māori proverb. The Māori are the indigenous Polynesian people of New Zealand. The quotation is from Lau (n.d.).

[3] The discussion in this section draws on Ascher (1978, 1981), Isserman (1984), and Wachs (2001).

[4] The discussion in the following section draws on Klosterman (1987) and Wachs (1989).

[5] The following discussion draws on Avin (2007), Hopkins and Zapata (2007), Isserman (1984, 2007) and Wachs (2001).

[6] Quoted in Arnstein (1969).

[7] The following discussion draws on Klein (2000) and Wachs (2001).

[8] The urban modeling literature is reviewed in Klosterman (1994). A particularly impressive example of urban theory is Michael Batty's *New science of cities* (2013). Large scale urban models are reviewed in Hunt (2005) and Wegener (2014). Paul Waddel's UrbanSim (Waddell 2002) is widely used in the United States.

[9] See, e.g., Harris (1989), Klosterman (1997), Brail and Klosterman (2001), Geertman and Stillwell (2003), Brail (2008), Geertman and Stillwell (2009), and Geertman et al. (2015).

[10] The acronym was coined by Kelly Johnson (1910–1990), lead engineer at the Lockheed Skunk Works, creators of the Lockheed U-2 and SR-71 Blackbird spy planes, among many others. While popular usage translates it as "Keep it simple, stupid," Johnson translated it as "Keep it simple [and] stupid." There was no implicit meaning that an engineer was stupid; just the opposite. The principle is best exemplified by the story of Johnson handing a team of design engineers a handful of tools, with the challenge that the jet aircraft they were designing must be repairable by an average mechanic in the field under combat conditions with only these tools. Hence, the "stupid" refers to the relationship between the way things break and the sophistication available to fix them.

[11] The following discussion draws on Klosterman (2012).

[12] The following discussion draws on Ascher (1981, 253–256), Alonso (1968), and Wachs (1982, 565–566).

[13] The following discussion draws on Isserman (1984, 2007).

Welcome to Decatur 2

This chapter introduces the City of Decatur, Georgia, and DeKalb County in which it is located, which will be considered in detail in the remainder of the book. The chapter contains eight major sections. The first section locates Decatur and DeKalb County and briefly describes the city's history. The second section describes the city's population and income and how they compare to DeKalb County and the twenty-eight county Atlanta region. The third section describes the city's employment and occupational structure, as they compare to the county and the Atlanta region. The fourth section describes the city's land uses and housing. The fifth section describes Decatur's block groups: their land uses, population, racial and ethnic composition, and household income. The final three sections describe three methods for describing and analyzing a community's population and income or wealth distribution: population pyramids, Gini coefficients, and Lorenz curves.

Background Information

Location

Decatur, Georgia, is named after Steven Decatur, a popular early-nineteenth-century American naval hero. The city is the county seat of DeKalb County, which is named after Baron Johann de Kalb, a German soldier who fought with the colonials in the American revolutionary war.[1]

As shown in map 2.1, DeKalb County is in the twenty-eight-county Atlanta-Sandy Springs-Roswell, GA, **Metropolitan Statistical Area** (MSA), which will be referred to as the Atlanta region in the remainder of this book. The Atlanta region contained 5.3 million residents in 2010, as many people as the state of Minnesota, and occupies 8,480 square miles (22,000 square kilometers), as much land as the state of Massachusetts.

As pointed out in appendix A, MSAs are geographic entities delineated by the US Office of Management and Budget that federal agencies use for collecting, tabulating, and publishing their statistics. An MSA consists of a core urban area of fifty thousand or more population and includes one or more counties that include the core urban area and any adjacent counties that have a high degree of social and economic integration (as measured by commuting to work) with the urban core.

As shown in map 2.2, the City of Decatur is immediately adjacent to the City of Atlanta, within the Interstate 285 corridor that encircles Atlanta. Decatur is surrounded by incorporated cities.

History

The first European settlers moved into the area that is now DeKalb County in the early 1820s. In 1822, the Georgia General Assembly established DeKalb County and designated a site where two Indian trails met as the location for the county courthouse, which still stands in Decatur's city center. The Georgia General Assembly incorporated the City of Decatur in 1823 as the state's second oldest incorporated city.

In the 1830s, the Western and Atlantic Railroad wanted to make Decatur the southernmost stop on its line. The citizens of Decatur did not want the noise, pollution,

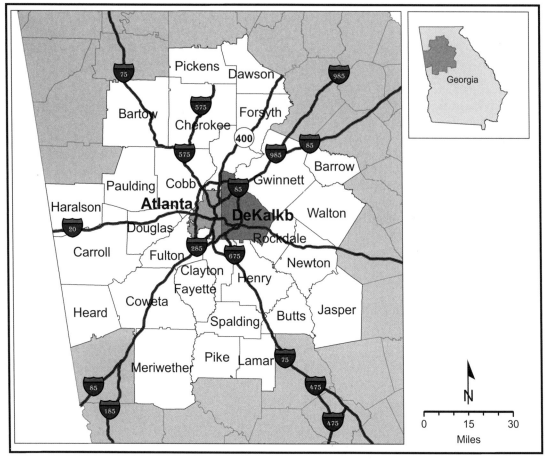

Map 2.1 Atlanta Region, 2010

and growth that would come with such a major terminal, so they rejected the proposal. In response, the railroad founded a new city to the west-southwest of Decatur for the terminal, which later became the City of Atlanta. By 1845, Atlanta had become the region's transportation center, and its growth soon eclipsed the small city of Decatur. Atlanta remained in DeKalb County until 1853 when Fulton County was created from the western half of DeKalb County.

Decatur became the commercial center of DeKalb County soon after its establishment but grew slowly in the late nineteenth and early twentieth centuries. The city's growth during this period was mainly along and near the numerous trolley lines that provided fast, clean, and convenient travel in an era of horse-drawn carriages. Many garden suburbs were developed in the city between 1910 and 1940 that provided larger lots with houses set back from the street and room for a garage and driveway. The city's development continued from the 1940s to the 1960s, featuring the then popular single-story ranch house. The city is now almost totally developed.

DeKalb County was primarily agricultural until the 1960s but became increasingly urbanized with the sprawl of metropolitan Atlanta. The completion of the eastern half of the Interstate 285 beltway in 1969 placed most of the county inside the region's perimeter highway.

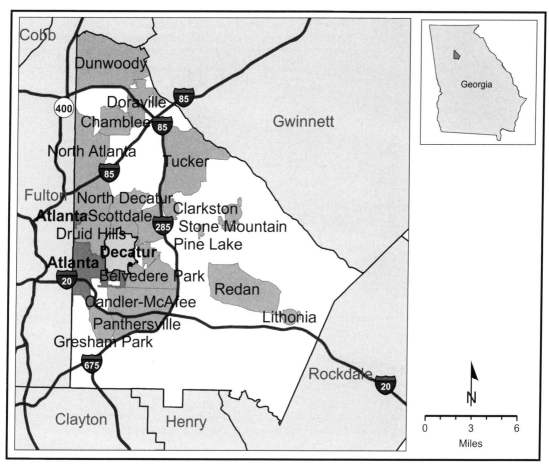

Map 2.2 DeKalb County, Georgia, 2010

Population and Housing Information

The City of Decatur had a population of 19,335 in 2010; 691,893 people lived in DeKalb County; 5.3 million people lived in the twenty-eight-county Atlanta region; and 9.7 million people lived in Georgia. Table 2.1 reports that 69.8 percent Decatur's residents are white, compared to 35.5 percent for DeKalb County and 56.1 percent for the Atlanta region. Correspondingly, Decatur has a substantially smaller proportion of people who are black, other races, or Hispanic than the comparison areas.

The population data in table 2.1 are for a five-year range (2010–2014), rather than for a single year such as the 2012 midpoint of this range. As noted in appendix B, these data from the US Bureau of the Census' **American Community Survey** (**ACS**) are estimates for a specified time period, five years in this case. This contrasts with point-in-time data from the decennial census of population and housing that counts the country's population on April 1 of every year ending in zero. The multi-year ACS population estimates have many advantages over the more familiar point-in-time estimates that are described in appendix B.

Table 2.2 reports that Decatur's population is older than the population of the county or region. DeKalb County's population distribution is roughly equal to that of the region.

| TABLE 2.1 | Population Breakdown by Race: City of Decatur, DeKalb County, and the Atlanta Region, 2010–2014 | | |

Population	City of Decatur	DeKalb County	Atlanta Region
(1)	(2)	(3)	(4)
White Alone	69.8%	35.5%	56.1%
Black Alone	22.2%	53.8%	32.9%
Other	8.0%	10.7%	11.1%
Hispanic	3.2%	9.3%	10.4%

Note: The Atlanta region corresponds to the twenty-eight county Atlanta-Sandy Springs-Roswell, GA Metropolitan Statistical Area.

Source: Computed by the authors from US Bureau of the Census (2016a).

| TABLE 2.2 | Population Breakdown by Age: City of Decatur, DeKalb County, and the Atlanta Region, 2010–2014 | | |

Population	City of Decatur	DeKalb County	Atlanta Region
(1)	(2)	(3)	(4)
19 and below	25.8%	26.3%	28.7%
20–44	36.1%	39.1%	35.8%
45–64	26.9%	25.0%	25.6%
65 and above	11.2%	9.5%	9.9%
Median age	38.1	34.9	35.4

Note: The Atlanta region corresponds to the twenty-eight county Atlanta-Sandy Springs-Roswell, GA Metropolitan Statistical Area.

Source: Computed by the authors from US Bureau of the Census (2016c).

Table 2.3 indicates that Decatur's mean (or average) household income is significantly higher than it is for the county or region. Half of the city's households earned more than $74,465 (the median) and more than 21 percent earned more than $150,000. Less than 9 percent of the city's households are below the poverty level. Nearly 16 percent of DeKalb County' households have incomes below the poverty line, significantly more than Decatur or the Atlanta region. DeKalb County's mean and median household incomes are also substantially lower than that of Decatur and the region.

Table 2.4 reports that the median value of housing units in Decatur is twice as high as in the three comparison regions. The city's **average household size** and **vacancy rate** are significantly lower than they are for the comparison areas.

This information suggests that Decatur is a well-established, relatively wealthy, largely residential inner-ring suburb of Atlanta. The city contains a substantially lower proportion of minority residents than the county and the region in which it is located. The next section examines the city's economy as it compares to the county and region.

TABLE 2.3	**Household Income Breakdown: City of Decatur, DeKalb County, and the Atlanta Region, 2010–2014**

Household Income	City of Decatur	DeKalb County	Atlanta Region
(1)	(2)	(3)	(4)
Less than $25,000	26.1%	24.6%	21.6%
$25,000–$74,999	24.3%	43.0%	42.2%
$75,000–$149,999	28.5%	22.3%	25.4%
$150,000 or more	21.1%	10.1%	10.7%
Median	$74,465	$49,510	$55,295
Mean	$99,594	$72,104	$76,582
Below poverty line	8.7%	15.8%	12.6%

Note: The Atlanta region corresponds to the twenty-eight county Atlanta-Sandy Springs-Roswell, GA Metropolitan Statistical Area.

Source: Computed by the authors from US Bureau of the Census (2016b).

TABLE 2.4	**Housing Statistics: City of Decatur, DeKalb County, and the Atlanta Region, 2010–2014**

Variable	City of Decatur		DeKalb County		Atlanta Region	
	Number	Percent	Number	Percent	Number	Percent
(1)	(2)	(3)	(4)	(5)	(6)	(7)
Owner-occupied housing units	5,172	60.2	145,729	55.2	38,288	63.8
Renter-occupied housing units	3,413	39.8	118,309	44.8	21,748	36.2
Occupied housing units	8,585	100.0	264,038	100.0	60,036	100.0
Average household size	2.22	—	2.62	—	2.78	—
Vacancy rate	7.3%	—	13.4%	—	11.7%	—
Median housing unit value	$335,800	—	$160,700	—	$162,800	—
Median rent	$866	—	$960	—	$946	—

Note: The Atlanta region corresponds to the twenty-eight county Atlanta-Sandy Springs-Roswell, GA Metropolitan Statistical Area.

Source: US Bureau of the Census (2016c).

Employment and Occupation Information

Table 2.5 records the employment breakdown for Decatur, DeKalb County, and the Atlanta region. The employment is recorded for the location in which people work, not where they live; these people may live in Decatur or commute into the city for work. It reveals that more than 55 percent of the 12,824 people who work in Decatur are employed in four employment sectors: Professional, Scientific, and

TABLE 2.5	Employment Breakdown by Major Employment Sector: City of Decatur, DeKalb County, and the Atlanta Region, 2013

NAICS Code	Industry	City of Decatur[a]	DeKalb County[b]	Atlanta Region[b]
(1)	(2)	(3)	(4)	(5)
11	Agriculture, Forestry, Fishing, and Hunting	0.0%	0.0%	0.0%
21	Mining, Quarrying, Oil, and Extraction	0.0%	0.0%	0.1%
22	Utilities	0.1%	0.5%	0.6%
23	Construction	2.1%	3.9%	4.7%
31	Manufacturing	2.3%	4.1%	6.3%
42	Wholesale Trade	1.2%	5.5%	6.7%
44	Retail Trade	11.6%	11.9%	12.2%
48	Transportation and Warehousing	1.2%	4.9%	5.3%
51	Information	1.1%	5.7%	4.8%
52	Finance and Insurance	2.6%	3.5%	5.3%
53	Real Estate and Rental and Leasing	1.5%	2.4%	1.9%
54	Professional, Scientific, and Technical Services	12.9%	7.3%	8.3%
55	Management of Companies and Enterprises	0.0%	4.1%	4.9%
56	Administrative and Support Services	8.5%	7.4%	8.5%
61	Educational Services	12.5%	8.6%	2.8%
62	Health Care and Social Assistance	22.7%	14.9%	11.4%
71	Arts, Entertainment, and Recreation	0.7%	1.0%	1.4%
72	Accommodation and Food Services	11.6%	9.5%	10.1%
81	Other Services (except Public Administration)	7.4%	4.9%	4.7%
99	Not classifiable	0.1%	0.0%	0.0%
—	**Total**	**100.0%**	**100.0%**	**100.0%**
—	**Total employment**	**12,824**	**260.079**	**2,003,708**

Note: The Atlanta region corresponds to the twenty-eight county Atlanta-Sandy Springs-Roswell, GA Metropolitan Statistical Area.

Source:
[a] Computed by the authors from US Bureau of the Census (2015b).
[b] Computed by the authors from US Bureau of the Census (2015a).

Technical Services (NAICS 54); Educational Services (NAICS 61); Health Care and Social Assistance (NAICS 62); and Other Services, except Public Administration (NAICS 81). More than one-fifth of the city's employment is in Health Care and Social Assistance. Employment in these sectors is substantially smaller for the county and the Atlanta region.

Table 2.6 reports the occupations for people who live in Decatur. They may work in Decatur or commute out of the city for work. It indicates that more than one-quarter of the people who live in Decatur work in Management, Business and Finance occupations. Another quarter of the city's residents are employed in two other occupations: (1) Computer, Engineering and Science; and (2) Education, Training and Library. Many Decatur residents work in three other occupations: (1) Healthcare, Practitioners and Technicians; (2) Sales and Related; and (3) Office and Administrative Support.

TABLE 2.6	Occupation Breakdown: City of Decatur, DeKalb County, and the Atlanta Region, 2013		
Occupational Category	**City of Decatur**	**DeKalb County**	**Atlanta Region**
(1)	(2)	(3)	(4)
Management, Business, and Finance	25.9%	16.9%	17.2%
Computer, Engineering, and Science	11.0%	6.6%	6.1%
Community and Social Service	2.2%	1.8%	1.4%
Legal	5.9%	1.9%	1.4%
Education, Training, and Library	14.2%	6.8%	6.4%
Arts, Design, Entertainment, Sports, and Media	4.8%	2.5%	2.1%
Healthcare Practitioners and Technicians	7.6%	5.4%	4.6%
Healthcare Support	0.2%	2.0%	1.7%
Protective Services	1.3%	2.1%	2.0%
Food Preparation and Serving	3.5%	6.2%	5.7%
Building and Grounds Services	0.1%	4.1%	3.8%
Personal Care	2.6%	2.8%	3.1%
Sales and Related	8.3%	10.0%	12.3%
Office and Administrative Support	9.4%	14.2%	13.3%
Farming, Fishing, and Forestry	0.0%	0.1%	0.2%
Construction and Extraction	1.2%	3.9%	4.6%
Installation, Maintenance, and Repair	0.0%	2.0%	3.2%
Production	0.6%	4.3%	4.5%
Transportation and Material Moving	1.2%	6.5%	6.5%
Total	**100.0%**	**100.0%**	**100.0%**
Total employed population, 16 and above	**10,100**	**332,738**	**2,514,806**

Note: The Atlanta region corresponds to the twenty-eight county Atlanta-Sandy Springs-Roswell, GA Metropolitan Statistical Area.

Source: Computed by the authors from US Bureau of the Census (2015d).

Together, Tables 2.5 and 2.6 support the previous analysis, which suggests that Decatur is an attractive place. Many of the people who live and work in Decatur are employed in management education, health care, and other technical areas. Many of the occupations in which Decatur is well-represented require higher education levels and better wages, which fits with the city's character as a relatively affluent suburban community.

Land Uses

Decatur occupies 4.4 square miles (2,820 acres or 1,100 hectares) of land. Map 2.3 shows the city's major features. The city center contains the Old Courthouse, the city and county administrative offices, and the library. The city has three underground connections to the Metropolitan Atlanta Rapid Transportation Authority system, providing convenient access to the entire Atlanta region. The city also has several parks scattered throughout the city.

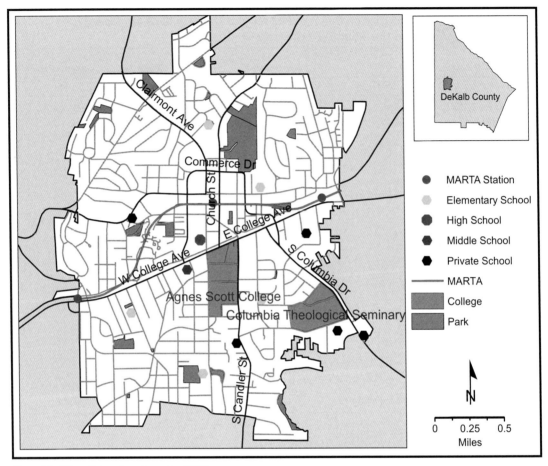

Map 2.3 City of Decatur, Georgia, 2010

Decatur has one high school, one middle school, five public elementary schools, and five private schools. Agnes Scott College, established in 1889 as the Decatur Female Academy, was the first school in Georgia to be fully accredited (in 1907) and is known for its high academic standards. The Columbia Theological Seminary is an educational institution of the Presbyterian Church (USA), which prepares women and men for leadership in the church's ordained and lay ministries.

Table 2.7 and map 2.4 indicate that Decatur is primarily a residential community. More than 59 percent of its land is residential, public and institutional land uses make up nearly 13 percent of the city's land area, and commercial land uses make up only 5 percent. The city has only thirty-one acres of vacant land and ten acres of industrial land. In contrast, less than 45 percent of the land in DeKalb County is residential, 5 percent is industrial, and more than 17 percent is vacant.

Decatur has much more residential land and much less commercial and industrial land than most cities in the United States. This reflects its long-standing position as a relatively small residential community for many people who leave the city to work elsewhere in the large and diverse Atlanta region.

TABLE 2.7					
Land Uses: City of Decatur and DeKalb County, 2014					

Land Use	City of Decatur		DeKalb County	
	Acres	**Percent**	**Acres**	**Percent**
(1)	**(2)**	**(3)**	**(4)**	**(5)**
Commercial	142	5.0	11,565	6.7
Industrial	10	0.4	9,253	5.3
Park/Recreation	127	4.5	5,026	2.9
Public/Institutional	358	12.7	15,894	9.2
Residential	1,679	59.4	77,380	44.6
Streets/Roads	480	17.0	23,753	13.7
Vacant	31	1.1	30,626	17.7
Total	**2,827**	**100.0**	**173,497**	**100.0**

Source: Computed by the authors from data provided by the City of Decatur Department of Community Development and the DeKalb County GIS Department.

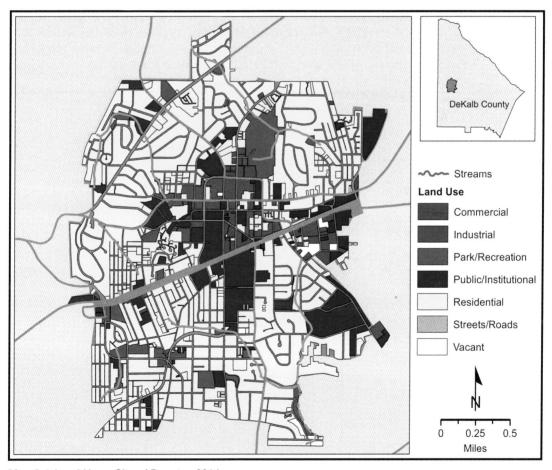

Map 2.4 Land Uses, City of Decatur, 2014

Source: Decatur Department of Community Development.

Decatur Block Groups

The preceding analysis looked at the City of Decatur and the county and metropolitan area in which it is located. The following analysis examines the city's thirteen block groups in 2010.

Decatur had four census tracts in 2010. The tract boundaries have changed very little over the last fifty years, allowing comparisons to be made easily for five census years: 1970, 1980, 1990, 2000, and 2010.

Map 2.5 shows Decatur's thirteen block groups in 2010; the city contained fifteen block groups in 2000. It had 303 census blocks in 2010 and 277 census blocks in 2000. Changes in the census boundaries make it difficult to compare data for these two years. Fortunately, the National Historical Geographic Information System at the University of Minnesota (Minnesota Population Center n.d.) provides 2000 census data for block groups, census tracts, and larger areas that have been adjusted to match the 2010 census boundaries.[2]

As pointed out in appendix A, census tracts, block groups, and census blocks are geographic units the US Bureau of the Census uses to collect and record its data. **Census tracts** are small, relatively permanent, statistical subdivisions of a county or equivalent entity. Census tracts generally have a population size between 1,200 and 8,000 people, with an optimum size of 4,000 people. **Block groups** are divisions of census tracts that generally contain between six hundred and three thousand people. **Census blocks** are subdivisions of block groups that are bounded by visible features, such as streets, roads,

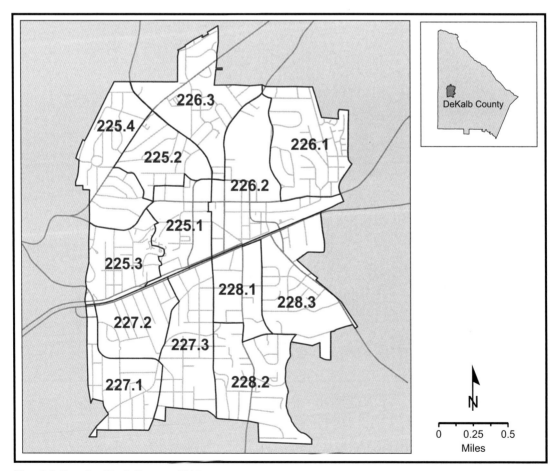

Map 2.5 Decatur Block Groups, 2010

Source: Decatur Department of Community Development.

streams, and railroad tracks, and by nonvisible political and administrative boundaries. Census blocks are generally small, for example a city block bounded on all sides by streets but may encompass hundreds of square miles in rural areas.

Land Use Information

Map 2.6 and table 2.8 reveal that residential land is distributed throughout Decatur with smaller amounts in Block Groups 225.1, 226.2, and 228.1 in the city center, highlighted in the table. Block Groups 225.1 and 226.2 contain nearly two-thirds of the city's commercial land and substantial public and institutional land. Public and institutional land uses are also concentrated in Block Group 228.1, which contains Agnes Scott College, and Block Group 228.3, which contains the Columbia Theological Seminary. A large public park is in the northern part of Block Group 226.2. The city's meager industrial land uses are in Block Group 228.3 on the city's eastern border. Nearly three-quarters of the city's vacant land is in Block Groups 228.2 and 228.3, on the city's southeastern border.

Population Density

Table 2.9 and map 2.7 report the residential population and density in Decatur's thirteen block groups. The residential density is highest in Block Groups 225.1 and 226.2 in the city center (highlighted in the table) that have relatively little

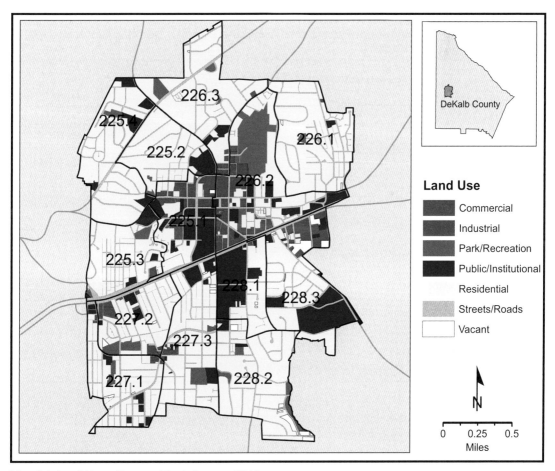

Map 2.6 Land Uses: Decatur Block Groups, 2014

Source: Decatur Department of Community Development.

TABLE 2.8   **Land Uses: Decatur Block Groups, 2014**

Block Group	Area (Acres)						
	Residential	Commercial	Public/ Institutional	Parks/ Recreation	Industrial	Streets/ Roads	Vacant
(1)	(2)	(3)	(4)	(5)	(6)	(7)	(8)
225.1	42.3	33.2	49.4	4.5	0.0	30.5	0.8
225.2	107.8	0.7	13.7	1.6	0.0	23.1	0.0
225.3	108.2	2.1	11.7	5.7	0.0	33.8	1.2
225.4	163.0	0.0	19.5	4.4	0.0	36.6	0.0
226.1	191.6	0.4	1.8	2.0	0.0	38.7	0.7
226.2	91.3	54.0	33.4	60.8	0.0	49.0	0.5
226.3	176.7	5.7	6.7	9.0	0.0	37.2	0.1
227.1	115.6	2.9	11.2	4.7	0.0	33.7	0.8
227.2	93.8	12.6	7.5	12.2	0.0	29.4	1.3
227.3	149.3	3.9	20.3	10.1	0.0	41.5	2.0
228.1	53.8	0.6	60.1	0.0	0.0	15.8	0.4
228.2	183.4	0.0	4.9	10.9	0.0	35.5	8.6
228.3	108.0	15.4	64.0	0.3	8.9	37.7	11.5
Total	1,584.8	131.5	304.2	126.2	8.9	442.5	27.9

Source: Computed by the authors from data provided by the City of Decatur, Department of Community Development.

TABLE 2.9   **Residential Population, Area, and Density: Decatur Block Groups, 2010–2014**

Block Group	Population (2010–2014)[a]	Residential Land Use (2014)	
		Area (acres)[b]	Density
(1)	(2)	(3)	(4)
225.1	2,065	42.3	48.8
225.2	842	107.8	7.8
225.3	1,218	108.2	11.3
225.4	1,540	163.0	9.4
226.1	1,858	191.6	9.7
226.2	2,015	91.3	22.1
226.3	1,866	176.7	10.6
227.1	1,978	115.6	17.1
227.2	970	93.8	10.3
227.3	1,569	149.3	10.5
228.1	958	53.8	17.8
228.2	1,688	183.4	9.2
228.3	1,225	108.0	11.3
Total	19,792	1,584.8	12.5

Source:
[a] US Bureau of the Census (2016d).
[b] Computed by the authors from data provided by the City of Decatur, Department of Community Development.

residential land. The residential density for Block Group 225.1 is nearly four times the city average, reflecting the large number of multi-story residential units in the city center.

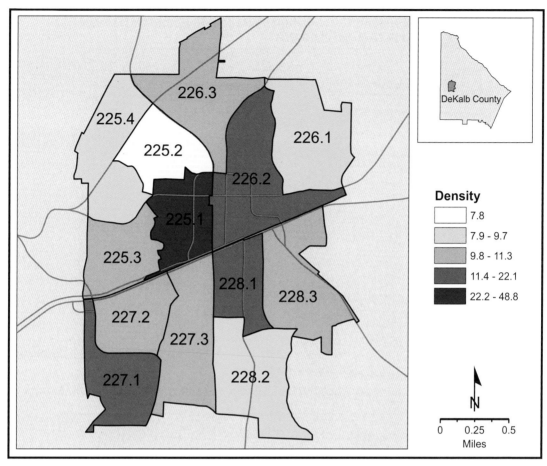

Map 2.7 Residential Density: Decatur Block Groups, 2010–2014

Racial and Ethnic Composition

Table 2.10 reports that Decatur's white-only population share ranges from more than 96 percent in Block Group 225.2 to less than 20 percent in Block Group 225.1, the city center (highlighted in the table). Map 2.8 indicates that the white-only population share is higher in the northern part of the city than it is in the southern part of the city. The only exception is Block Group 225.1 that has the smallest white-only population share. The city's small Hispanic population is located primarily in the northern portion of the city.

Household Income

Table 2.11 reports median household incomes that range from more than $110,000 in Block Groups 225.2 and 225.3 adjacent to Atlanta to roughly $9,000 in Block Group 225.1, the city center. A **household** contains one or more people living in a housing unit. A **housing unit** is a house, apartment, mobile home, group of rooms, or a single room that is occupied (or if vacant, is intended for occupancy) as separate living quarters.

| TABLE 2.10 | Population Composition by Race and Ethnicity: Decatur Block Groups, 2010–2014 | | | |

Block Group	Percentage			
	White Alone[a]	Black Alone[a]	Other Races[a]	Hispanic[b]
(1)	(2)	(3)	(4)	(5)
225.1	19.5	80.0	0.5	0.6
225.2	96.2	1.3	2.5	5.9
225.3	85.3	13.5	1.2	8.7
225.4	80.8	0.8	18.3	0.8
226.1	82.5	7.1	10.4	5.6
226.2	75.9	17.6	6.5	5.7
226.3	86.7	4.2	9.1	2.2
227.1	66.5	28.8	4.7	1.4
227.2	71.4	20.9	7.6	0.9
227.3	72.2	17.0	10.8	4.9
228.1	60.8	20.5	18.8	2.8
228.2	53.6	36.7	9.7	3.0
228.3	80.8	12.8	6.4	0.0
Total	**69.7**	**22.3**	**8.0**	**3.2**

Note: The Hispanic population can be of any race.

Source:
[a] Computed by the authors from US Bureau of the Census (2016e).
[b] Computed by the authors from US Bureau of the Census (2016f).

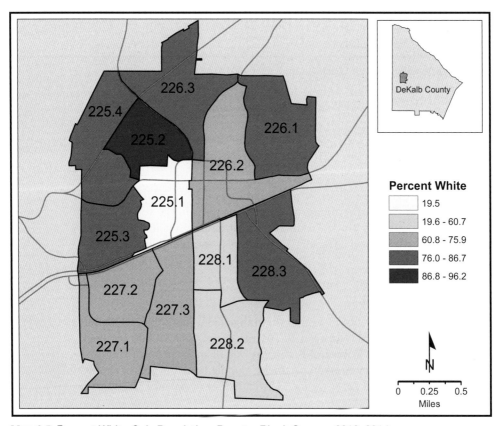

Map 2.8 Percent White-Only Population: Decatur Block Groups, 2010–2014

Block Group	Median Income ($)
(1)	(2)
225.1	9,008
225.2	56,250
225.3	111,300
225.4	110,435
226.1	98,625
226.2	64,861
226.3	71,354
227.1	103,021
227.2	95,347
227.3	73,900
228.1	87,143
228.2	86,458
228.3	97,216
Total	**77,202**

TABLE 2.11 Median Household Income: DeKalb County Block Groups, 2010–2014

Source: US Bureau of the Census (2016g).

Block Group 225.1 has the lowest white-only population share and the lowest median household income. Map 2.9 reveals that several of the block groups in the southern part of the city have relatively low white-only population shares and high household incomes; this is particularly true of Block Groups 227.1 and 227.2 adjacent to Atlanta. Conversely, Block Groups 225.2, 226.3, and 226.3 in the northern part of the city have relatively large white-only population shares and some of the lowest household incomes.

The preceding analysis reveals that the City of Decatur includes a range of neighborhoods with substantially different land uses and populations. Block Groups 225.1 and 226.2, which contain the city center, contain most of the city's commercial land and its highest population densities. Block Group 225.1, the city center, has by far the smallest white-only population share and the lowest household incomes. The block groups in the northern part of the city outside of the city center have the highest white population shares; the southern part of the city contains lower white-only population shares. The block groups along the city's western edge, nearest to the City of Atlanta, have the highest median incomes.

Population Pyramids

The preceding analyses used readily available census and GIS data to analyze Decatur, the city's block groups, and DeKalb County. The following sections describe three widely used ways to represent and analyze an area's population: population pyramids, Gini coefficients, and Lorenz curves.

Population pyramids display the different age and sex segments, or cohorts, of a population clearly and understandably. Population pyramids are horizontally stacked histograms that plot a population's age groups on the vertical axis and the number of people or the percentage of the population in each age group on the horizontal axis. By convention, the male population is plotted to the left of the vertical axis and the female population is plotted to the right.[3]

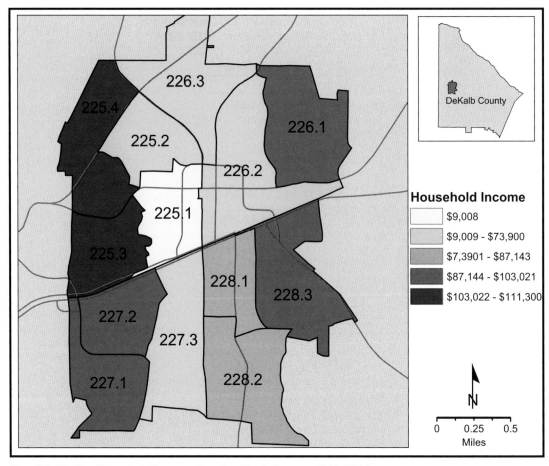

Map 2.9 Median Household Income: Decatur Block Groups, 2010–2014

Figure 2.1 illustrates three types of population pyramids. The population pyramid for Bangladesh is an **expansive population pyramid** that has many more people in younger age groups than in older age groups. This means that a large share of the population will age to produce their own children, causing future population growth. Expansive pyramids are generally found in Third World countries with growing populations, high fertility rates, and low life expectancies.

The population pyramid for Germany is a **constrictive population pyramid** that has fewer people in the younger age groups than in the older age groups. This means that a smaller part of the population will have children, causing the population to decline. This population distribution is found in many developed societies with shrinking populations, low fertility and mortality rates, and relatively older populations.

The population pyramid for the United States is a **stationary or nearly stationary population pyramid** that has roughly equal population numbers for most age groups. In these cases, the population is not growing or declining rapidly, and the graph looks more like a column than a pyramid, except at the very top, where the size of older age groups naturally diminishes.

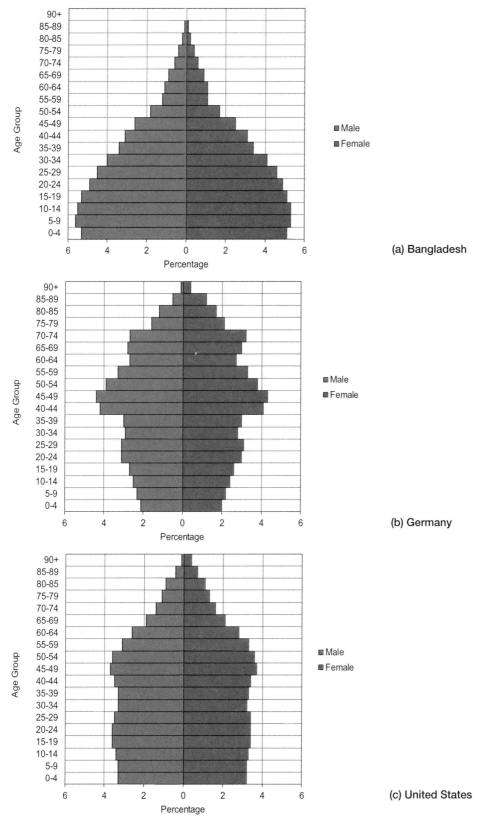

(a) Bangladesh

(b) Germany

(c) United States

Figure 2.1 Population Pyramids for Bangladesh, Germany, and the United States, 2010.

Source: Anonymous (2016).

The population pyramid for Georgia in figure 2.2 is stationary like the pyramid for the United States in figure 2.1.

DeKalb County's population pyramid in figure 2.3 is roughly stationary with slightly fewer people below the age of twenty-five. Decatur's population pyramid in figure 2.4 is a constrictive population pyramid that has a dramatic population deficit in the male population between the ages of fifteen and thirty. The implications that these population distributions may have on the city and county's future population growth will be analyzed in chapter 5.

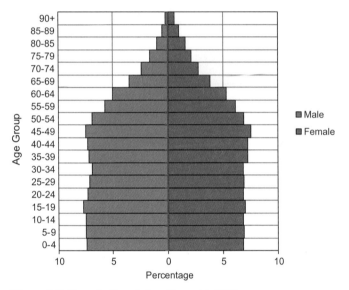

Figure 2.2 Georgia Population Pyramid, 2010

Source: Computed by the authors from data in US Bureau of the Census (2013c).

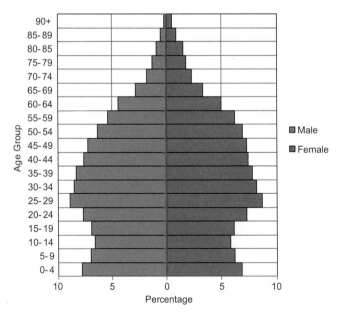

Figure 2.3 DeKalb County Population Pyramid, 2010

Source: Computed by the authors from data in US Bureau of the Census (2013c).

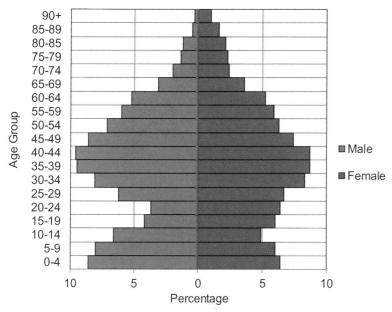

Figure 2.4 Decatur Population Pyramid, 2010

Source: Computed by the authors from data in US Bureau of the Census (2013c).

Lorenz Curves

Lorenz curves visualize the distribution of income, wealth, or other variables for one or more populations. They were developed by in 1905 by Max Lorenz, a graduate student at the University of Wisconsin. They are generally used to represent the inequality of a population's income or wealth distribution; they can also be used to describe the distribution of other variables such as the number of species in an ecology or the number of loan delinquencies for different credit scores.

When used with data on income or wealth, Lorenz curves plot the cumulative share of an area's population on the horizontal axis and the cumulative share of the area's income or wealth on the vertical axis. The more even the income distribution, the closer the curve is to a diagonal line.

Figure 2.5 displays the Lorenz for three hypothetical populations. The blue line is for a population with a perfectly equal income distribution. In this case, the population's cumulative share of the area's income is equal to its share of the population. Thus, for this example, the bottom 40 percent of the population has 40 percent of the income, the bottom 80 percent of the population has 80 percent of the income, and so on.

The orange line in figure 2.5 corresponds to a perfectly unequal distribution. The highest category has all the income and the rest of the population has nothing. In this case, the population's income share is zero for all but the highest category, whose income share is 100 percent.

The red line in figure 2.5 is for a distribution that lies between these two extremes. The lowest 40 percent of the population has 5 percent of the income, the lowest 80 percent of the population has 35 percent of the income, and the top 20 percent of the population has the remaining 65 percent of the income.

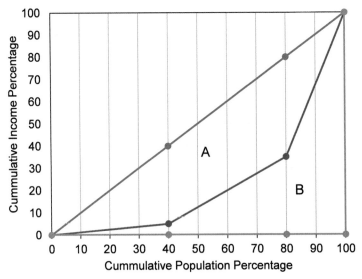

Figure 2.5 Sample Lorenz Curves

TABLE 2.12 **Computing Lorenz Curve Values**

Household Income	Households			Income			
	Number	Percent	Cumulative Percent	Average ($)	Total ($)	Percent	Cumulative Percent
(1)	(2)	(3)	(4)	(5)	(6)	(7)	(8)
Less than $20,000	4,000	40	40	10,000	40,000,000	5	5
$20,000 to $99,999	4,000	40	80	60,000	240,000,000	30	35
$100,000 or more	2,000	20	100	—	520,000,000	65	100
Total	10,000	100	—	—	800,000,000	100	—
Mean income	$80,000	—	—	—	—	—	—

Table 2.12 computes the plotted values for the intermediate distribution repre-sented by the red line in figure 2.5. Columns one and two record the number and percentage of households in each income category. The cumulative percentage values in column four are plotted on the horizontal axis of figure 2.5.

Column five records the assumed average income for each income category. The total income in column six is equal to the number of people in each category in column two multiplied by the average income in column five. The average income cannot be computed for the top income category because the top of its income range isn't reported, as is the case for the US census data. However, the population's total income ($800,000,000) is equal to the number of households (10,000) multiplied by the mean (or average) income ($80,000). The total income for the highest cat-egory ($520,000) is the population's total income ($800,000,000) minus the total income in the first two categories ($280,000,000). The cumulative income percent-ages in column eight are plotted on the vertical axis of figure 2.5.

Figure 2.6 displays the Lorenz curves for Decatur, DeKalb County, the State of Georgia, and the United States. It reveals that all four areas have roughly equivalent household income distributions.

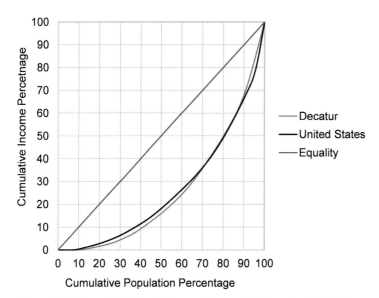

Figure 2.6 Household Wealth Lorenz Curves for Decatur, the State of Georgia, and the United States, 2010

Source: Computed by the authors from data in US Bureau of the Census (2015c).

Gini Coefficients

Gini coefficients provide a single measure of the income or wealth distribution displayed in a Lorenz curve. They were invented in 1912 by Corrado Gini, an influential Italian statistician and demographer. The Gini coefficient is the ratio between: (1) the area between a Lorenz curve and the diagonal line (area A in figure 2.5) and (2) the area below the diagonal line (area A plus area B in figure 2.5).

The Lorenz curve for a perfectly equal distribution corresponding to the blue line in figure 2.5, which lies along the diagonal line. As a result, the area of polygon A for this curve is 0 and the Gini coefficient for a perfectly equal distribution is 0.

The Lorenz curve for a perfectly unequal distribution represented by the orange line in Figure 2.5 lies along the **X-axis**. As a result, the area of polygon A for this curve is equal to total area below the diagonal line and the Gini coefficient is 1 for a perfectly unequal distribution. The Gini coefficient for a real population is generally between 0 and 1.[4] The closer the Gini coefficient is to 1, the greater the inequality.

Table 2.13 computes the Gini coefficient for the intermediate income distribution represented by the red line in figure 2.5. The table computes the area for polygon B in figure 2.5. This area is subtracted from the total area below the diagonal line to determine the area in polygon A, which is divided by the total area under the diagonal line to compute the Gini coefficient.

TABLE 2.13   **Computing a Gini Coefficient**

Polygon	Household Percent		Income Percent		Polygon		
	Low	High	Low	High	Base	Average Height	Area
(1)	(2)	(3)	(4)	(5)	(6)	(7)	(8)
1	0	40	0	5	40	2.5	100.0
2	40	80	5	35	40	20.0	800.0
3	80	100	35	100	20	67.5	1,350.0
Total	—	—	—	—	—	—	2,250.0

| TABLE 2.14 | Gini Coefficients for Decatur, DeKalb County, the State of Georgia, and the United States, 2010 |

Area	Gini Coefficient
Decatur	0.476
DeKalb County	0.475
Georgia	0.464
United States	0.453

Source: Computed by the authors from data in US Bureau of the Census (2015c).

Polygon B is first divided into three polygons: (1) polygon one, containing the lowest 40 percent of the population; (2) polygon two, containing the population between 40 and 80 percent of the total population, and (3) polygon three, containing the top 20 percent of the population. Each polygon's area is computed by multiplying its base by the average of its sides, as shown below.

Columns two and three of table 2.13 record the low and high household income values for each polygon, as plotted in figure 2.5. Columns four and five record the low- and high-income percentages for each polygon plotted in figure 2.5. Column five, the base for each polygon, is equal to the high-income percentage in column three minus the low household percentage in column two. Column seven, the average height of each polygon, is the average of the income percentage values in columns four and five. The polygon areas in column eight are equal to the polygon base in column six multiplied by the average height in column seven. The sum of these values, 2,250, is the area of polygon B in figure 2.5.

The area under the diagonal line in figure 2.5 is a right triangle; its area is equal to one half of the base (100/2) multiplied by the height (100), or 5,000. The area of polygon A is equal to the area under the diagonal line (5,000) minus the area of polygon B (2,250) or 2,750. The Gini coefficient is equal to the area of polygon A (2,750) divided by the area under the diagonal line (5,000) or 0.550.

Table 2.14 records the Gini coefficient for Decatur, DeKalb County, the State of Georgia, and the United States. It reveals that Decatur's household income distribution is roughly equal to DeKalb County and Georgia and slightly less equal than that of the United States. Georgia's Gini coefficient is tenth largest of the fifty states in the United States. Gini coefficients for United States reveal that it has the most unequal income distribution in the developed world and has become increasingly unequal over the last forty years.[5]

Notes

[1] The following information on the City of Decatur and DeKalb County is taken from http://www.decaturga.com/, http://www.co.dekalb.ga.us/, and http://en.wikipedia.org/wiki/DeKalb_County,_Georgia, accessed on December 1, 2014.

[2] The spatial interpolation methods described in chapter 6 use the data for one set of boundaries (e.g., the 2000 block group boundaries) to estimate the data for a second set of boundaries (the 2010 block group boundaries in this case).

[3] Center for Family and Demographic Research (n.d.) provides directions for preparing population pyramids in Microsoft Excel.

[4] Gini coefficients can exceed one if some members of the population have negative income or wealth.

[5] https://en.wikipedia.org/wiki/List_of_U.S._states_by_Gini_coefficient; https://en.wikipedia.org/wiki/List_of_countries_by_income_equality.

Trend Projection Methods 3

Chapter 2 described the past and existing conditions for Decatur and DeKalb County, Georgia. This description is a good place to start, but planning is about the future: What is likely to happen, and how can purposeful and practical action now change the likely future into something that citizens and their elected officials consider more desirable? This chapter and chapter 4 present several techniques that use historical data to project the future.

An overwhelming array of methods is described in the literature for using data from the past and present to consider what the future may be. The methods range from the simple methods considered in this chapter and chapter 4 to complex structural models incorporating dozens of equations that attempt to represent the complex interplay among household demographics, the price and supply of land and buildings, local and regional economic conditions, local land-use regulations, and how these variables interact over time and space.[1] This chapter introduces trend projection methods and uses them to project the 2040 population for the City of Decatur and DeKalb County, Georgia, described in chapter 2. Chapter 4 introduces share projection methods and uses them to project the 2040 population of Decatur and its block groups.

This chapter contains three major sections. The first section defines some of the concepts and terms that will be used in this chapter and the remainder of this book. The second section introduces trend projection methods and four widely used trend projection curves. The final section uses the trend projection methods to project the 2040 population of Decatur and DeKalb County and illustrates procedures for selecting the most appropriate projection, or projections.

Concepts and Terminology

Many different terms can be used to describe attempts to use information from the past and present to consider what the future may be. Although the terms are often used interchangeably, it is important to define them carefully to avoid confusion. The following definitions will be used in this book.

Estimates, Projections, and Forecasts

An **estimate** is an indirect measure of a present or past condition that could be measured directly but isn't because doing so would be too difficult or expensive given the limitations of time and other resources.[2] For example, the current population of a city could be estimated from voter registration records when a complete population count is impractical. The US Bureau of the Census's Population Estimates Program produces annual population estimates for the United States, its states, counties, cities, and towns, and for the Commonwealth of Puerto Rico and its *municipios*. The estimates are used to allocate billions of federal dollars to states and sub-state areas.[3] The US Bureau of the Census (or "Census Bureau") also produces a wide array of multi-year population and housing estimates through the American Community Survey described in appendix B.

A **projection** is a conditional ("what if") statement about the future. Projections calculate *what* the future will be *if* their underlying assumptions about the future are correct. For example, an analyst may state that "if current birth, death and migration rates continue, the city's population will exceed 250,000 in the year 2040." In making this projection, the analyst is not claiming that current rates will continue or that the county's population will exceed 250,000 in 2040. She is not predicting *what* will happen, only stating what would happen *if* the underlying projection assumptions—stated and unstated—prove to be right. If the projected future does not occur, it is not because the analyst made a mistake, it is because the projections underlying assumptions were incorrect. Since the analyst never claimed that the assumptions would be right, she has made no error if the projected future does not arrive.

A **forecast** is a judgmental statement of what the analyst believes to be the most likely future. Unlike merely stating what would happen if a set of assumptions are satisfied, forecasters accept the responsibility for evaluating alternative "ifs" and selecting those that are most likely to occur. Given this information, forecasters state what they believe the future will be.

The distinction between conditional projections of what the future may be, given a set of assumptions, and judgmental forecasts of what the future will be is often ignored in practice. Planners and their clients routinely treat the projections prepared by others as forecasts without understanding their conditional nature and the need to evaluate their underlying assumptions. More troubling, analysts may prepare projections knowing that they will be used as forecasts, without evaluating their underlying assumptions to determine whether their projections are reasonable. This book will consider methods for preparing projections, not forecasts.

Projection Terminology

Figure 3.1 illustrates the key terms that will be used to describe the analysis and projection methods in this book. The figure illustrates the graphical projection method in which an analyst plots the historical data, visually fits a line to the data, and extends the line by hand to project future values. For example, in figure 3.1 the observed population for DeKalb County, Georgia, for 1950 through 2010 has been plotted

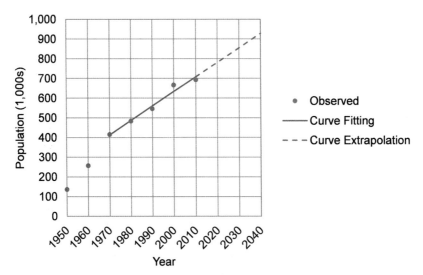

Figure 3.1 Graphical Projection: DeKalb County Population, 2040

Source: Minnesota Population Center (n.d.).

and a solid red line has been drawn that appears to fit the data for 1970 through 2010. The dashed red line extends this curve to project the county's population for 2020, 2030, and 2040.

Graphic conventions assume that the **independent variable** is plotted on the horizontal (or X) axis and the **dependent variable** is plotted on the vertical (or Y) axis. This suggests in figure 3.1 that the county's population is dependent on (or "caused by") time. Clearly, this is not the case. Population change reflects the aggregate effects of three other factors—births, deaths, and migration—that are time-related and are caused, in turn, by other time-related factors such as changes in an area's health levels and economic conditions. As a result, the time dimension serves as a composite variable reflecting the net effect of many unmeasured events and causal processes.

Population Value. An area's population in year t can be represented as P^t. For example, in figure 3.1 the population of DeKalb County in 1950, P^{1950}, is approximately 140,000.

Observation Period. The observation period is the time period between the first and last years for which data are available. For example, in figure 3.1 the observation period is 1950–2010.

Base Year. The base year, b, is the first year used to project future values. For example, in figure 3.1 the base year is 1970 and the base year population, P^{1970}, is roughly 410,000. Figure 3.1 demonstrates that the base year is not always the beginning of the observation period.

Launch Year. The launch year, l, is the last year used to project the future population. For example, in figure 3.1, the launch year is 2010 and the launch year population, P^{2010}, is approximately 690,000. In this case, the launch year is equal to the last year of the observation period but that isn't always the case.

Target Year. The target year, ta, is the last projection year, that is, 2040 in figure 3.1.

Fit Period. The fit period is the time period used to fit the trend curve, that is, the period between the launch year and the target year. For example, in figure 3.1 the fit period is 1970–2010.

Projection Period. The projection period is the time period for which projected values are being computed, that is, the period between the launch year and the target year. For example, in figure 3.1 the projection period is 2010–2040.

Observation Interval. The observation interval is the time interval between the dates in the observation period, for example, ten years in figure 3.1.

Projection Interval. The projection interval is the time interval between the dates in the projection period, for example, ten years in figure 3.1. The projection interval may not equal the observation interval. For example, information collected every ten years by the Census Bureau's decennial population and housing census can be used project the population at one- or five-year intervals.

The trend projection methods in this chapter use explicit mathematical procedures to identify and extend the curve, or curves, that best fit the past and suggests what the future may be. These methods introduce the following concepts and terms.

Observed Values. Y is the observed value for the dependent variable when the independent variable is equal to X. For example, figure 3.2 and table 3.1 contain two linear curves with the general form

$$Y_c = a + bX. \hspace{3cm} 3.1$$

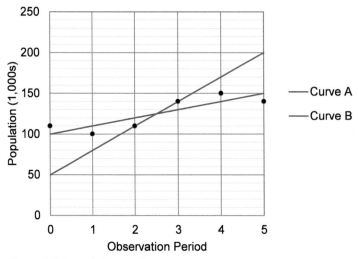

Figure 3.2 Sample Linear Curves

TABLE 3.1	Observed and Computed Values for Sample Linear Curves		

Observation Period	Observed Value	Computed Value	Residual Error
X	Y	Y_c	$Y - Y_c$
(1)	(2)	(3)	(4)
Curve A: $Y_c = 100 + 10X$			
0	110	100	10
1	100	110	−10
2	110	120	−10
3	140	130	10
4	150	140	10
5	140	150	−10
Curve B: $Y_c = 50 + 30X$			
0	110	50	60
1	100	80	20
2	110	110	0
3	140	140	0
4	150	170	−20
5	140	200	−60

Figure 3.2 and table 3.1 report that the observed population, Y, is 110 when X is 0; Y is 140 when X is 3.

Computed Values. Y_c is the computed value for a curve when the independent variable is equal to X. For example, for Curve A in figure 3.2 and table 3.1 the computed value, Y_c, is 100 when the observation period X is 0; Y_c is 130 when X is 3. For Curve B, the computed value is 50 when X is 0 and Y_c is 140 when X is 3.

Residual Error. The residual error is the difference between the observed value, Y, and the computed value, Y_c, for a given value of X. For example, for Curve A, the observed value, Y, is 110 when X is 0 and the computed value, Y_c, is 100; the residual error ($Y - Y_c$) is $110 - 100$ or 10.

Variables and Parameters. X and Y_c are variables that assume different values for a particular curve. Parameters remain fixed for a given curve and assume different values for different curves.

For example, the X and Y_c values for Curve A in figure 3.2 and table 3.1 are related by the linear relationship $Y_c = 100 + 10X$ in equation 3.1. Substituting different values for X into this equation yields Y_c values that lie along this curve. For example, when $X = 2$, $Y_c = 100 + 10(2) = 120$. The X and Y_c values for Curve B define a second linear curve, $Y_c = 50 + 30X$. Substituting X values into this equation yields Y_c values that are different from the values for Curve A. For example, for Curve B, when $X = 2$, $Y_c = 50 + 30(2) = 110$.

Y-Intercept. The Y-intercept is the value of the dependent variable, Y_c, when the independent variable, X, is 0. The Y-intercept for Curve A is 100; the Y-intercept for Curve B is 50.

Growth Increment. The growth increment is the change in the dependent variable, ΔY_c, for a unit change in the independent variable X. For example, Y_c is 100 when X is 0, and Y_c is 110 when X is 1; the growth increment (ΔY_c) is $110 - 100$ or 10.

Slope. The slope of a curve is the change in the dependent variable Y_c (ΔY_c) over a period divided by the change in the independent variable X (ΔX) over that period. That is,

$$\text{Slope} = \frac{\Delta Y_c}{\Delta X}. \qquad\qquad 3.2$$

For example, for Curve A, Y_c is 100 when X is 0; Y_c is 110 when X is 1. Substituting these values into equation 3.2,

$$\text{Slope} = \frac{\Delta Y_c}{\Delta X} = \frac{110 - 100}{1 - 0} = \frac{10}{1} = 10$$

Trend and Share Projection Methods

The trend and share projection methods provide two strategies for projecting the population, employment, or other time-related characteristic of a county, city, neighborhood, or other area (the **study area**) located in a larger area (the **context area**). Trend projection methods assume that the study area's past and future growth are so different from the context area that the study area can be treated as an isolated, unique entity. Given this assumption, the trend projection methods use the study area's past growth trends to project its future, without considering the context area in which it is located.

The share projection methods described in chapter 4 assume that the study area's future growth is closely tied to the growth of the context area. Given this assumption, they project the study area's future by relating it to the context area's projected growth and the observed relationship between the two areas' growth. The share projection methods recognize that projections for large areas such as a county, region, or state are generally more detailed and more reliable than projections for smaller areas. They are particularly useful when large-area projections are available and small-area projections are not.

The trend and share projection methods are not the only—and often not the preferred—projection technique. The cohort component technique described in chapter 5 is preferred for projecting an area's population by age, sex, and race. The economic projection methods described in chapter 6 can be used to project an area's economic activity by sector. Many other, more elaborate, projection techniques have been developed but they are rarely used by planners and other public policy analysts, and will not be discussed in this text.

Trend Projection Methods

Trend projection methods are the easiest and the most widely used methods for projecting the future. They will be used in this chapter to project the population of the City of Decatur and DeKalb County in 2040. However, they can also be used to project any variable for which consistent data are available at uniform points in time—employment, transit ridership, or fiscal expenditures—for any area—states, counties, cities, or neighborhoods. They are also widely used as components of more complex methods, for example, used to project mortality, fertility, and migration rates for the cohort-component projection methods described in chapter 5.

Trend projections are theoretically less appealing than projection techniques such as the cohort component methods that divide an area's population or economy into subcomponents and attempt to represent the underlying causes of past and future trends for each subcomponent. More complex projection methods also provide detailed answers to a wide range of policy questions and explicitly express causal relationships that are ignored by trend methods. However, complex projection methods often require information that is not available at the appropriate spatial or sectoral levels for a study area. As a result, trend projections are particularly useful when time and other resources are limited and for small areas where disaggregated population and employment data are not available. Equally importantly, an extensive body of research demonstrates convincingly that simple projection models are just as accurate as, and often preferable to, more complex models (Klosterman 2012).

The four trend curves described in the following section are more formal, explicit, and reproducible versions of the graphical projection method illustrated in figure 3.1. Both approaches are based on a simple two-stage process. The first stage, **curve fitting**, analyzes past data to identify an overall trend of growth or decline. The second stage, **curve extrapolation**, extends the past trend to project the future. The underlying procedure is identical—a curve is selected that most closely corresponds to the past data and is extended to project future values. However, the trend projection methods described in this chapter replace the intuitive process of visually fitting a curve with explicit quantitative criteria for identifying the best fitting curve. They also replace the manual extension of the trend line with formal computational procedures for continuing past trends. The most widely used criterion for identifying the best fitting line will be described after the linear curve is introduced.

Linear Curve

The simplest and most widely used trend curve is the **linear curve** described in equation 3.2:

$$Y_c = a + bX.$$

As shown in figure 3.3, the curve is linear because it plots as a straight line. The a parameter for the linear curve is equal to the Y-intercept, the value of Y_c when X is equal to 0.[4] For example, for Curve A in figure 3.2 and table 3.1, when $X = 0$, then $Y_c = 100 + 10(0) = 100$, the a parameter.

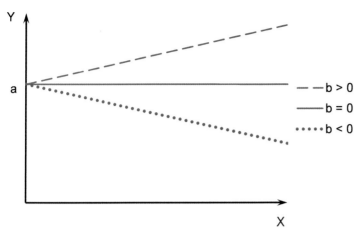

Figure 3.3 Alternate Forms of the Linear Curve

The b parameter for the linear curve is the slope. For example, for Curve A in figure 3.2 and table 3.1,

when $X = 0$, $Y_c = 100 + 10(0) = 100$;

when $X = 1$, $Y_c = 100 + 10(1) = 110$; and

when $X = 2$, $Y_c = 100 + 10(2) = 120$.

Substituting into equation 3.2, the slope for the first growth increment is equal to

$$\text{Slope} = \frac{\Delta Y_c}{\Delta X} = \frac{110 - 100}{1 - 0} = \frac{10}{1} = 10.$$

Similarly, the slope for the second growth increment is equal to

$$\text{Slope} = \frac{\Delta Y_c}{\Delta X} = \frac{120 - 110}{2 - 1} = \frac{10}{1} = 10.$$

That is, the linear curve has a constant slope equal to the b parameter.

The linear curve plots as a straight line because it has a constant slope. As figure 3.3 illustrates, it increases without limit if the b parameter is greater than 0 and decreases without limit if the b parameter is less than 0. If the b parameter is equal to 0 the curve is a horizontal line with a constant value equal to the a parameter.[5]

The constant incremental growth assumed by the linear curve is rarely appropriate for demographic and economic phenomena. Newly developing regions generally grow rapidly with increasing large incremental changes, but this growth eventually levels off as the region matures, leading to declining incremental changes. Nevertheless, Isserman (1977) suggests that the linear curve provides the most accurate projections for small, slow-growing areas. In any case, analysts cannot merely assume that the linear curve is the most appropriate model without considering the other curves described in this chapter.

Identifying the Best Fitting Trend Curve

An area's growth rarely fits a linear curve (or any of the trend curves) precisely. As a result, a decision rule must be used to identify the curve parameters that best fit a data set. The most widely used decision rule is the **least squares criterion** that minimizes the **sum of squared residual errors**, $(Y - Y_c)^2$. Squared residual errors are used because the residual errors, $Y - Y_c$, will be positive in some cases and negative in others. Adding these positive and negative values may result in cancellations that mask the extent of the actual deviations.

TABLE 3.2		Error of Estimates for Sample Linear Curves		
Observation Period	Observed Value	Computed Value	Residual Error	Squared Residual Error
X	Y	Y_c	$Y - Y_c$	$(Y - Y_c)^2$
(1)	(2)	(3)	(4)	(5)
Curve A: $Y_c = 100 + 10X$				
0	110	100	10	100
1	100	110	−10	100
2	110	120	−10	100
3	140	130	10	100
4	150	140	10	100
5	140	150	−10	100
Sum	—	—	0	600
Curve B: $Y_c = 50 + 30X$				
0	110	50	60	3,600
1	100	80	20	400
2	110	110	0	0
3	140	140	0	0
4	150	170	−20	400
5	140	200	−60	3,600
Sum	—	—	0	8,000

For example, table 3.2 reports the observed values, computed values, and residual errors for the two linear curves in figure 3.2. While Curve A clearly provides the better fit, the sum of the residual errors in column four is 0 for both curves, suggesting that they are equally appropriate. The squared residual errors in column five eliminate all negative numbers, ensuring that the total error between the observed and estimated values is as small as possible. For example, the sum of squared residual errors in column five is dramatically smaller for Curve A than for Curve B, indicating that it is clearly the better curve—as visual inspection of figure 3.2 suggests.

Geometric Curve

The **geometric curve** equation is

$$Y_c = ab^X.$$ 3.3

Geometric curves describe the compound growth of phenomena measured in discrete time intervals, for example, an area's population or employment reported by month or year. Geometric curves are similar to exponential curves that have the general form

$$Y_c = ae^{bX},$$

where e is a natural number equal to 2.71828. Exponential curves describe the instantaneous rate of compound growth that many natural processes approach as the intervals between the X variables become increasingly smaller. Because planners generally deal with data measured at discrete points in time, only the geometric curves will be considered here.

Geometric curves describe phenomena that grow by a constant **growth rate**, r. The growth rate is defined as the change in the dependent variable, Y, for a time period divided by the Y value at the beginning of the period. That is, the growth rate, $r^{t,t+1}$, for any period from time t to time $t+1$ is equal to the incremental growth for the period, $Y^{t+1} - Y^{t}$, divided by the initial value, Y^{t}, or

$$r^{t,t+1} = \frac{Y^{t+1} - Y^{t}}{Y^{t}}. \qquad\qquad 3.4$$

As figure 3.4 indicates, the a parameter for the geometric curve is equal to the Y-intercept.[6] The b parameter is equal to 1 plus the growth rate, r. As a result, equation 3.3 can be rewritten as equation 3.5 where r is the growth rate.

$$Y_{c} = a\,(1+r)^{X}. \qquad\qquad 3.5$$

As figure 3.4 shows, the geometric curve increases without limit if b is greater than 1. If b is between 0 and 1, the geometric curve approaches a Y value of 0 as X increases. And if b is equal to 1, the geometric curve is equal to a for all values of X.[7]

Table 3.3 illustrates the difference between the constant incremental growth of the linear curve and the constant growth rate of the geometric curve. The two curves have the same a parameter and identical values when X is 0 or 1. However, the linear curve has equal growth increments (column three) and decreasing growth rates (column four) for all values of X. The geometric curve has a constant growth rate (column four) and increasing growth increments (column three) for all values of X.

Examples of geometric growth include money deposited in a savings account yielding a constant interest rate and biological populations whose growth is not limited by resource constraints. For example, the current value of $100 deposited in a savings account yielding 6 percent interest per year can be expressed as equation 3.5 by substituting the initial value ($100) for the a parameter and the interest rate expressed as a decimal for r. That is,

$$Y_{c} = a\,(1+r)^{X},$$

$$Y_{c} = \$100(1.00 + 0.06)^{X},$$

or

$$Y_{c} = \$100(1.06)^{X}.$$

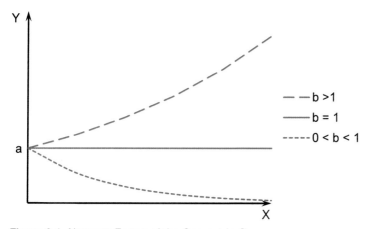

Figure 3.4 Alternate Forms of the Geometric Curve

TABLE 3.3	Comparison of the Linear and Geometric Curves

Linear Curve: $Y_c = 100 + 20X$

Observation Period	Linear Value	Growth Increment	Growth Rate
X	Y_c^t	$Y_c^{t+1} - Y_c^t$	$\left(Y_c^{t+1} - Y_c^t\right) / Y_c^t$
(1)	(2)	(3)	(4)
0	100	—	—
1	120	20	0.2
2	140	20	0.167
3	160	20	0.143
4	180	20	0.125
5	200	20	0.111

Geometric Curve: $Y_c = 100(1.2)^X$

Observation Period	Geometric Value	Growth Increment	Growth Rate
X	Y_c^t	$Y_c^{t+1} - Y_c^t$	$\left(Y_c^{t+1} - Y_c^t\right) / Y_c^t$
(1)	(2)	(3)	(4)
0	100	—	—
1	120	20	0.2
2	144	24	0.2
3	172.8	28.8	0.2
4	207.4	34.6	0.2
5	248.8	41.4	0.2

The saving account's current value at any point in time can be computed by substituting the appropriate X value into this equation. For the initial deposit,

$$X = 0 \text{ and } Y_c = \$100(1.06)^0 = \$100.00.$$

After the first year,

$$X = 1 \text{ and } Y_c = \$100(1.06)^1 = \$106.00.$$

After the second year,

$$X = 2 \text{ and } Y_c = \$100(1.06)^2 = \$112.36; \text{ and so on.}$$

The growth rate r for the first year is equal to

$$r = \frac{(\$106.00 - \$100.00)}{\$100.00} = \frac{\$6.00}{\$100.00} = 0.06 \text{ or } 6\%$$

and the growth rate for the second year is

$$r = \frac{(\$112.36 - \$106.00)}{\$106.00} = \frac{\$6.36}{\$106.00} = 0.06 \text{ or } 6\%.$$

That is, the money is growing at a constant growth rate of 6 percent, equal to the interest rate.

Geometric projections assume that growth or decline will continue at a constant rate. This assumption is often reasonable because demographic processes generally

Box 3.1

Malthus's Theory of Population Growth

Geometric growth is an essential part of the influential theory of population growth developed by Thomas Malthus (1766–1834). Writing in response to the belief that the human condition could be perfected, Malthus argued that unrestrained populations increase at a geometric rate while the ability of the Earth to support them will grow at a linear rate. As figures 3.3 and 3.4 show, the geometric growth of an increasing population inevitably outstrips linear growth. For Malthus, this implied that sustained population growth eventually

harmed society, particularly its least well-off members. Malthus argued that population growth could only be restrained by "positive" checks such as hunger, disease, and war that raised the death rate, and "preventative" checks such as abortion, birth control, prostitution, and celibacy that lowered the birth rate. Fortunately, Malthus seriously underestimated the ability of technological advances to support much higher population and living levels than could be imagined nearly three centuries ago.

tend to approach a fixed or intrinsic growth rate. However, because geometric growth continues by increasingly larger increments if b is greater than 1, the geometric curve may not account for the fact that growth will eventually be restricted by congestion and resource constraints. Thus, while possibly appropriate for short-term projections for rapidly growing regions, the geometric curve often produces unrealistically high projections for the long term.

Parabolic Curve

The **parabolic curve** has the general formula:

$$Y_c = a + bX + cX^2. \qquad 3.6$$

The parabolic curve is a member of a larger family of polynomial curves that share the general form:

$$Y_c = a(X)^0 + b(X)^1 + c(X)^2 + d(X)^3 + \cdots + m(X)^n.$$

The linear curve (or first degree) curve has the general formula:

$$Y_c = a(X)^0 + b(X)^1 \text{ or}$$

$$Y_c = a + bX.$$

The parabolic (or second-degree) curve has the general formula:

$$Y_c = a(X)^0 + b(X)^1 + c(X)^2 \text{ or}$$

$$Y_c = a + bX + cX^2.$$

The third-degree (or cubic) curve has the general formula:

$$Y_c = a(X)^0 + b(X)^1 + c(X)^2 + d(X)^3 \text{ or}$$

$$Y_c = a + bX + cX^2 + dX^3; \text{ and so on.}$$

The "degree" of a polynomial curve is equal to the highest exponent on the independent variable, X, and the number of parameters is equal to one plus the curve degree.

As figure 3.5 indicates, the parabolic curve has a constantly changing slope and one bend. Given a sufficient range of X values, the parabolic curve is positively inclined (or "upward sloping") in one section and negatively inclined (or "downward sloping") in another section.

The a parameter for the parabolic curve is equal to the Y-intercept.[8] The b parameter is equal to the slope at the Y-intercept. When c is positive, the curve is concave upward. When c is negative, the curve is concave downward. And when c is 0 the parabolic curve becomes a linear curve with a Y-intercept of a and a slope of b.[9]

As the top portion of table 3.4 indicates, when the c parameter for the parabolic curve is greater than 1, the growth increments increase and the growth rates decrease

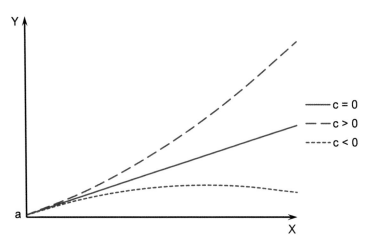

Figure 3.5 Alternate Forms of the Parabolic Curve

TABLE 3.4	Sample Parabolic Curves		
Observation Period	Parabolic Value	Growth Increment	Growth Rate
X	Y_c^t	$Y_c^{t+1} - Y_c^t$	$\left(Y_c^{t+1} - Y_c^t\right) / Y_c^t$
(1)	(2)	(3)	(4)
$Y_c = 100 + 20X + 1.2X^2$			
0	100.0	—	—
1	121.2	21.2	0.2120
2	144.8	23.6	0.1947
3	170.8	26.0	0.1796
4	199.2	28.4	0.1663
5	230.0	30.8	0.1546
$Y_c = 100 + 20X + 0.8X^2$			
0	100.0	—	—
1	119.2	19.2	0.1920
2	136.8	17.6	0.1477
3	152.8	16.0	0.1170
4	167.2	14.4	0.0942
5	180.0	12.8	0.0766

as X increases. In this case, the parabolic curve grows less rapidly than the geometric curve, which has a constant growth rate. As the lower part of table 3.4 indicates, if the c parameter is less than 1, the growth increments and growth rates both decrease as X increases. As a result, this version of the curve may be appropriate for slowly growing areas. The parabolic curve perfectly fits two or three observations; just as the linear curve perfectly fits two observations.

Gompertz Curve

The linear, geometric, and parabolic curves assume that growth or decline will continue without limit. While these trends may continue for some time, it is extremely unlikely that they will continue forever. At some point an area's growth will be impeded by a lack of adequate resources, public facilities, and other amenities. Declining communities rarely disappear entirely; rather they decline only until they reach population and employment levels appropriate to their current position in the regional or national economy. In both cases it seems reasonable to assume that long-term growth or decline will continue at a decreasing rate until the region's population or employment approaches an upper or lower limit.

The **Gompertz curve**, developed by Benjamin Gompertz (1779–1865), an English actuary and mathematician, is an **asymptotic growth curve** that recognizes that a region's population or employment will eventually approach an upper or lower growth limit or **asymptote**.[10] The Gompertz curve equation is

$$Y_c = ca^{bX}. \qquad\qquad 3.7$$

That is, the dependent variable, Y_c, is equal to the c parameter multiplied by the a parameter raised to the b^X power.

As shown in figure 3.6, the Gompertz curve assumes different shapes depending on the values of the a and b parameters. When the a parameter is less than 1 and the b parameter is between 0 and 1, the Gompertz curve is S-shaped with an upper limit of c for large positive values of X. For these parameter values, the Gompertz curve begins growing slowly, increases for a period, and then tapers off to approach an asymptotic upper limit. If a is greater than 1 and b is between 0 and 1, the Gompertz curve approaches an asymptotic lower limit for large values of X. The Y-intercept for the Gompertz curve is ca.[11]

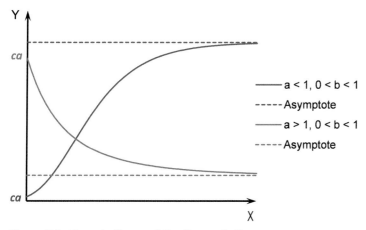

Figure 3.6 Alternate Forms of the Gompertz Curve

Using Trend Curves

Table 3.5 contains 16 different projections for DeKalb County's 2040 population that were computed with the 4 trend curves, 4 different fit periods, and a projected growth limit of 960,000 residents. The extremely low parabolic projection for the 1990–2010 fit period is clearly unlikely and can be ignored. The other projected values range from 758,993 to 1,076,160, a difference of more than 40 percent. All the projections are mathematically correct, that is, computed with the best fitting trend curves for each set of observed population values. Readily available trend projection software can easily generate many other projections by using other fit periods and deleting or adjusting questionable population values.[12]

As this example illustrates, the trend projection methods often require analysts to select the preferred projection, or projections, from perhaps dozens of computationally correct alternatives. It is unreasonable to expect them to know what the future will be, but their projections should be based on more than mere whim, wishful thinking, or arbitrary choice. Instead, planners' projections should incorporate their considered judgment on what the future is likely to be. Equally importantly, they should use generally accepted procedures to select and defend their projections.

Fortunately, several procedures can be used to select the most appropriate trend projection or set of projections. The procedures do not provide simple, clear-cut tests for identifying the single "best" projection. Instead, they consider different factors, examine different information, and often recommend different projections. However, they help guide the selection process and provide the basis for informed and defensible decision making about the unknowable future—which is all one can expect.

Five selection procedures will be considered: (1) plotting the observed data, (2) plotting the projected values, (3) defining a growth limit, (4) combining projections, and (5) evaluating the **goodness of fit** statistics.

Plotting the Observed Data

Plotting the observed data is the first, simplest, and most important procedure for preparing trend projections. Plots of the observed data are particularly useful for selecting the most appropriate fit period and for identifying trends and anomalies in the data that are extremely difficult to identify in a table of numbers.

For example, figure 3.7 shows that DeKalb County's population has grown steadily for more than one hundred years, as one would expect in a state that has grown consistently over this period. However, the fit period to be used in projecting the future isn't clear. Table 3.5 identifies four possible fit periods: 1970–2010,

| TABLE 3.5 | **Projected Population for Different Trend Curves and Fit Periods: DeKalb County, 2040** |

Trend Curve	Projected Population for the Fit Period (P^{2040})			
	1970–2010	1980–2010	1990–2010	2000–2010
(1)	(2)	(3)	(4)	(5)
Linear	928,328	932,641	926,644	769,977
Geometric	1,076,160	1,048,708	1,014,179	776,242
Parabolic	885,608	1,057,238	205,977	769,977
Gompertz	827,053	834,149	837,014	758,993

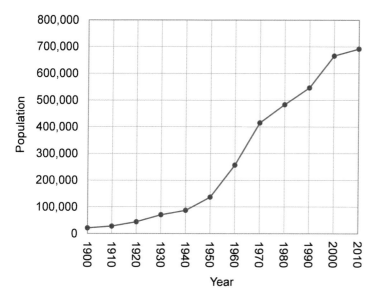

Figure 3.7 DeKalb County Population, 1900–2010

Source: Minnesota Population Center (n.d.).

1980–2010, 1990–2010, and 2000–2010. The projections for the 1970–2010 fit period in column two range from 827,053 for the Gompertz curve to 1,076,160 for the geometric curve. The projections for the 2000–2010 fit period in column five range from 758,993 for the Gompertz curve to 776,242 for the geometric curve. The projections for the other fit periods lie between these two extremes.

DeKalb County's reduced population growth between 2000 and 2010 is due in part to the global "Great Recession" of 2007–2009. The recession in the United States was the longest and most severe in sixty years. From the recession's peak in 2007 to its trough in 2009, the United States' gross domestic product fell by 4.3 percent; the country's unemployment rate doubled from 5 to 10 percent; home prices fell by approximately 30 percent on average, and the net worth of US households fell from approximately $69 trillion in 2007 to roughly $55 trillion in 2009 (Rich 2013). The dramatic decline in the national economy was reflected in substantial declines in birth rates and migration throughout the country and in DeKalb County's reduced population growth for the decade.

Figure 3.8 reveals somewhat surprisingly that Decatur's population was larger in 1950 through 1970 than it is today. This is remarkable because the city is in a county, region, and state that have grown steadily over the last century. This may be due in part to changes in the city's boundaries that often make it difficult to compare population values over long periods of time. However, Decatur has been surrounded by other municipalities for many years so its boundaries have changed little since the 1950s. As a result, other factors may account for Decatur's unusual population trends.

Decatur's surprising growth pattern can be explained in part by applying the **housing unit method** for estimating an area's current or past population.[13] The housing unit method is expressed in equation 3.8:

$$P^t = (\mathrm{HU}^t \times \mathrm{AHS}^t) + \mathrm{GQ}^t \qquad\qquad 3.8$$

where P^t = study area population in year t, HU^t = number of housing units in year t, AHS^t = average household size in year t, and GQ^t = group quarter population in year t.

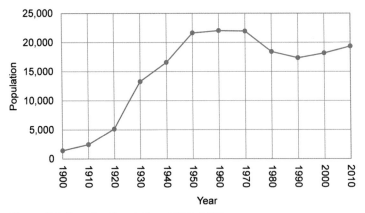

Figure 3.8 Decatur Population, 1900–2010

Source: Minnesota Population Center (n.d.).

A **housing unit** is a house, apartment, mobile home or trailer, a group of rooms, or a single room that is occupied as separate living quarters, or if vacant, is intended for occupancy as separate living quarters. The **household population** includes everyone who uses a housing unit as their usual place of residence. It is equal to the total population minus the **group quarters population,** which includes people residing in college dormitories, military quarters, nursing facilities, and correctional institutions. The **average household size** is the average number of persons per household, that is, the household population divided by the number of households.

Table 3.6 uses equation 3.8 to analyze Decatur's population trend from 1970 to 2010. Subtracting the group quarters population in the second row from the total population in the first row yields the household population values in the third row. Dividing the household population in the third row by the number of households in the fourth row yields the average household size in the fifth row.

Table 3.6 indicates that the number of households in Decatur has grown steadily over the past forty years and the group quarters population has fluctuated. This suggests that the city's population decline from 1970 to 1990 and its growth since 1990 are due in part to substantial changes in the average household size. The city's population declined between 1970 and 1990 primarily because the average household size declined by over 40 percent during this period, and the number of households grew by only 10 percent. This suggests that Decatur's future population may be determined partially by changes in its average household size.

TABLE 3.6 **Selected Population and Household Statistics: Decatur, 1970–2010**

Variable	Year				
	1970[a]	1980[a]	1990[a]	2000[b]	2010[c]
(1)	(2)	(3)	(4)	(5)	(6)
Total population	21,942	18,404	17,336	18,147	19,335
Group quarters population	919	703	856	989	691
Household population	21,023	17,701	16,480	17,158	18,644
Number of households	7,479	7,893	8,230	8,051	8,599
Average household size	2.81	2.24	2.00	2.13	2.17

Source:
[a] Minnesota Population Center (n.d.).
[b] US Bureau of the Census (2003b).
[c] US Bureau of the Census (2013c).

Plotting the Projected Values

Plotting the projected values is helpful for identifying curves that are inappropriate for a given data set and for determining whether a curve reasonably extends past trends. For example, as pointed out earlier, the parabolic curve always perfectly fits three observation values. As a result, as figure 3.9 illustrates, the DeKalb County parabolic population projections for the 1990–2010 fit period correspond exactly to the observed values. However, the graph also demonstrates that the parabolic curve is clearly inappropriate for this case because it projects a declining population from 2010 onward.

Extending trend curves beyond the target year is also useful for identifying long-term trends that are unreasonable in the light of past trends. For example, figure 3.10 extends the linear, geometric, and Gompertz DeKalb County population projections for the 1990–2010 fit period to the year 2070. It suggests that the geometric curve may be unacceptable because it projects more than a doubling of the county's population by 2070. The liner curve projection is likewise suspect in projecting that the county's population will grow by two-thirds over the next sixty years. The Gompertz curve is more reasonable in projecting that the county's population will grow by roughly a third over this period.

Defining Growth Limits

As has been pointed out earlier, the Gompertz curve assumes an asymptotic upper or lower limit that the projected values approach but never reach. As a result, projections using this curve must assume what the limit growth will be. The City of Decatur and DeKalb County illustrate two approaches for doing this.

Chapter 2 reported that Decatur had only thirty-one acres of vacant land, suggesting that the city can accommodate little additional development. The preceding analysis also reveals that the city's past growth has been dependent largely on changes in the number of households and the average household size. This suggests that the city's future growth may be determined largely by changes in these two factors.

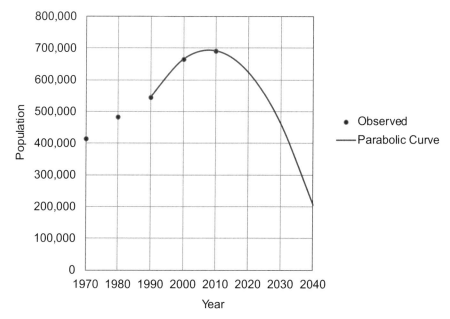

Figure 3.9 Parabolic Trend Projections: DeKalb County, 1990–2010

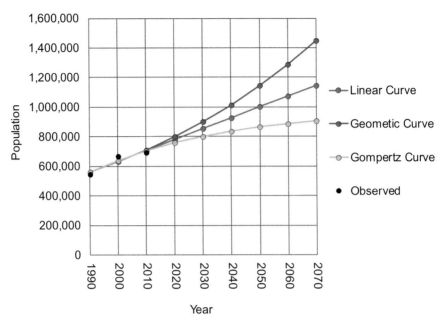

Figure 3.10 Trend Projections: DeKalb County Population, 1990–2070

Table 3.6 indicates that the average household size has never exceeded 2.81 and the group quarters population has consistently been below one thousand. Trend projects for the number of households suggest that the city will have less than ten thousand households in 2040. These values can be substituted into the housing unit population (equation 3.8) to project the upper limit for Decatur's 2040 population. That is,

$$P^{2040} = (HU^{2040} \times AHS^{2040}) + GQ^{2040}$$
$$= (10,000 \times 2.81) + 1,000 = 29,100.$$

This suggests that 29,100 is a reasonable upper limit for Decatur's future population growth.

Chapter 2 reported that DeKalb County had 77,380 acres of residential land and 30,626 acres of vacant land in 2010. Figure 3.7 reports that the county's residential population in that year was roughly 690,000. Dividing the residential population (690,000) by the residential land area (77,380) yields a residential density of 8.9 persons per residential acre. Assuming the county's vacant land is totally developed at this density, multiplying the currently vacant land (30,626 acres) by 8.9 persons per residential acre yields a projected upper limit for the residential population on currently vacant land of roughly 270,000. Adding this value to the county's current population of 690,000 suggests that the county's population will not exceed 960,000 people. This limit was used to compute the Gompertz projections in table 3.5.

Combining Projections

The proceeding discussion followed general practice by considering the trend curves in isolation. However, the forecasting literature suggests that combined projections are particularly useful for projecting county- and subcounty-level populations.[14] Combined projections are often less biased because they combine the data and assumptions of several projection methods, offsetting the errors in individual projections. Three methods for combining trend projections will be considered here: average projections, trimmed average projections, and composite projections.

Average Projections. As their name suggests, **average projections** average a number of individual projections. The averages can include any number of projections, different trend curves, both trend and share projections, or different fit periods. The projections may be weighted but equally weighted projections are generally used.

Trimmed Average Projections. **Trimmed average projections** compute an average projection that excludes questionable projections. For example, the trimmed average may exclude the highest and lowest projections or questionable projections.

Composite Projections. **Composite projections** use a combination of projection methods that have been found to perform better for areas with particular characteristics. Their use assumes that consistent patterns can be identified in the performance of different projection models under different situations, allowing the projections for a particular place to utilize the model, or models, that are most accurate for places with its characteristics. For example, Isserman (1977) compared models that projected the population of 1,777 subcounty units in Illinois and Indiana. Similarly, Rayer (2008) evaluated five projection methods that projected the population for the 2,482 counties in the United States with consistent boundaries from 1900 to 2000, utilizing sixty-three projection horizon/base period/projection period combinations.

It must be remembered that the performance of the projection methods in these comparative tests is not strictly applicable to other regions or different time periods. Nevertheless, this research suggests the following:[15]

- Averages generally produce more precise and less biased projections than individual trend methods.
- Trimmed average and composite methods are generally more accurate than individual or average projections. However, further research is required to identify methods that perform particularly well (or poorly) for places with particular characteristics and the best ways to combine these methods.
- There is a tendency to under-project in areas with declining populations and over-project in areas with increasing populations, which suggests that population changes tend to moderate over time, that is, regress toward the mean. Smith (1987, 999–1000) provides a number of reasons why this may occur.

Table 3.7 applies the average and trimmed average methods to the DeKalb County trend projections in table 3.5. The trimmed averages in column five are the

TABLE 3.7	Average and Trimmed Average Population Projections: DeKalb County, 2040				
Trend Curve	Projected Population for the Fit Period (P^{2040})				
	1970–2010	1980–2010	1990–2010	Trimmed Average	2000–2010
(1)	(2)	(3)	(4)	(5)	(6)
Linear	928,328	932,641	926,644	929,204	769,977
Geometric	1,076,160	1,048,708	1,014,179	–	776,242
Parabolic	885,608	1,057,238	205,977	–	769,977
Gompertz	827,053	834,149	837,014	832,739	758,993
Average	929,287	968,184	745,954	–	768,797
Trimmed average	880,330	883,395	881,829	881,851	768,797

averages of the values in columns two, three, and four. The trimmed average for the geometric curve is not reported because it exceeds the assumed upper limit of 960,000. The trimmed average is not reported for the parabolic curve because it includes the obviously incorrect projection for the 1990–2010 fit period.

Averages of the four trend curves for each fit period are recorded at the bottom of each column. The trimmed average for the 1970–2010 fit period does not include the geometric projection that exceeds the assumed upper limit. The trimmed average for the 1980–2010 fit period does not include the geometric and the parabolic projections that exceed the assumed upper limit. The trimmed average for the 1990–2010 fit periods does not include the geometric projection that exceeds the assumed upper limit and the obviously incorrect parabolic projection. The average of the trimmed averages in column five is the average of the trimmed averages for the 1970–2010, 1980–2010, and 1990–2010 fit periods. All the projected values are included in the trimmed average for the 2000–2010 fit period.

The trimmed average projections are preferable to the average values because they eliminate highly questionable projection values. The three trimmed averages for the 1970–2010, 1980–2010, and 1990–2010 fit periods are very similar, suggesting that if any of these trends continue, the county's 2040 population will be slightly more than 880,000. The projections for the 2000–2010 fit period consistently project a smaller population value of less than 770,000, reflecting the slower growth for this period, as revealed in figure 3.7.

Table 3.8 computes the average and trimmed average projections for Decatur's 2040 population using the 1990–2010 fit period and an assumed growth limit of 29,100. The projections are approximately equal for the linear, geometric, and Gompertz curves and substantially larger for the parabolic curve. The trimmed average excludes the highest and lowest projections, that is, it is the average of the linear and geometric trend values. The trimmed average projection suggests that Decatur's 2030 population will be approximately 22,500, 10.6 percent larger than its 2010 population.

Evaluating Goodness of Fit Statistics

Another approach evaluates the extent to which the different trend curves fit the observed values, on the assumption that the curve that best fits past trends will most accurately predict future trends. This assumption does not always hold but this criterion provides a reasonable basis for evaluating different trend curves.[16]

| TABLE 3.8 | Projected Population for Four Trend Curves and the 1990–2010 Fit Period: Decatur, 2040 |

Trend Curve	2040 Population (P^{2040})
(1)	(2)
Linear	22,271
Geometric	22,707
Parabolic	25,161
Gompertz	21,793
Average	**22,983**
Trimmed average	**22,489**

A surprisingly large number of error statistics are described in the forecasting literature.[17] Four widely used statistics will be described here: (1) the mean error, (2) the mean absolute error, (3) the mean percentage error, and (4) the mean absolute percentage error.

Mean error. As its name suggests, the **mean error** (**ME**) is the average residual error, recognizing positive and negative residual errors. That is,

$$ME = \frac{\sum(Y - Y_c)}{N},$$

3.9

where Y = observed value, Y_c = estimated value, and N = number of observations.

Table 3.9 computes the ME and the other goodness of fit measures for Decatur's linear curve population estimates for the 1990–2010 fit period. The residual error terms in column four are the observed values in column two minus the estimated values in column three. The ME is the sum of these values divided by the number of observations. That is,

$$ME = \frac{\sum(Y - Y_c)}{N} = \frac{0}{3} = 0.00.$$

As this example illustrates, positive and negative residual errors may cancel each other, yielding an ME of zero. As a result, the ME is a poor measure of the total deviation between the observed and estimated values. However, it is a good measure of **bias**, the extent to which the curve estimates are, on average, larger or smaller than the observed values. The zero ME value indicates that, on average, the linear curve does not over- or under-estimate the observed population values.

Mean absolute error. The **mean absolute error** (**MAE**) is the average error, ignoring signs. That is,

$$MAE = \frac{\sum(|Y - Y_c|)}{N},$$

3.10

where the difference between the observed and estimated values is expressed as an absolute (or positive) value.

The absolute error terms in column five of table 3.9 are the absolute values of the error terms in column four. The MAE is the average of these values:

$$MAE = \frac{\sum(|Y - Y_c|)}{N} = \frac{252}{3} = 84.00.$$

TABLE 3.9	Computing Goodness of Fit Statistics: Decatur Linear Population Estimates for the 1990–2010 Fit Period

Year	Observed Population	Linear Estimate	Residual Error	Absolute Error	Percentage Error	Absolute Percentage Error
X	Y	Y_c	$Y - Y_c$	$|Y - Y_c|$	$(Y_c - Y)/Y$	$|Y_c - Y|/Y$
(1)	(2)	(3)	(4)	(5)	(6)	(7)
1990	17,336	17,273	63	63	0.36%	0.36%
2000	18,147	18,273	−126	126	−0.69%	0.69%
2010	19,335	19,272	63	63	0.33%	0.33%
Sum	—	—	0	252	0.00%	1.38%

The MAE evaluates the total estimation error, regardless of whether the estimates are too large or too small. As a result, it provides a good measure of the total variation between the observed and estimated values.

The ME and the MAE consider the total difference between the observed and estimated values. They do not consider the how large the errors are, relative to the values being considered, which can be very important. For example, an error of 1,000 is much more important when the observed value is 5,000 (a 20 percent error) than it is when the observed value is 500,000 (a 0.2 percent error). The MPE and the MAPE both consider the size of the estimation error.

Mean percentage error. The **mean percentage error** (**MPE**) is the average percentage error, considering signs. That is,

$$\text{MPE} = \frac{\sum\left(\frac{Y - Y_c}{Y} \times 100\right)}{N},$$
3.11

where the numerator is the sum of the percentage errors.

The percentage error terms in column six of table 3.9 are the residual error terms in column four divided by the observed values in column two, multiplied by 100. For example, the percentage error for 1990 is:

$$\frac{(Y - Y_c)}{Y} \times 100 = \frac{(17,336 - 17,273)}{17,336} \times 100 = 0.0036 \times 100 = 0.36\%.$$

The MPE is the average of the percentage errors. That is,

$$\text{MPE} = \frac{\sum\left(\frac{Y - Y_c}{Y} \times 100\right)}{N} = \frac{0.00\%}{3} = 0.00\%.$$

This indicates that, on average, the linear curve estimates are neither larger nor smaller than the observed values.

The MPE allows positive and negative errors to offset each other, which means that it provides a poor measure of the total error. However, like the ME, it provides a good measure of bias, that is, whether, on average, the estimates are larger or smaller than the observed values. It is particularly useful because it is dimensionless and is not affected by the number of observations, making it appropriate for comparing estimates for different data sets and different numbers of observations.

Mean absolute percentage error. The **mean absolute percentage error** (**MAPE**) is the average percentage error, ignoring signs. That is,

$$\text{MAPE} = \frac{\sum\left(\frac{|Y - Y_c|}{Y} \times 100\right)}{N}.$$
3.12

The absolute percentage error terms in column seven of table 3.9 are the absolute values of the percentage error terms in column six. The MAPE is the average of the absolute percentage errors. That is,

$$\text{MAPE} = \frac{\sum\left(\frac{|Y - Y_c|}{Y} \times 100\right)}{N} = \frac{1.38\%}{3} = 0.46\%.$$

This indicates that, on average, the linear curve estimates are slightly larger than the observed values.

The MAPE evaluates the total estimation error, regardless of sign, providing a good measure of the total variation between the observed and estimated values. It is particularly useful because it is dimensionless and is not affected by the number of observations.

Evaluating the DeKalb County and Decatur Trend Projections

Figure 3.7 suggests four possible fit periods for projecting DeKalb County's population: 1970–2010, 1980–2010, 1990–2010, and 2000–2010. Table 3.7 compares the county's projected 2040 population for the four trend curves and the average and trimmed average projections. The average of the trimmed average values suggests that a projected value of approximately 882,000 for the 1970–2010, 1980–2010, and 1990–2010 fit periods and roughly 769,000 if the 2000–2010 trend continues.

Figure 3.8 suggests that 1990–2010 provides the most appropriate fit period for projecting Decatur's population. Figure 3.11 plots the linear, geometric, parabolic, Gompertz projections for Decatur's population to the year 2070 for this fit period. It reveals that the parabolic projection exceeds the assumed growth limit of 29,100 by the year 2060, suggesting that it may not be appropriate for projecting the 2040 population. It also reveals that all four curves project a 2050 population larger than the city's historic maximum, 22,026.

Table 3.10 records the goodness of fit statistics for the Decatur projections for the 1990–2010 fit period plotted in figure 3.11 and reported in table 3.8. The average statistics are computed for the average of the four trend curves. The trimmed averages statistics are computed for the average of the linear, geometric, and Gompertz curves; the parabolic curve is not included because it exceeds the assumed upper limit in 2060, as figure 3.11 indicates.

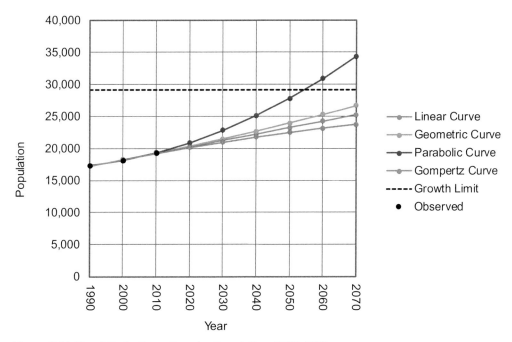

Figure 3.11 Trend Projections: Decatur Population, 1990–2070

TABLE 3.10	Goodness of Fit Statistics: Decatur Population Estimates for the 1990–2010 Fit Period

Projection	Goodness of Fit Statistic			
	ME	MAE	MPE	MAPE
(1)	(2)	(3)	(4)	(5)
Linear	0.00	84.00	0.00%	0.46%
Geometric	0.33	71.67	0.00%	0.39%
Parabolic	0.00	0.00	0.00%	0.00%
Gompertz	−0.33	97.67	0.00%	0.54%
Average	**0.00**	**63.33**	**0.00%**	**0.35%**
Trimmed average	**0.00**	**84.44**	**0.00%**	**0.46%**

Note: ME, mean error; MAE, mean absolute error; MPE, mean percentage error; MAPE, mean absolute percentage error.

The goodness of fit statistics are the lowest or second lowest for the average projections. However, these values are distorted by the questionable parabolic estimates that exactly fit the observed values for three observations. As a result, figure 3.11 and table 3.10 suggest that the trimmed average projection of roughly 24,500 may be the preferred projection for Decatur's 2040 population.

Notes

[1] See, e.g., De Gooijer and Hyndman (2006), Wilson and Rees (2005), and Miller et al. (1999).

[2] The following discussion draws on Isserman (1984, 208–209).

[3] See US Bureau of the Census (2017) for further information on the Census Bureau's annual population and housing unit estimates.

[4] This can be demonstrated by setting X to 0 in the linear curve equation, i.e., $Y_c = a + b(0)$, or a for all values of b.

[5] This can be demonstrated by setting b to 0 in the linear curve equation, i.e., $Y_c = a + (0)X$, or a, for all values of X.

[6] This can be shown by setting X equal to 0 in the geometric curve to yield $Y_c = ab^0$. Any number raised to the 0 power is equal to 1 so $b^0 = 1$ and $Y_c = a(1)$ or a for any value of the b parameter.

[7] The geometric curve is undefined for negative values of the b parameter because the curve is not continuous in this case. If b is negative, odd values of X yield negative geometric values because negative numbers raised to an odd power are negative; e.g., if $b = -1$, $b^3 = (-1)^3$ or -1. If b is negative, even values of X yield positive geometric values because negative numbers raised to an even power are positive; e.g., if $b = -1$, $b^2 = (-1)^2$ or 1. As a result, the geometric curve values flip flops around the X-axis when b is negative.

[8] This can be demonstrated by substituting an X value of 0 into the parabolic curve equation to yield $Y_c = a + b(0) + c(0)^2$ or a for all values of the a, b, and c parameters.

[9] This can be demonstrated by substituting a value of 0 for the c parameter in the parabolic curve to yield $Y_c = a + bX + 0X^2$ or $Y_c = a + bX$, the equation for a linear curve with a slope of b.

[10] The logistic curve is another widely used asymptotic growth curve. However, unlike the Gompertz curve, it assumes that declining series always approach zero, limiting its usefulness for declining areas. The logistic curve and the modified exponential curve, another asymptotic growth curve, are described in Klosterman (1990, 19–29).

[11] The Gompertz curve $Y_c = ca^{b^X}$ approaches an asymptotic value of c when b is less than 1, because: (1) b^X approaches 0 for increasingly large values of X; as a result, (2) a^{b^X} approaches a^0 or 1; and, therefore, (3) ca^{b^X} approaches $c(1)$ or c. If a is less than 1, a^{b^X} approaches 1 from below and ca^{b^X} approaches c as an upper limit. If a is greater than 1, a^{b^X} approaches 1 from above and ca^{b^X} approaches c as a lower limit.

[12] See, e.g., the free, open source models available at PlanningSupport.org.

13 A comprehensive review and evaluation of the housing unit method of population estimation is provided in Smith (1986).

14 The following discussion draws on Smith et al. (2001, 328–331) and Rayer and Smith (2010, 154–155).

15 These results are drawn from Smith et al. (2001, 307–331) and Rayer (2008).

16 The forecasting literature suggests that the ability to fit historical data provides a poor basis for selecting variable, specifying relationships, or selecting the functional forms for complex forecasting models (Armstrong 2001a, 460–462).

17 See, e.g., Armstrong and Collopy (1992) and Armstrong (2001b). The discussion here draws on Smith et al. (2001, 302–307).

Share Projection Methods 4

The trend projection methods described in chapter 3 assume that a study area's past and future growth or decline are so different from the **context area** in which it is located that the study area can be treated as an isolated entity. Given this assumption, the trend projection methods use the study area's past growth trends to project its future, without considering the context area in which it is located. The share projection methods described in this chapter assume that the study area's future growth or decline will be closely tied to the growth or decline of the context area in which it is located. Given this assumption, they project the study area's future by relating it to the context area's projected change and the observed relationship between the two areas' growth trends.

Share methods are easy to apply and have limited data requirements, which makes them particularly useful for projecting the growth or decline of small areas and the components of a population or an economy. The methods recognize that projections for large areas such as a county, region, or state are generally more detailed and more reliable than projections for smaller areas. As a result, they are particularly useful when large-area projections are available and small-area projections are not.

This chapter has three sections. The first section uses four share projection methods and the projected population of DeKalb County to project Decatur's 2040 population. The second section uses four share projection methods and the projected population for the City of Decatur to project the 2040 population of Decatur's block groups. The final section evaluates the share projection methods.

Share Projections for One Area

This section of the chapter uses four share projection methods—constant-share, shift-share, share-of-change, and share trend—to project Decatur's 2040 population, given the trimmed average population projection for DeKalb County prepared in chapter 3.

Constant-Share Method

The **constant-share method** assumes that a study area's share of the population or employment a larger context area in which it is located is constant over time. For example, to project an area's population, the method assumes that

$$s^t = \frac{p_i^t}{P^t} = \text{constant for all years } t, \qquad 4.1$$

where s^t = study area population share for any time t, p^t = study area population for any time t, and P^t = context area population at time t.
Given this assumption,

$$s^l = \frac{p_i^l}{P^l} = \frac{p_i^{ta}}{P^{ta}}, \qquad 4.2$$

where l = launch year when population data are available for the study area and the context area, ta = target year for projecting the study area population, s^l = study area population share in launch year l, p_i^l = study area population in launch year l, P^l = context area population in launch year l, p_i^{ta} = projected study area population in target year ta, and P^{ta} = projected context area population in target year ta.

Rearranging terms in equation 4.2 yields the constant-share projection equation for a single study area:

$$p^{ta} = s^l \times P^{ta}.$$ 4.3

Equation 4.3 says that a study area's projected population in the target year (p^{ta}) is equal to the context area's projected population in the target year (P^{ta}) multiplied by the study area's population share in the launch year (s^l). This equation reflects the constant-share method's assumption that the study area will have a constant share of the context area's future growth. This assumption will be considered in conjunction with the shift-share method discussed in the next section.

Table 4.1 uses equation 4.3 to compute Decatur's 2040 population, given the trimmed average projection for DeKalb County's 2040 population in table 3.7. The 2010 share in column three (0.0279) is equal to Decatur's 2010 population (19,335) divided by DeKalb County's 2040 population in column two (691,893). Decatur's projected 2040 population is computed by applying equation 4.3:

$$p^{2040} = s^{2010} \times P^{2040} = 0.0279 \times 881,851 = 24,604.$$

Shift-Share Method

The constant-share projection method requires data from only a single point in time. This makes it very useful when political or enumeration area boundaries change and consistent data for the past are not available. However, it assumes that the study area and the context area will grow at the same rate, which is highly unlikely. In some cases, a study area will grow more rapidly than the context area, causing its population share to increase. In other cases, a study area will grow more slowly than the context area, causing its population share to decline. Rarely will their growth be the same, as the constant-share projection technique assumes.

The **shift-share method** recognizes that the constant share assumption is rarely correct. It modifies the constant-share projection formula by adding a "shift" term that accounts for the observed difference between the growth rates of a larger context area and a smaller study area.

The shift-share equation for projecting a single area's population is:

$$p^{ta} = \left(s^l + \Delta s^{l,ta} \right) P^{ta},$$ 4.4

where p^{ta} = study area population in target year ta, s^l = study area population share in launch year l, $\Delta s^{l,ta}$ = projected population shift between launch year l and target year ta, and P^{ta} = projected context area population in the target year ta.

TABLE 4.1   **Constant-Share Population Projections: Decatur, 2040**

2010 Population[a]		2010 Share	2040 Population	
Decatur	DeKalb County		DeKalb County[b]	Decatur
p^{2010}	P^{2010}	s^{2010}	P^{2040}	p^{2040}
(1)	(2)	(3)	(4)	(5)
19,335	691,893	0.0279	881,851	24,604

Source:
[a] US Bureau of the Census (2013a).
[b] Table 3.7.

The projected population shift is computed as follows:

$$\Delta s^{l,ta} = \left(\frac{ta - l}{l - b}\right)\left(s^l - s^b\right),\qquad 4.5$$

where ta = target year for projecting the study area population, l = launch year when population data are available for the study area and reference area, b = base year when population data are available for the study area and reference area, $\Delta s^{l,ta}$ = projected study area population shift between launch year l and target year ta, s^l = study area population share in launch year l, and s^b = study area population share in base year b.

Decatur's 2000 population share in table 4.2 is equal to the city's 2000 population (18,147) divided by DeKalb County's 2000 population (665,865) or 0.0273. Decatur's population share in 2010 (0.0279) is computed in table 4.1. Substituting these values into equation 4.5 yields the projected shift in column five:

$$\Delta s^{2010,2040} = \left(\frac{2040 - 2010}{2010 - 2000}\right)(0.0279 - 0.0273) = 3 \times 0.0006 = 0.0018.$$

Table 4.3 uses equation 4.4 to project Decatur's 2040 population, given the trimmed average projection for DeKalb County's 2040 population:

$$p^{2040} = \left(s^{2010} + \Delta s^{2010,2040}\right)P^{2040} = (0.0279 + 0.0018)881,851 = 26,191.$$

Share-of-Change Method

The constant-share method assumes that an area's future growth or decline is directly proportional to its *size*, that is, that larger areas will grow or decline more rapidly than smaller areas. The shift-share method accounts for differences between the study area

TABLE 4.2   Projected Population Shift: Decatur, 2010–2040

2000 Population[a]		2000 Share	2010 Share[b]	Projected Shift
Decatur p^{2000}	DeKalb County P^{2000}	s^{2000}	s^{2010}	$\Delta s^{2010,2040}$
(1)	(2)	(3)	(4)	(5)
18,147	665,865	0.0273	0.0279	0.0018

Source:
[a] US Bureau of the Census (2003b).
[b] Table 4.1.

TABLE 4.3   Shift-Share Population Projections: Decatur, 2040

2010 Share[a]	Projected Shift[b]	2040 Population	
		DeKalb County[c]	Decatur
s^{2000}	$\Delta s^{2010,2040}$	P^{2040}	p^{2040}
(1)	(2)	(3)	(4)
0.0279	0.0018	881,851	26,191

Source:
[a] Table 4.1.
[b] Table 4.2.
[c] Table 3.7.

and context area's past growth but the study area's size is still the most important factor affecting its future growth or decline. The **share-of-change method** assumes that an area's future growth or decline is proportional to its *change* over the fit period, that is, that areas that grew (or declined) more rapidly than the context area in the past will grow (or decline) more rapidly in the future.[1] This assumption is expressed in the share-of-change projection equation for a single area:

$$p^{ta} = p^{l} + \left(c^{b,l} \times \Delta P^{l,ta}\right),$$ 4.6

where p^{ta} = projected study area population in target year ta, p^{l} = study area population in launch year l, $c^{b,l}$ = observed share of population change between the base year b and launch year l, and $\Delta P^{l,ta}$ = projected context area population change between launch year l and target year ta.

The observed share of population change share, $c^{b,l}$, is computed by applying equation 4.7:

$$c^{b,l} = \frac{\Delta p^{b,l}}{\Delta P^{b,l}},$$ 4.7

where $c^{b,l}$ = observed share of population change between base year b and launch year l, $\Delta p^{b,l}$ = study area population change between base year b and launch year l, and $\Delta P^{b,l}$ = context area population change between base year b launch year l.

Table 4.4 uses equation 4.7 to compute Decatur's share of DeKalb County's population change between 2000 and 2010. The share of population change in column five is computed by dividing Decatur's population change (1,188) by the county's population change (26,028).

Table 4.5 computes the share-of-change projections for Decatur's 2040 population, given the trimmed average projection for DeKalb County's 2040 population.

TABLE 4.4   **Share of Population Change: Decatur, 2000–2010**

Area	Population		Population Change	Share of Change
	2000[a]	2010[b]		($c^{2000,2010}$)
(1)	(2)	(3)	(4)	(5)
Decatur	18,147	19,335	1,188	0.0456
DeKalb County	665,865	691,893	26,028	—

Source:
[a] US Bureau of the Census (2003b).
[b] US Bureau of the Census (2013a).

TABLE 4.5   **Share-of-Change Population Projections: Decatur, 2040**

DeKalb County Population			Share of Change[c]	Decatur Population	
2010[a]	2040[b]	Change		2010[a]	2040
P^{2010}	P^{2040}	$\Delta P^{2010,2040}$	$c^{2010,2040}$	p^{2010}	p^{2040}
(1)	(2)	(3)	(4)	(5)	(6)
691,893	881,851	189,958	0.0456	19,335	27,997

Source:
[a] US Bureau of the Census (2013a).
[b] Table 3.7.
[c] Table 4.4.

The projected change in column three is computed by subtracting DeKalb County's observed population in 2010 from its projected 2040 population. Decatur's projected 2040 population is computed by applying equation 4.6:

$$p^{2040} = p^{2010} + \left(c^{2000,2010} \times \Delta P^{2010,2040}\right) = 19,335 + (0.0456 \times 189,958) = 27,997.$$

Share-Trend Method

The **share-trend method** uses information on the *trend* in the study area's share of the context area's population to project the study area's future population. The share-trend method is expressed in equation 4.8:

$$p^{ta} = s^{ta} \times P^{ta}, \tag{4.8}$$

where p^{ta} = projected study area population in target year *ta*, s^{ta} = projected study area population share in target area *ta*, and P^{ta} = projected context area population in target year *ta*.

Figure 4.1 shows a consistent decline in Decatur's share of DeKalb County's population between 2000 and 2010. The share-trend method assumes that this trend will continue in the future and uses Decatur's projected share of DeKalb County's population to project the city's future population.

The projected share in column two of table 4.6 was computed by applying the trend projection methods described in chapter 3 to the observed share values displayed in figure 4.1. Table 4.6 uses equation 4.8 to project Decatur's 2040 population, given the trimmed average projection for DeKalb County's 2040 population. That is,

$$p^{2040} = s^{2040} \times P^{2040} = 0.0229 \times 881,851 = 20,194.$$

Comparing the Decatur Share Projections

Table 4.7 reports five share projections for Decatur's 2040 population given the trimmed average projection for DeKalb County's 2040 population: (1) the constant-share projections computed in table 4.1, (2) the shift-share projections computed in

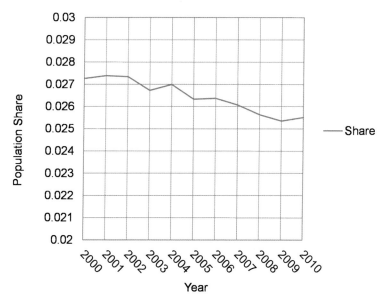

Figure 4.1 Observed Population Share: Decatur and DeKalb County, 2000–2010

TABLE 4.6	Share-Trend Population Projection: Decatur, 2040

DeKalb County 2040 Population[a] P^{2040}	Projected Share[b] s^{2040}	Decatur 2040 Population p^{2040}
(1)	(2)	(3)
881,851	0.0229	20,194

Source:
[a] Table 3.7.
[b] Computed by the authors.

TABLE 4.7	Share Population Projections: Decatur, 2040

Share Projection	DeKalb County			Decatur		
	2010 Population P^{2010}	2040 Population P^{2040}	Growth Rate $R^{2010,2040}$	2010 Population p^{2010}	2040 Population p^{2040}	Growth Rate $r^{2010,2040}$
(1)	(2)	(3)	(4)	(5)	(6)	(7)
Constant-share	691,893	881,851	27.5%	19,335	24,604	27.2%
Shift-share	691,893	881,851	27.5%	19,335	26,191	35.5%
Share-of-change	691,893	881,851	27.5%	19,335	27,997	44.8%
Share-trend	691,893	881,851	27.5%	19,335	20,194	4.4%
Average	**691,893**	**881,851**	**27.5%**	**19,335**	**24,747**	**28.0%**

table 4.3, (3) the share-of-change projections computed in table 4.5, (4) the trend-share projections in table 4.6, and (5) the average of the four share projections.

Decatur's projected growth rate for the constant-share projections are identical to DeKalb County's growth rate, as the method assumes; the slight difference is due to rounding error. The growth rates for the shift-share method are larger than the constant-share rates and smaller than the share-of-change rates. The share-trend projects less than a 5 percent growth in the city's population over the thirty-year projection interval. The average projection (24,747) is 10 percent larger than the trimmed average trend projection, 22,489, reported in table 3.8.

Share Projections for Multiple Areas

This section uses four share projection methods—constant-share, shift-share, share-of-change, and adjusted-share-of-change—to project the 2040 population of Decatur's block groups. The share-trend method for projecting the population of a single area cannot be used for multiple areas and the adjusted-share-of-change method cannot be used for a single area.

Constant-Share Method

The constant-share method for projecting multiple areas assumes that each study area's share of a larger context area in which it is located is constant over time. That is, the method assumes that

$$s_i^t = \frac{p_i^t}{P^t} = \text{constant for all years } t, \qquad 4.9$$

where s_i^t = population share for study area i for any time t, p_i^t = population of study area i for any time t, and P^t = population of the context area at time t.

Given this assumption, the constant share equation for projecting the population of study area i is

$$p_i^{ta} = s_i^l \times P^{ta}, \qquad\qquad 4.10$$

where l = launch year when population data are available for study area i and the context area, ta = target year for projecting the population of study area i, p_i^{ta} = projected population of study area i in target year ta,
s_i^l = population share for study area i in launch year l, and
P^{ta} = projected population of the context area in target year ta.

Equation 4.10 says that the projected population of study area i in the target year (p_i^{ta}) is equal to the context area's projected population in the target year (P^{ta}) multiplied by the study area's population share in the launch year (s_i^l).

Table 4.8 uses the constant-share method to project the population of Decatur's thirteen block groups, mapped on maps 2.5 and 2.6. Column three records the block group and the total population in 2010, the launch year.[2] The population shares in column four are computed by dividing each block group's 2010 population by the total population in 2010.

Column five uses equation 4.10 to compute the projected block group population for an assumed 2040 population of 24,500, the preferred projection for Decatur's 2040 population computed in chapter 3. For example, the 2040 population of study area one is:

$$p_1^{2040} = s_1^{2010} \times P^{2040} = 0.0866 \times 24{,}500 = 2{,}122.$$

The sum of the projected block group population values does not equal the projected total because of rounding error. The growth rates are approximately equal,

TABLE 4.8 **Constant-Share Population Projections: Decatur Block Groups, 2040**

Study Area	Block Group	2010 Population[a]	2010 Share	2040 Population	Growth Rate
i		p_i^{2010}	s_i^{2010}	p_i^{2040}	$r_i^{2010,2040}$
(1)	(2)	(3)	(4)	(5)	(6)
1	225.1	1,665	0.0866	2,122	27.4%
2	225.2	1,046	0.0544	1,333	27.4%
3	224.5	1,232	0.0640	1,568	27.3%
4	225.4	1,418	0.0737	1,806	27.4%
5	226.1	2,001	0.1040	2,548	27.3%
6	226.2	1,738	0.0903	2,212	27.3%
7	226.3	2,056	0.1069	2,619	27.4%
8	227.1	1,532	0.0796	1,950	27.3%
9	227.2	1,151	0.0598	1,465	27.3%
10	227.3	1,750	0.0910	2,230	27.4%
11	228.1	871	0.0453	1,110	27.4%
12	228.2	1,481	0.0770	1,887	27.4%
13	228.3	1,296	0.0674	1,651	27.4%
Total	—	**19,237**	**1.0000**	**24,501**	**27.4%**

Source:
[a] US Bureau of the Census (2013a).

reflecting the constant-share method's assumption that all the block groups will grow by the same rate. The slight differences in these values are due to rounding error.

Shift-Share Method

The shift-share equation for projecting the population of study area i is:

$$p_i^{ta} = \left(s_i^l + \Delta s_i^{l,ta}\right) P^{ta},$$ 4.11

where p_i^{ta} = projected population of area i in target year ta, s_i^l = population share for area i in launch year l, $\Delta s_i^{l,ta}$ = projected population shift for area i between launch year l and target year ta, and P^{ta} = projected population of the context area in target year ta.

The projected shift term is computed as follows:

$$\Delta s_i^{l,ta} = \left(\frac{ta - l}{l - b}\right)\left(s_i^l - s_i^b\right),$$ 4.12

where b = base year when study area and context area population are available, $\Delta s_i^{l,ta}$ = projected population shift for study area i between launch year l and target year ta, s_i^l = population share for study area i in launch year l, and s_i^b = population share for study area i in base year b.[3]

That is, the projected shift for area i between the launch year l and the target year ta ($\Delta s_i^{l,ta}$) is equal to: (1) the observed difference between the area's population share in the base year (s_i^l) and its population share in the target year (s_i^l), multiplied by (2) the length of the projection period ($ta - l$), divided by (3) the length of the fit period ($l - b$).

Table 4.9 computes Decatur's block groups' projected shifts. The shift-share method requires identical enumeration boundaries for the study area and context area

TABLE 4.9	Projected Population Shift: Decatur Block Groups, 2010–2040					
Study Area	Block Group	2000 Population[a]	2000 Share	2010 Population[b]	2010 Share	Projected Shift
i		p_i^{2000}	s_i^{2000}	p_i^{2010}	s_i^{2010}	$\Delta s_i^{2010,2040}$
(1)	(2)	(3)	(4)	(5)	(6)	(7)
1	225.1	1,559	0.0862	1,665	0.0866	0.0012
2	225.2	1,061	0.0586	1,046	0.0544	−0.0126
3	224.5	1,221	0.0675	1,232	0.0640	−0.0105
4	225.4	1,353	0.0748	1,418	0.0737	−0.0033
5	226.1	1,876	0.1037	2,001	0.1040	0.0009
6	226.2	1,269	0.0701	1,738	0.0903	0.0606
7	226.3	1,936	0.1070	2,056	0.1069	−0.0003
8	227.1	1,469	0.0812	1,532	0.0796	−0.0048
9	227.2	1,187	0.0656	1,151	0.0598	−0.0174
10	227.3	1,424	0.0787	1,750	0.0910	0.0369
11	228.1	1,091	0.0603	871	0.0453	−0.0450
12	228.2	1,504	0.0831	1,481	0.0770	−0.0183
13	228.3	1,143	0.0632	1,296	0.0674	0.0126
Total	–	18,093	1.0000	19,237	1.0000	0.0000

Source.
[a] Minnesota Population Center (n.d.).
[b] US Bureau of the Census (2013a).

in the base year, b (2000 in this example), the launch year, l (2010 in this case), and the target year (2040 in this example). The 2000 population values for Decatur's block groups in table 4.9 that match the 2010 census boundaries were obtained from the Minnesota Population Center at the University of Minnesota (Minnesota Population Center n.d.). The 2040 projections are for the 2010 block group boundaries. The block groups' 2000 share values in column four are computed by dividing each block group's 2000 population by the total 2010 population. The block groups' 2010 shares in column six are the block group 2010 population divided by the total 2010 population.

The projected shift values in column seven are computed by applying equation 4.12. For example, for study area one:

$$\Delta s_1^{2010-2040} = \left(\frac{2040-2010}{2010-2000}\right)\left(s_1^{2010} - s_1^{2000}\right) = \left(\frac{30}{10}\right)(0.0866 - 0.0862)$$
$$= 3 \times 0.0004 = 0.0012$$

Table 4.10 uses equation 4.11 and the projected shift values in table 4.9 to project the 2040 block group population values. For example, assuming Decatur's 2040 population is 24,500, the 2040 population in study area one is computed as follows:

$$p_1^{2040} = \left(s_1^{2010} + \Delta s_1^{2010-2040}\right)P^{2040} = (0.0866 + 0.0012)24,500 = 2,151.$$

As was true for the constant-share projections, the sum of the block group projections is not equal to the projected total population due to rounding error.

Table 4.11 reports the observed and projected growth rates for the shift-share projections reported in table 4.10. It indicates that Decatur's total population grew by 6.3 percent between 2000 and 2010 and is projected to grow by 27.4 percent between 2010 and 2040. The shift-share method projects dramatic growth changes for different block groups. For example, the population of Block Group 226.2 grew by 37 percent between 2000 and 2010 and is projected to more than double over

TABLE 4.10　　**Shift-Share Population Projections: Decatur Block Groups, 2040**

Study Area	Block Group	2010 Population[a]	2010 Share	Projected Shift[b]	2040 Population
i		p_i^{2010}	s_i^{2010}	$\Delta s_i^{2010,2040}$	p_i^{2040}
(1)	(2)	(3)	(4)	(5)	(6)
1	225.1	1,665	0.0866	0.0012	2,151
2	225.2	1,046	0.0544	−0.0126	1,024
3	224.5	1,232	0.0640	−0.0105	1,311
4	225.4	1,418	0.0737	−0.0033	1,725
5	226.1	2,001	0.1040	0.0009	2,570
6	226.2	1,738	0.0903	0.0606	3,697
7	226.3	2,056	0.1069	−0.0003	2,612
8	227.1	1,532	0.0796	−0.0048	1,833
9	227.2	1,151	0.0598	−0.0174	1,039
10	227.3	1,750	0.0910	0.0369	3,134
11	228.1	871	0.0453	−0.0450	7
12	228.2	1,481	0.0770	−0.0183	1,438
13	228.3	1,296	0.0674	0.0126	1,960
Total	–	19,237	1.0000	0.0000	24,501

Source:
[a] US Bureau of the Census (2013a).
[b] Table 4.9.

TABLE 4.11		Projected Shift-Share Growth Rates: Decatur, 2000–2040				

Study Area	Block Group	2000 Population[a]	2010 Population[b]	2040 Population[c]	Growth Rate	
i		p_i^{2000}	p_i^{2010}	p_i^{2040}	$r_i^{2000,2010}$	$r_i^{2010,2040}$
(1)	(2)	(3)	(4)	(5)	(7)	(8)
1	225.1	1,559	1,665	2,151	6.8%	29.2%
2	225.2	1,061	1,046	1,024	−1.4%	−2.1%
3	224.5	1,221	1,232	1,311	0.9%	6.4%
4	225.4	1,353	1,418	1,725	4.8%	21.7%
5	226.1	1,876	2,001	2,570	6.7%	28.4%
6	226.2	1,269	1,738	3,697	37.0%	112.7%
7	226.3	1,936	2,056	2,612	6.2%	27.0%
8	227.1	1,469	1,532	1,833	4.3%	19.6%
9	227.2	1,187	1,151	1,039	−3.0%	−9.7%
10	227.3	1,424	1,750	3,134	22.9%	79.1%
11	228.1	1,091	871	7	−20.2%	−99.2%
12	228.2	1,504	1,481	1,438	−1.5%	−2.9%
13	228.3	1,143	1,296	1,960	13.4%	51.2%
Total	—	18,093	19,237	24,501	6.3%	27.4%

Source:
[a] Minnesota Population Center (n.d.).
[b] US Bureau of the Census (2013a).
[c] Table 4.10.

the thirty-year projection period. Conversely, the population of Block Group 228.1 declined by 20 percent between 2000 and 2010 and is projected to decline by 99 percent by 2040.

The dramatic differences in the projected values reflect the fact that the shift-share projection method assumes that an area's growth is directly related to its observed population share, the first term in equation 4.11, and its projected population shift, the second term in equation 4.11. Table 4.10 reports that Block Group 226.2 has the fourth largest population share (column four) and the largest projected population shift (column five), which yield a large population growth. Conversely, Block Group 228.1 has the smallest population share and the largest negative projected population shift, which means that its projected population declines dramatically.

These results may be mathematically correct, but it is doubtful that a block group's population will decline by more than 99 percent if the city's population is growing. It is more reasonable to assume that block groups in a growing area will decline less (not more) than they did in the past. Fortunately, the adjusted-share-of-change method to be discussed below provides this result.

Share-of-Change Method

The **share-of-change method** assumes that an area's future growth or decline is proportional to its *change* over the base period, that is, that areas which grew (or declined) more rapidly than the context area in the past will grow (or decline) more rapidly in the future. This assumption is expressed in the share-of-change projection equation for study area i:

$$p_i^{ta} = p_i^l + \left(\Delta s_i^{b,l} \times \Delta P^{l,ta} \right), \qquad\qquad 4.13$$

where p_i^{ta} = projected population of study area i in target year ta, p_i^l = population of study area i in launch year l, $\Delta s_i^{b,l}$ = observed share of population change for study area i between base year b and launch year l, and $\Delta P^{l,ta}$ = projected population change for the context area between launch year l and target year ta.

The observed share of population change is computed as follows:

$$\Delta s_i^{b,l} = \frac{\Delta p_i^{b,l}}{\Delta P^{b,l}}, \qquad\qquad 4.14$$

where $\Delta s_i^{b,l}$ = observed change of population share for study area i between base year b and launch year l, $\Delta p_i^{b,l}$ = observed population change for study area i between base year b and launch year l, and $\Delta P^{b,l}$ = observed population change for the context area between base year b and launch year l.

Table 4.12 computes the observed share of population change for the Decatur block groups. The population change values in column five are the difference between the 2010 and 2000 populations. The share of change values in column six are the population change values divided by the total population change. For example, the share of population change for study area one is computed as follows:

$$\Delta s_1^{2000,2010} = \frac{\Delta p_1^{2000,2010}}{\Delta P^{2000,2010}} = \frac{106}{1,144} = 0.0927.$$

Table 4.13 uses equation 4.13 to project the 2040 population for the Decatur block groups. Assuming the city's 2040 population will be 24,500, the projected total population change ($\Delta P^{2010-2040}$) is equal to the projected population (24,500) minus the observed population (19,237) or 5,263. The projected population change for study area one in column five (488) is equal to its share of change in column four

TABLE 4.12	Observed Share of Change: Decatur Block Groups, 2000–2010				

Study Area	Block Group	2000 Population[a]	2010 Population[b]	Population Change	Share of Change
i		p_i^{2000}	p_i^{2010}	$\Delta p_i^{2000,2010}$	$\Delta s_i^{2000,2010}$
(1)	(2)	(3)	(4)	(5)	(6)
1	225.1	1,559	1,665	106	0.0927
2	225.2	1,061	1,046	−15	−0.0131
3	224.5	1,221	1,232	11	0.0096
4	225.4	1,353	1,418	65	0.0568
5	226.1	1,876	2,001	125	0.1093
6	226.2	1,269	1,738	469	0.4100
7	226.3	1,936	2,056	120	0.1049
8	227.1	1,469	1,532	63	0.0551
9	227.2	1,187	1,151	−36	−0.0315
10	227.3	1,424	1,750	326	0.2850
11	228.1	1,091	871	−220	−0.1923
12	228.2	1,504	1,481	−23	−0.0201
13	228.3	1,143	1,296	153	0.1337
Total	–	**18,093**	**19,237**	**1,144**	**1.0000**

Source:
[a] Minnesota Population Center (n.d.).
[b] US Bureau of the Census (2013a).

TABLE 4.13	Share-of-Change Population Projections: Decatur Block Groups, 2040					
Study Area	Block Group	2010 Population[a]	Share of Change[b]	Projected Change	2040 Population	Growth Rate
i		p_i^{2010}	$\Delta s_i^{2000,2010}$	$\Delta p_i^{2010,2040}$	p_i^{2040}	$r_i^{2010,2040}$
(1)	(2)	(3)	(4)	(5)	(6)	(7)
1	225.1	1,665	0.0927	488	2,153	29.3%
2	225.2	1,046	−0.0131	−69	977	−6.6%
3	224.5	1,232	0.0096	51	1,283	4.1%
4	225.4	1,418	0.0568	299	1,717	21.1%
5	226.1	2,001	0.1093	575	2,576	28.7%
6	226.2	1,738	0.4100	2,158	3,896	124.2%
7	226.3	2,056	0.1049	552	2,608	26.8%
8	227.1	1,532	0.0551	290	1,822	18.9%
9	227.2	1,151	−0.0315	−166	985	−14.4%
10	227.3	1,750	0.2850	1,500	3,250	85.7%
11	228.1	871	−0.1923	−1,012	−141	−116.2%
12	228.2	1,481	−0.0201	−106	1,375	−7.2%
13	228.3	1,296	0.1337	704	2,000	54.3%
Total	–	19,237	1.0001	5,263	24,501	27.4%

Source:
[a] US Bureau of the Census (2013a).
[b] Table 4.12.

(0.0927) multiplied by the projected total population change (5,263). Study area one's 2040 population in column six is equal to it 2010 population (1,665) plus the projected change (488) or 2,153. As before, the block group total does not match the projected population due to rounding error.

Table 4.13 contains some surprising projections for individual block groups. The population of Block Group 226.2 (study area six) more than doubles over the thirty-year projection period. More surprisingly, the projected population for Block Group 228.1 (study area eleven) in column five is negative, which is obviously impossible. These results reflect the fact that the projected population changes in column five of table 4.13 are more than four times larger than the observed population changes in column four of table 4.12. This means that the population of block groups such as Block Group 228.1 that declined between 2000 and 2010 are projected to shrink more than four times as much between 2010 and 2040. This is obviously not a reasonable assumption because the city's total population is growing. In this case, it seems more appropriate to assume that (1) block groups whose population declined in the past will decline less, not more, and maybe even grow slightly in the future and (2) block groups that grew in the past will continue to grow in the future. Fortunately, the adjusted-share-of-change method expresses this assumption.

Adjusted-Share-of-Change Method

The **adjusted-share-of-change method** computes the projected population in an area by adding the area's projected population growth to its current population. That is, the adjusted-share-of-change projection for study area i is:

$$p_i^{ta} = p_i^{l} + \Delta p_i^{l,ta}, \qquad 4.15$$

where p_i^{ta} = projected population for study area i in target year ta, p_i^l = observed population for study area i in launch year l, and $\Delta p_i^{l,ta}$ = projected population change for study area i between launch year l and target year ta.

The area's projected population change is computed by applying a *plus-minus adjustment procedure* (Shryock et al. 1975, 705–706) to the area's observed population change in the fit period. As its name suggests, the plus-minus adjustment applies different adjustments to the positive and negative components of a total. The adjustments are:

$$f_1 = \frac{\sum |n_i| + (N - n)}{\sum |n_i|} \quad (n_i > 0)$$ 4.16

and

$$f_1 = \frac{\sum |n_i| - (N - n)}{\sum |n_i|} \quad (n_i < 0),$$ 4.17

where f_1 = adjustment for positive values, f_2 = adjustment for negative values, n_i = observed value, n = sum of observed values recognizing signs, N = sum of adjusted values recognizing signs, and $\sum |n_i|$ = sum of absolute values.

Table 4.14 applies the plus-minus adjustment factors to the observed population change values in column five of table 4.12. The sum of observed changes recognizing signs (n), 1,144, is computed in column three. The sum of the absolute values $(\sum |n_i|$ or 1,732) in column four is the sum of the absolute values of the observed changes values. The sum of the adjusted change values recognizing signs (N) is the projected total population change between 2010 and 2040 (24,500 minus 19,237

TABLE 4.14	**Plus-Minus Adjustment for Share-of-Change Population Projections: Decatur Block Groups, 2010–2040**

Study Area	Block Group	Population Change[a]	Absolute Change	Adjusted Change
i		$\Delta p_i^{2000,2010}$	$\lvert \Delta p_i^{2000,2010} \rvert$	$\Delta p_i^{2010,2040}$
(1)	(2)	(3)	(4)	(5)
1	225.1	106	106	358
2	225.2	−15	15	21
3	224.5	11	11	37
4	225.4	65	65	220
5	226.1	125	125	422
6	226.2	469	469	1,584
7	226.3	120	120	405
8	227.1	63	63	213
9	227.2	−36	36	50
10	227.3	326	326	1,101
11	228.1	−220	220	303
12	228.2	−23	23	32
13	228.3	153	153	517
Total	–	**1,144**	**1,732**	**5,263**

Source:
[a] Table 4.12.

or 5,263). Substituting these values in equations 4.16 and 4.17 yield the following values for the f_1 and f_2 adjustment factors:

$$f_1 = \frac{1,732 + (5,263 - 1,144)}{1,732} = 3.3782, \text{ and}$$

$$f_2 = \frac{1,732 - (5,263 - 1,144)}{1,732} = -1.3782.$$

The sum of the adjustment factors is 2.0, as it always is.

The adjustments are applied to the observed population changes in column three to compute the adjusted population changes in column five. The observed population change for study area one is positive so the f_1 adjustment factor is used to compute its adjusted change; that is, 106×3.3782 equals 358. The observed population change for study area two is negative so the f_2 adjustment factor is used to compute its adjusted change; that is, -15×-1.3782 equals 21.

Table 4.15 uses equation 4.15, the observed values in column three, and the adjusted change values from column five of table 4.14 to project the 2040 population for Decatur's block groups. For example, the projected population for study area one is equal to:

$$p_1^{2040} = p_1^{2010} + \Delta p_1^{2010,2040} = 1,665 + 358 = 2,023.$$

The adjusted change of share projections in column four of table 4.15 are all positive, as one would expect in a growing city. Comparing the adjusted change values in column four of table 4.15 to the unadjusted projected change values in column five of table 4.13 reveals that (1) the projected change for study area six (Block

TABLE 4.15   **Adjusted-Share-of-Change Population Projections: Decatur Block Groups, 2040**

Study Area	Block Group	2010 Population[a]	Adjusted Change[b]	2040 Population	Growth Rate
i		p_i^{2010}	$\Delta p_i^{2010,2040}$	p_i^{2040}	$r_i^{2010,2040}$
(1)	(2)	(3)	(4)	(5)	(6)
1	225.1	1,665	358	2,023	21.5%
2	225.2	1,046	21	1,067	2.0%
3	224.5	1,232	37	1,269	3.0%
4	225.4	1,418	220	1,638	15.5%
5	226.1	2,001	422	2,423	21.1%
6	226.2	1,738	1,584	3,322	91.1%
7	226.3	2,056	405	2,461	19.7%
8	227.1	1,532	213	1,745	13.9%
9	227.2	1,151	50	1,201	4.3%
10	227.3	1,750	1,101	2,851	62.9%
11	228.1	871	303	1,174	34.8%
12	228.2	1,481	32	1,513	2.2%
13	228.3	1,296	517	1,813	39.9%
Total	–	19,237	5,263	24,500	27.4%

Source:
[a] US. Bureau of the Census (2013a).
[b] Table 4.14.

Group 226.2) is reduced by roughly 27 percent and (2) study area eleven (Block Group 228.1) and the other block groups whose populations declined between 2000 and 2010 are projected to grow, not decline.

Comparing the Decatur Block Group Share Projections

The four share projection methods generate very different projections for Decatur's block groups. The projected changes for constant-share method projections in table 4.8 are equal to roughly 27.4 percent for all the block groups. The projected changes for the shift-share projections in table 4.11 range from a high of 112.7 percent to a low of –99.2 percent. The projected changes for the share-of-change method in table 4.13 range from a high of 124.2 percent to a low of –116.2 percent. And the projected changes for the adjusted-share-of-change projections in table 4.15 range from a high of 91.1 percent to a low of 2.0 percent. The adjusted-share-of-change projections seem to be the most reasonable for this situation. However, this does not mean that this method is always the most appropriate choice.

Figure 4.2 graphically displays the substantial projected population growth of Block Groups 226.1, 226.2, and 227.3. Map 4.1 reveals that the largest population growth will occur in Block Group 226.2 that includes the city center and most of the city's commercial development.

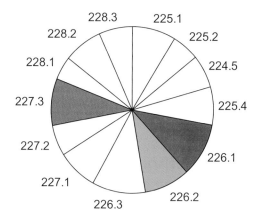

(a) 2010 Population

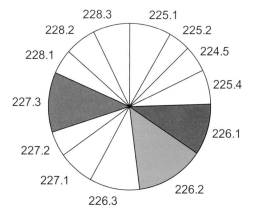

(b) 2040 Population

Figure 4.2 Observed and Adjusted-Share-of-Change Projected Population: Decatur, 2010 and 2040.

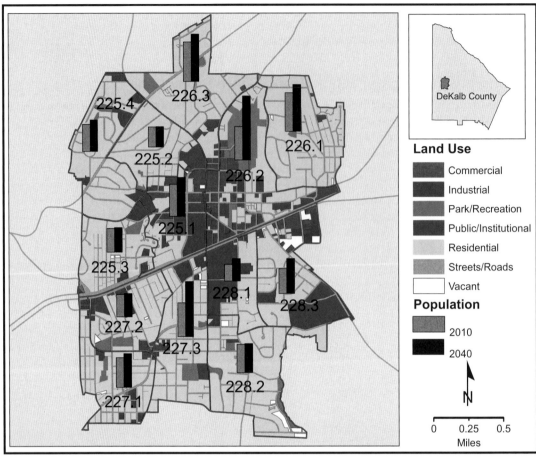

Map 4.1 Observed and Adjusted-Share-of-Change Population Projections and Land Uses: Decatur Block Groups, 2010 and 2040

Evaluating the Share Projection Methods

Several studies have evaluated the accuracy of different share projection methods.[4] Their results are inconclusive because they examined different projection models, data sets, geographic areas, and levels of sectoral aggregation. However, they generally support the constant-share model, particularly for small areas with limited data.

The constant-share method is questioned by its assumption that the different parts of a population or economy will all grow or decline the same.[5] The shift-share and share-of-change methods avoid this limitation by assuming that areas which grew or declined more rapidly than the context area in the past will continue to do so in the future. The shift-share and share-of-change methods implicitly assume that an area which grew faster (or slower) than the context area did so because it had a comparative advantage (or disadvantage) compared to other parts of the context area. Given this assumption, they further assume that this advantage (or disadvantage) will continue for the forecast period. Together these assumptions imply that population or employment shifts that occurred in the past will continue in the future.

The first assumption of the shift-share and share-of-change methods is reasonable because an area's past growth differs from the context area's growth at least in part because local conditions have stimulated its growth or decline. The second assumption is questionable, however. In some cases, positive shifts will reduce an area's locational advantage by, for example, raising its housing and labor costs. In these situations, positive shifts in the past may be reversed in the future. In other cases, past growth may increase an area's locational advantage, increasing its ability to stimulate future growth. In these situations, positive shifts in the past can be expected to be larger in the future. In both cases, there is no reason to assume that past population and employment changes will continue.

These observations suggest that none of the share projection techniques can be justified on theoretical grounds. Nevertheless, like the trend projections described in chapter 3, the share methods' computational simplicity and modest data requirements justify using them to consider what the future may be.

Notes

[1] This method is often called the share-of-growth method. However, it will be referred to as the share-of-change method here to make clear that it can be applied in both growing and declining areas.

[2] The total block group population in table 4.8 is not equal to the total Decatur population because the Census Bureau's block group boundaries do not correspond precisely with the city's political boundaries.

[3] This version of the shift-share method assumes that an area's share of the context area's population changes by the constant increments of a linear curve. Many formulations of the shift-share model assume the constant rate of the geometric curve (e.g., Klosterman 1990, 175–184) but it is equally reasonable, and computationally easier, to assume linear change.

[4] These studies generally evaluate alternate share models for projecting an area's employment. However, their conclusions are equally appropriate for using these methods to project an area's population.

[5] The following discussion draws on Stevens and Moore (1980).

Cohort-Component Methods 5

Perhaps no factor is more important for local government planning than the size and composition of an area's population and how it may change in the future. Changes in an area's population size, distribution, and composition place new demands on educational facilities, transportation systems, health and welfare services, and employment and recreational facilities. Conversely, efforts to encourage or direct growth help determine future population patterns. Even when the total population remains constant, changes in its composition can fundamentally alter the need for public facilities and services. As a result, the consideration of an area's current and future demographic characteristics is essential for determining how many and what kinds of people will be affected by—and affect—planners' proposals.

The trend and share projection methods described in the previous two chapters are extremely useful when time, resources, and information are limited. However, they project an area's future population by extending past trends without considering the factors that caused these trends to occur and offer little assurance that the trends will continue in the future. As a result, it seems that improved projections can be obtained by considering the three components of population change—fertility, mortality, and migration. Together these factors determine an area's overall rates of population change because a population can only change when someone is born, dies, or moves into or out of the area.

This basic demographic fact is expressed as:

$$P^{t+1} = P^t + B^{t,t+1} - D^{t,t+1} + (\text{IM}^{t,t+1} - \text{OM}^{t,t+1}), \qquad 5.1$$

where P^{t+1} = an area's future population at time $t + 1$, P^t = the area's population at time t, $B^{t,t+1}$ = the area's births between t and $t + 1$, $D^{t,t+1}$ = the area's deaths between t and $t + 1$, $\text{IM}^{t,t+1}$ = the number of **in-migrants** into the area between t and $t + 1$, and $\text{OM}^{t,t+1}$ = the number of **out-migrants** out of the area between t and $t + 1$.

Demographic Trends

The three components of population change vary substantially by age, sex, and race. They are also influenced by short-term, cyclical events, such as business cycles, and long-term trends driven by new technologies and changing societal norms and behavior.

Mortality

As figure 5.1 indicates, death rates have declined steadily in the United States over the past seventy years due, in part, to advances in medical technologies and changing life styles. The US mortality rate in 2010 was less than half of what it was in 1940, even accounting for the nation's changing age profile. Women lived significantly longer (81.0 years) than men (76.2 years) in 2010. The life expectancy for women increased much more rapidly than for men for most of the twentieth century, although the gender gap has narrowed since its peak in 1970 (Murphy et al. 2013).

The probability of death is extremely low for people below the age of forty. It increases gradually from forty to seventy and dramatically after that. Death rates are much higher in the first year of life (623.4 deaths per 100,000) than for children between the ages of one and four (26.5 deaths per 100,000). Infant mortality rates declined sharply over the twentieth century and continue to improve.

Survival rates tend to be similar from place to place in the United States; regional differences often reflect differences in the racial, ethnic, or socioeconomic profile of a region's population. As figure 5.2 indicates, African Americans have notably higher

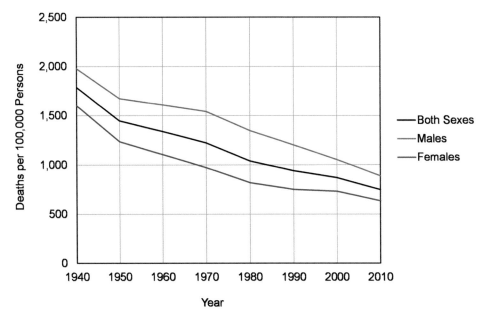

Figure 5.1 Age-Adjusted Mortality Rates, United States, 1940–2010

Source: Murphy et al. (2013).

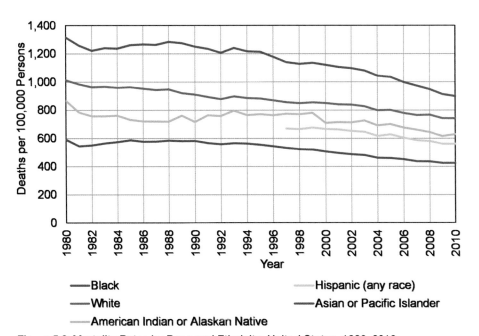

Figure 5.2 Mortality Rates by Race and Ethnicity: United States, 1980–2010

Source: National Center for Health Statistics (n.d.d).

mortality rates than whites and Asians have notably lower mortality rates than other races. The mortality rates for all major racial groups have declined over the past thirty years, showing some degree of convergence.

Fertility

Births are determined by the number of women in the child-bearing age groups and by prevailing **fertility rates**—the likelihood that a woman will give birth in a specified time period. As figure 5.3 indicates, fertility rates in the United States dropped substantially during the 1960s and early 1970s and have declined gradually since then.

There is considerable variation in fertility rates between age cohorts. Women in their late twenties and early thirties have notably higher fertility rates than other cohorts. However, there have recently been fewer births to younger women and more births for women in their late thirties and early forties as medical advances and changes in family-career choices have led many women to delay childbirth. As a result, the age progression of population bubbles (such as baby boomers and millennials) can have a huge influence on projected births.

Age-specific fertility rates vary by race, ethnicity, nationality, educational attainment, and income levels, among other things. As figure 5.4 indicates, Hispanic and black women typically have the highest fertility rate among the major racial groups, although both have declined substantially since 2006. Fertility rates are also affected by cyclical events. For example, figure 5.4 suggests that the fertility rate for Hispanic women was particularly sensitive to the 2008 recession.

Migration

Migration is by far the most volatile component of population change and typically distinguishes growing areas from those that are stable or declining.[1] Migration rates have fallen over the past several decades in the United States. As shown in figure 5.5, the proportion of people moving in 2012–2013 was at a near record low of 11.7 percent, down from the 1984 peak of 20.2 percent. The reasons for this downward trend are not fully understood.

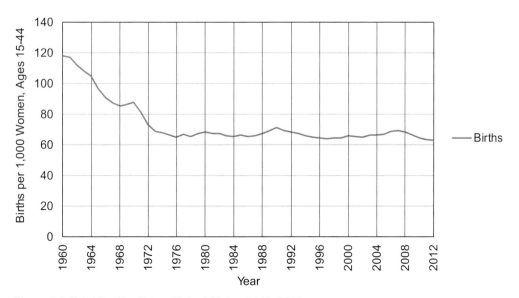

Figure 5.3 Total Fertility Rates: United States, 1960–2012

Source: Martin et al. (2013).

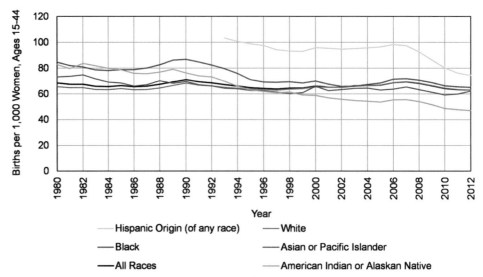

Figure 5.4 Total Fertility Rates by Race: United States, 1980–2012

Source: Martin et al. (2013).

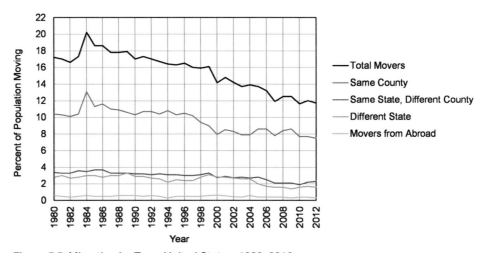

Figure 5.5 Migration by Type: United States, 1980–2012

Source: US Bureau of the Census (2013e).

When people move, they generally do not move very far. Most moves are to a different residence within the same county, as shown in figure 5.5. Short-distance moves are often related to a change in housing; such as moving into a different home or apartment. Longer-distance moves are more typically job-related. Migration tends to drop in recessionary periods and pick up during economic recoveries.

Migration is highly age-dependent. As figure 5.6 indicates, migration peaks for people in their twenties and thirties and declines thereafter as people settle into their jobs, families, and communities. Young children are also highly mobile, but they generally move with their parents. The age profile of long-distance migrants generally mirrors that of short-distance migrants. However, when retirees and elderly move, they are more likely to move to another county than other age groups.

Migration is also influenced by many socioeconomic factors. People with higher education levels tend to be more mobile. Homeownership, stable employment, and

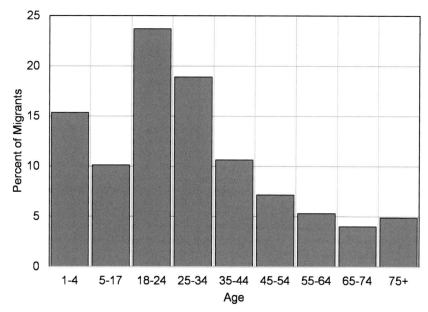

Figure 5.6 Age of Migrants: United States, 2012

Source: Computed by the authors from US Bureau of the Census (2013d).

the presence of children at home all tend to reduce mobility. Migration rates are similar for men and women, although men are more likely to move for job-related reasons. There are also racial differences in migration, although much of these differences are explained by age differences, homeownership patterns, family status, and education. Blacks and Hispanics are more mobile than either whites or Asians but are more likely to move locally for housing-related reasons and not for work opportunities (Ihrke 2014). Asians and whites are more likely to move longer distances because they are more likely to move for job-related reasons.

Cohort-Component Methods

The significant differences in mortality, fertility, and migration rates for different segments of the population suggest that population projections can be improved by dealing separately with the three components of change. Projection methods that do this are referred to as **cohort-component** methods: "cohort" methods because they disaggregate a population into age, sex, and racial groups or cohorts and "component" methods because they deal separately with the three components of population change—fertility, mortality, and migration.

   Whelpton and Thompson (1933) were the first to analyze and project the three components of population change separately and combine them to project population change. Dublin and Lotka (1930) refined the technique by dividing the population into age, sex, and race cohorts and applying the three components of population change to each cohort. Since then, cohort-component methods have become the most widely used technique for projecting the population of states, metropolitan areas, and counties in the United States (Smith 1986a, 127).

   An extensive body of research demonstrates that simple trend projections for an area's total population are generally as accurate as cohort-component projections (Smith 1987; Smith and Shahidullah 1995; Long 1995). This somewhat surprising finding reflects the fact that it is often as difficult—or even more difficult—to project an area's mortality, fertility, and migration rates than it is to project its total population because the required data may be unavailable or poorly measured. Nevertheless, cohort-component methods continue to be extremely useful because they project an area's population by age, sex, and race, which is required to project school enrollments and the demand for medical facilities and other services that meet the needs of specialized populations. Cohort-component methods also project the number of births, deaths, and migrants that are useful for health planning and other areas.

Hamilton-Perry Method

The concepts behind cohort-component projections are illustrated by a simplified cohort-component model that projects the aging of a population over time without directly considering the three components of population change. This approach is known as the **cohort-survival method** or the **Hamilton-Perry method** (Hamilton and Perry 1962). The Hamilton-Perry method requires only data on an area's population by age and sex for two time periods. The method's modest data requirements make it particularly useful for projecting the population of subcounty areas or for many areas such as all the municipalities in a state.

Data on the population by age, sex, and race for years ending in 0 are readily available in the United States for counties, county subdivisions, cities, and census tracts from the decennial census of population and housing. These data are only available for subcounty areas from the decennial census. Like other cohort-component methods, the Hamilton-Perry method projects an area's population at intervals equal to the length of the age cohorts; this allows the population in each cohort to move completely into the next higher cohort in a projection period. As a result, ten-year age cohorts are required to project Decatur's population in the ten-year increments between the decennial population censuses.

Projecting the Female Population

Table 5.1 uses the Hamilton-Perry method to project Decatur's female population for ten-year age cohorts in 2040. The female population in cohort n at time t can be expressed as PF_n^t. Thus, for example, Decatur's female population in the first cohort (ages 0–9) in 2000, PF_1^{2000}, is 1,009. In ten years some of the women in a cohort will die, some will move away, and others will move in. The net effect of these changes constitutes the female population in the next higher cohort in the second observation year (PF_{n+1}^{t+1}). Thus, for example, the 1,190 girls remaining from the first cohort in 2000 make up the second cohort in 2010 (PF_2^{2010}).

The Hamilton-Perry method does not consider separately the mortality, fertility, and migration processes that determine an area's future population. Instead, it incorporates the combined effect of these processes in a **Cohort Change Ratio** (CCR). The CCR for females and ten-year age is defined as follows:

$$CCRF_{n,n+1} = \frac{PF_{n+1}^{t+10}}{PF_n^t}, \qquad 5.2$$

where $CCRF_{n,n+1}$ = cohort change ratio for females from cohort n to cohort $n + 1$, PF_{n+1}^{t+10} = female population in cohort $n + 1$ in year $t + 10$, and PF_n^t = female population in cohort n in year t.

TABLE 5.1		Hamilton-Perry Population Projections: Decatur Females, 2020–2040					
Cohort	Ages	Observed Female Population[a]		Transition Ratio	Projected Female Population		
		2000	2010		2020	2030	2040
n		PF_n^{2000}	FP_n^{2010}		PF_n^{2020}	PF_n^{2030}	PF_m^{2040}
(1)	(2)	(3)	(4)	(5)	(6)	(7)	(8)
0	—	—	—	0.279	—	—	—
1	0–9	1,009	1,347	1.179	1,245	1,275	1,368
2	10–19	1,205	1,190	1.185	1,588	1,468	1,503
3	20–29	1,677	1,428	1.101	1,410	1,882	1,740
4	30–39	1,939	1,846	0.907	1,572	1,552	2,072
5	40–49	1,542	1,759	0.863	1,674	1,426	1,408
6	50–59	1,058	1,331	0.914	1,518	1,445	1,231
7	60–69	581	967	0.886	1,217	1,387	1,321
8	70–79	727	515	0.361	857	1,078	1,229
9	80+	715	521	—	374	444	550
	Total	10,453	10,904	—	11,455	11,957	12,422

Source:
[a] US Bureau of the Census (2003b, 2013c).

For example, the CCRF from cohort one (ages 0–9) to cohort two (ages 10–19) in table 5.1 is computed as follows:

$$\text{CCRF}_{1,2} = \frac{PF_2^{2010}}{PF_1^{2000}} = \frac{1,190}{1,009} = 1.179.$$

That is, there were 1,009 females between the ages of zero and nine in Decatur in 2000. Ten years later the city had 1,190 females between the ages of ten and nineteen. These data suggest that more girls moved into Decatur than died or moved out, resulting in a CCRF value greater than one.

The 2020 female population in all but the first and last cohorts can be projected by multiplying a cohort's 2010 female population from column four of table 5.1 by the cohort's CCRF value in column five. That is,

$$PF_{n+1}^{t+10} = P_n^t \times \text{CCRF}_{n,n+1}. \qquad 5.3$$

For example, the projected population in the second cohort in 2020 is computed as follows:

$$PF_2^{2020} = PF_1^{2010} \times \text{CCRF}_{1,2} = 1,347 \times 1.179 = 1,588.$$

A different process is used to project the female population of the first and last age cohorts. The projected female population in the first cohort, PF_1^{t+10}, is born between year t and year $t + 10$. A cohort change ratio cannot be computed for this cohort because its population was not born in 2000. Instead, the **child-to-woman ratio** for female births (CWRF) is computed by dividing the number of females born over the decade by the number of females of child-bearing age (cohorts two, three, and four in this example) at the beginning of the decade. This is:

$$\text{CWRF} = \frac{\text{PF}_1^{t+10}}{\text{PF}_2^{t} + \text{PF}_3^{t} + \text{PF}_4^{t}}. \qquad 5.4$$

where CWRF = child to women ratio for female births.

For example, given the data in table 5.1, the CWRF is computed as follows:

$$\text{CWRF} = \frac{\text{PF}_1^{2010}}{\text{PF}_2^{2000} + \text{PF}_3^{2000} + \text{PF}_4^{2000}} = \frac{1,347}{1,205 + 1,667 + 1,939} = 0.279.$$

The projected female population in the first cohort is computed by applying equation 5.5:

$$\text{PF}_1^{t+10} = \left(\text{PF}_2^{t} + \text{PF}_3^{t} + \text{PF}_4^{t}\right) \times \text{CWRF}. \qquad 5.5$$

For example, given the data in table 5.1, the 2020 female population in cohort one is computed as follows:

$$\text{PF}_1^{2020} = \left(\text{PF}_2^{2010} + \text{PF}_3^{2010} + \text{PF}_4^{2010}\right) \times \text{CWRF}$$
$$= (1,190 + 1,428 + 1,846) \times 0.279 = 1,245.$$

The final age cohort (80+) is open ended and some of its members will live beyond age 90 in the projection year, $t + 10$. As a result, the projected female population in the final cohort (cohort nine in this example) in year $t + 10$ includes women who were in cohort eight and in cohort nine in year t. As a result, the female CCR for the final cohort is:

$$\text{CCRF}_{8+9,9} = \frac{\text{PF}_9^{t+10}}{\text{PF}_8^{t} + \text{PF}_9^{t}} \qquad 5.6$$

For example, the CCRF for the final cohort in table 5.1 is:

$$\text{CCRF}_{8+9,9} = \frac{\text{PF}_9^{2010}}{\text{PF}_8^{2000} + \text{PF}_9^{2000}} = \frac{521}{727 + 715} = 0.361.$$

The projected female population in the final cohort is computed by applying equation 5.7:

$$\text{PF}_9^{t+10} = \text{CCRF}_{8+9,9} \times \left(\text{PF}_8^{t} + \text{PF}_9^{t}\right). \qquad 5.7$$

For example, given the data in table 5.1, the 2020 female population in the final age cohort is computed as follows:

$$P_9^{2020} = \text{CCRF}_{8+9,9} \times \left(\text{PF}_8^{2010} + \text{PF}_9^{2010}\right) = 0.361 \times (515 + 521) = 374.$$

The projections for Decatur's female population in 2030 and 2040 in table 5.1 are prepared in the same way. The 2030 population is projected by applying the CCRF and CWRF ratios in column five to the 2020 population in column six. The 2040 population is projected by applying the transition ratios in column five to the 2030 population in column seven.

Projecting the Male Population

The Hamilton-Perry method projects the male population with a nearly identical process that replaces the current and projected female population values in equations

5.2 through 5.7 with the equivalent male populations, PM_n^t and PM_{m+1}^{t+1}. For example, the CCR for females in equation 5.2 is converted to the CCR for males and ten-year age cohorts as follows:

$$\text{CCRM}_{n,n+1} = \frac{\text{PM}_{n+1}^{t+10}}{\text{PM}_n^t},\qquad 5.8$$

where $\text{CCRM}_{n,n+1}$ = cohort change ratio for males from cohort n to cohort $n + 1$, PM_{n+1}^{t+10} = male population in cohort $n + 1$ in year $t + 10$, and PM_n^t = male population in cohort n in year t.

Thus, in table 5.2 the CCRM for males from cohort one to cohort two is computed as follows:

$$\text{CCRM}_{1,2} = \frac{\text{PM}_2^{2010}}{\text{PM}_1^{2000}} = \frac{913}{1,024} = 0.892.$$

The procedure for projecting the male population in table 5.2 is identical to the process for projecting the female population in table 5.1 for the second through the ninth, open-ended, cohort. The only difference is projecting the male population in the first cohort. Because males do not give birth, the child-to-woman ratio for males (CWRM) is computed by dividing the 2010 male population in the first cohort in table 5.2 by the 2000 female population in the child-bearing years (cohorts two, three, and four) in table 5.1. That is,

$$\text{CWRM} = \frac{\text{PM}_1^{t+10}}{\text{PF}_2^t + \text{PF}_3^t + \text{PF}_4^t}.\qquad 5.9$$

where CWRM = child to women ratio for male births, PM_1^{t+10} = male population in the first cohort in time $t + 10$, and PF_n^t = female population in cohort n in time t.

TABLE 5.2		Hamilton-Perry Population Projections: Decatur Males, 2020–2040					
Cohort	Ages	Male Population[a]		Transition Ratio	Male Population		
		2000	2010		2020	2030	2040
n		PM_n^{2000}	PM_n^{2010}		PM_n^{2020}	PM_n^{2030}	PM_n^{2040}
(1)	(2)	(3)	(4)	(5)	(6)	(7)	(8)
0	—	—	—	0.290	—	—	—
1	0–9	1,024	1,399	0.892	1,295	1,325	1,422
2	10–19	915	913	0.913	1,248	1,155	1,182
3	20–29	1,026	835	1.447	834	1,139	1,055
4	30–39	1,681	1,485	0.915	1,208	1,207	1,648
5	40–49	1,325	1,538	0.835	1,359	1,105	1,104
6	50–59	838	1,106	0.834	1,284	1,135	923
7	60–69	377	699	0.764	922	1,071	947
8	70–79	287	288	0.331	534	704	818
9	80+	221	168	—	151	227	308
	Total	7,694	8,431	—	8,835	9,068	9,407

Source:
[a] US Bureau of the Census (2003b, 2013c).

For example, given the data in tables 5.1 and 5.2, the CWRM is computed as follows:

$$\text{CWRM} = \frac{\text{PM}_1^{2010}}{\text{PF}_2^{2000} + \text{PF}_3^{2000} + \text{PF}_4^{2000}} = \frac{1{,}399}{1{,}205 + 1{,}667 + 1{,}939} = 0.290.$$

The CWRM is applied to the female population in the child-bearing years and not to the male population. The projected male population in the first cohort is computed by applying equation 5.10:

$$\text{PM}_1^{t+10} = \left(\text{PF}_2^t + \text{PF}_3^t + \text{PF}_4^t\right) \times \text{CWRM}. \qquad 5.10$$

For example, given the data in tables 5.1 and 5.2,

$$\text{PM}_1^{2020} = \left(\text{PF}_2^{2010} + \text{PF}_3^{2010} + \text{PF}_4^{2010}\right) \times \text{CWRM}$$
$$= \left(1{,}190 + 1{,}428 + 1{,}846\right) \times 0.290 = 1{,}295.$$

Decatur Hamilton-Perry Population Projections Table 5.3 records the Hamilton-Perry projections for Decatur's total population by age (PT_n^t) in 2020, 2030, and 2040. The city's population is projected to grow by 13 percent between 2010 and 2040. The projections includes dramatic changes in some of the city's age groups. For example, the population between seventy and seventy-nine is projected to grow by more than 150 percent, the population between the ages of ten and thirty is projected to grow by 25 percent, and the population between forty and sixty is projected to decline by 18 percent.

Table 5.3 illustrates the movement of age cohorts through a community over time. For example, the large population below the age of ten in 2010 causes consistent growths of more than 34 percent as it ages. Conversely, the comparatively small population between the ages of twenty and thirty in 2010, causes 16 percent population declines as it ages.

TABLE 5.3 **Hamilton-Perry Population Projections: Decatur Total Population, 2020–2040**

Cohort	Ages	Total Population[a]				Percent Change			
		2010	2020	2030	2040	2010–2020	2020–2030	2030–2040	2010–2014
n		P_n^{2010}	P_n^{2020}	P_n^{2030}	P_n^{2040}				
(1)	(2)	(3)	(4)	(5)	(6)	(7)	(8)	(9)	(10)
1	0–9	2,746	2,540	2,600	2,790	−7.5	2.4	7.3	1.6
2	10–19	2,103	2,836	2,623	2,685	34.9	−7.5	2.4	27.7
3	20–29	2,263	2,244	3,021	2,795	−0.8	34.6	−7.5	23.5
4	30–39	3,331	2,780	2,759	3,720	−16.5	−0.8	34.8	11.7
5	40–49	3,297	3,033	2,531	2,512	−8.0	−16.6	−0.8	−23.8
6	50–59	2,437	2,802	2,580	2,154	15.0	−7.9	−16.5	−11.6
7	60–69	1,666	2,139	2,458	2,268	28.4	14.9	−7.7	36.1
8	70–79	803	1,391	1,782	2,047	73.2	28.1	14.9	154.9
9	80+	689	525	671	858	−23.8	27.8	27.8	24.5
	Total	19,335	20,290	21,025	21,829	4.9	3.6	3.8	12.9

Source:
[a] Tables 5.1 and 5.2.

The population pyramids in figure 5.7 illustrate the dramatic changes in Decatur' projected population composition between 2010 and 2040. The projected population has more females for all but the youngest age group. It has a larger male and female population between the ages of ten and nineteen, a smaller population between the ages of forty and sixty, and a dramatically larger population over the age of sixty. These differences suggest that Decatur should think carefully about the educational and social services it provides in its future.

Evaluating the Hamilton-Perry Method

The Hamilton-Perry method is useful because it is easy, quick, and requires only readily available data. This method has been found to be at least as accurate as more complicated cohort-component models for projecting the population of states in the United States (Swanson and Tayman 2013). However, combining the three components of

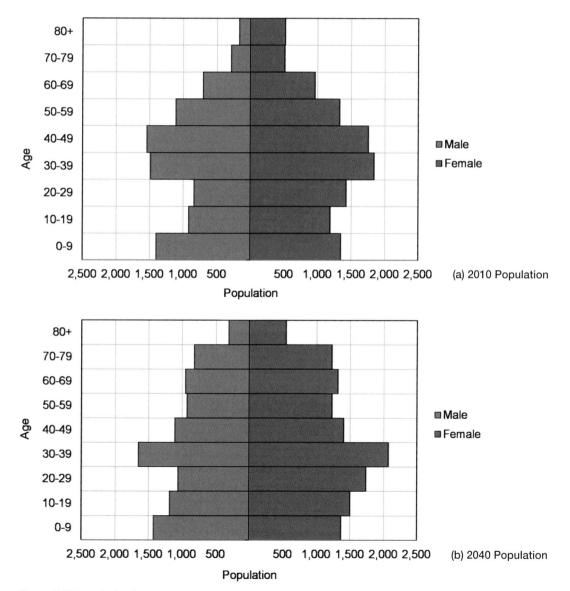

Figure 5.7 Population Pyramids: Decatur 2010 and 2040.

population into a single cohort-change ratio reveals little about the underlying causes of past population changes and makes it difficult to determine whether past trends will continue in the future. The Hamilton-Perry method also does not allow planners to take advantage of local knowledge and expert judgment on anticipated trends in migration, births, and deaths that can be used to develop alternate population projections.

As a result, it is generally preferable to consider the three components of population change—deaths, births, and migration—separately. This allows planners to evaluate the applicability of past trends in each cohort on future growth and choose rates that more closely match their understanding of local conditions. However, considering the components of change separately for multiple age/sex/race cohorts greatly increases the complexity of the analysis and the accompanying data requirements. These data requirements mean that the full cohort-component model typically cannot be used for small areas such as neighborhoods and cities. Instead, it is generally used to project the population for nations, states, multi-county metropolitan areas, counties, and larger cities.

The cohort-component method encourages analysts to understand the demographic forces that shaped an area's past and will determine its future. It recognizes that demographic projections are determined largely by the projected trends for the individual components, requiring analysts to consider the three components of population change individually. The following section begins this process by considering the mortality component of population change.

Mortality Component

Mortality is incorporated in cohort-component models as **cohort survival rates**—the probability that the members of a cohort will survive and move into the next higher age cohort. Two methods can be used to compute a community's survival rates. The **life table method** derives survival rates from national or state life tables. The **mortality count method** calculates survival rates directly from the study area's population counts and recorded number of deaths.

The mortality count method computes survival rates directly from mortality and population data specific to the study area. There are several reasons for using this approach. National life tables are often available only after a considerable time lag and many states do not prepare or regularly update their life tables. And while survival rates do not vary as much as fertility and migration rates, some regions have unique demographic or cultural characteristics that might not be reflected in national or state life tables.

The data needed to compute local survival rates—the number of deaths and the population size in a year for each age, sex, and racial or ethnic cohort—are readily available from most state vital statistics offices; many states distribute these data on the web. Annual county- and state-level vital statistics records are also reported by the US Centers for Disease Control and Prevention (2016a). The vital statistics data are derived from death certificates that record deaths by the place of residence of the deceased. These data are generally very accurate for nations with well-developed health and human services institutions.

Nevertheless, published life tables for the nation or state are generally adequate for estimating local mortality rates. Local mortality rates rarely differ substantially from national trends, especially in countries with well-established medical institutions. Most of the variation in mortality rates reflects differences in age structure or racial and ethnic distribution that can be captured by properly subdividing an area's population.

National survival rates also tend to be less erratic than local rates. Deaths are rare events in all but the oldest age cohorts and a local tragedy can temporarily skew the mortality rates for small areas. National data also provide more detailed breakdowns by age, race, and ethnicity than are available for most local areas. It must also be remembered that data from the recent past are being used to project population trends for twenty, thirty, or more years in the future. This means that recent history may not reflect the direction a community is heading and it may be appropriate to assume that diverse communities will become more like the rest of the nation in the future (Franklin 2014). As a result, only the life method will be considered in the following discussion.

Life Tables

Life tables are summary statistical models reporting the mortality and survivorship rates and life expectations of a population.[2] They were invented by Edmund Halley, the seventeenth century mathematician and astronomer who also discovered the comet that bears his name. They are particularly useful for preparing projections because they provide summary mortality measures that are independent of the age distribution of a population.

Cohort or **generation life tables** are based on the life experience of a real population (e.g., females born in the United States in 1990) from birth until all its members die. These life tables literally take lifetimes to build and are rarely used for population forecasting. Instead, most cohort-component models use **current** or **period life tables** that are based on a snapshot of the mortality experiences of the different components of an existing population over a short period of time, generally one to three years. Current life tables summarize the mortality experiences of a hypothetical population of one hundred thousand live births subjected throughout its life to the mortality rates experienced by a real population over a short time period. The term "life table" in the remainder of this discussion refers only to current life tables.

Life tables for the United States and states are published in the United States by the Centers for Disease Control's National Vital Statistics System.[3] Life tables are also available online for many states. These tables routinely report the population by sex, race, and ethnicity.

Table 5.4 displays a portion of the life table for America's female population in 2010. The first column, "Age (years)," refers to the time interval between the exact ages (or birth anniversaries) indicated. For example, the 0–1 age interval is the one-year period between the population's birth and its first birthday.

The second column, "Probability of dying between ages x and aged $x + 1$" (q_x), records the proportion of the population that is alive at the beginning of an age interval and dies before the interval ends. For example, in table 5.4, the proportion dying for the 0–1 age interval (.00555) indicates that 555 of the 100,000 live births will die before they reach their first birthday. The proportion dying values are computed from the observed mortality rates for a real population and are used to derive the remainder of the life table.

The third column, "Number surviving to age x" (l_x), records the number of person who survived to the beginning of an age interval. For example, in table 5.4, 99,445 children will survive to their first birthday, 99,404 survive to their second birthday, and so on.

The fourth column, "Number dying between ages x and $x + 1$" (d_x), reports the number of persons who die during the age interval x to $x + 1$. The number dying is equal to number living at the beginning of an age interval, l_x, multiplied by the proportion dying during that interval, q_x. For example, the number dying in the first age interval, 555, is equal to the number living at the beginning of the interval, 100,000,

TABLE 5.4 **Life Table for Females: United States, 2010**

Age (years)	Probability of Dying between Ages x and x + 1	Number Surviving to Age x	Number Dying between Ages x and x + 1	Person-Years Lived between Ages x and x + 1	Total Number of Person-Years Lived above Age x	Expectation of Life at Age x
	q_x	l_x	d_x	L_x	T_x	e_x
(1)	(2)	(3)	(4)	(5)	(6)	(7)
0–1	0.00555	100,000	555	99,514	8,104,166	81.0
1–2	0.00041	99,445	41	99,424	8,004,653	80.5
2–3	0.00023	99,404	22	99,393	7,905,228	79.5
3–4	0.00017	99,382	17	99,373	7,805,835	78.5
4–5	0.00014	99,365	14	99,358	7,706,462	77.6
5–6	0.00012	99,351	12	99,345	7,607,104	76.6
6–7	0.00011	99,339	11	99,334	7,507,759	75.6
7–8	0.00010	99,328	10	99,323	7,408,425	74.6
8–9	0.00009	99,319	9	99,314	7,309,102	73.6
9–10	0.00009	99,310	9	99,305	7,209,788	72.6
10–11	0.00009	99,301	9	99,297	7,110,483	71.6
...	...	...	...	...	...	...
80–81	0.04295	63,820	2,741	62,450	617,151	9.7
81–82	0.04767	61,079	2,912	59,623	554,702	9.1
82–83	0.05304	58,167	3,085	56,625	495,078	8.5
83–84	0.05933	55,082	3,268	53,448	438,454	8.0
84–85	0.06696	51,814	3,470	50,079	385,006	7.4
85–86	0.07556	48,344	3,653	46,518	334,927	6.9
86–87	0.08478	44,691	3,789	42,797	288,409	6.5
87–88	0.09493	40,903	3,883	38,961	245,613	6.0
88–89	0.10609	37,019	3,927	35,056	206,652	5.6
89–90	0.11828	33,092	3,914	31,135	171,596	5.2
...	...	...	...	...	...	...

Source: Arias (2014).

multiplied by the proportion dying (.00555). Forty-one people die between their first and second birthday, and so on.

The l_x values for an age group in column three are computed by subtracting the d_x for the proceeding cohort from the preceding cohort's l_x value. For example, the l_x value for the first age interval (ages 1–2), 99,455, is equal to the l_x value for cohort one, 100,000, minus the d_x value for cohort one, 555.

The fifth and six columns measure the total number of person-years lived in each age interval; they are also measures of the **stationary population**. The stationary population is a hypothetical population in which one hundred thousand persons are born and die each year with the proportion dying in each age interval corresponding to the q_x values in column two. If there is no migration and births and deaths are evenly distributed over the year, the population will be "stationary" because the total population and the number of persons in each age interval do not change. When a person dies or enters the next higher age interval, their place is immediately taken by someone entering from the next lower age interval and the number of persons in the age interval remains the same.

The l_x values in column three report the *number of persons* who are alive on the anniversary of their birth; the L_x values in column five are the *average population* between two birth anniversaries, considering the distribution of deaths throughout the year. For example, for the second age interval (ages 1–2), 99,445 individuals are alive at the start of the interval and 99,404 are alive at the end of the interval. The L_x value for this age interval is the average of these values, 99,424, because deaths within this interval are assumed to be evenly distributed over time. The L_x value for the first age interval, 99,514, is less than the average of the cohort's initial and final populations because a large proportion of the infant deaths occur shortly after birth, reducing the average population in this age interval.

The sixth column, "Total number of person-years lived above age x" (T_x), records the cumulative number of person-years that will be lived by the people in each age interval. The T_x values are equal to the cumulative sum of the L_x values in an age interval and all higher age intervals.

The final column, "Expectation of life at age x" (e_x), is the average remaining life-time (or life expectancy) for the people in an age interval. For example, the one hundred thousand live female births in table 5.4 can expect to live eighty-one years. The e_x values are computed by dividing the L_x values in column six by the l_x values in column three.

Computing One-Year Life Table Survival Rates

One-year life table survival rates can be computed for all but the first and last age intervals by substituting the L_x values from a one-year life table into equation 5.11:

$$s_{n,n+1}^{t,t+1} = \frac{L_{n+1}}{L_n}, \qquad\qquad 5.11$$

where $s_{n,n+1}^{t,t+1}$ = one-year survival rate from cohort n to cohort $n + 1$, L_{n+1} = life table person-years lived in age interval $n + 1$, and L_n = life table person-years lived in age interval n.

For example, given the values in table 5.4, the one-year survival rate from the first age interval (ages 0–1) to the second age interval (ages 1–2) is computed as follows:

$$s_{1,2}^{2010,2001} = \frac{L_2}{L_1} = \frac{99,424}{99,514} = 0.9991.$$

One-year survival rates from birth into the first cohort are computed by dividing the L_x value for the first cohort by 100,000, the number of births in a year. That is,

$$s_{0,1}^{t,t+1} = \frac{L_1}{100,000}. \qquad\qquad 5.12$$

Thus, for the life table in table 5.4, the one-year survival rate for births is equal to

$$s_{0,1}^{2010,2011} = \frac{99,514}{100,000} = 0.9951.$$

One-year survival rates for the final, open-ended cohort are computed from the total number of person-years lived (T_x) in column six of table 5.4 and not from the number of person-years lived within the age interval (L_x) in column five. Assume, for example, that the final cohort is made up of the population eighty-five and above. In one year this population will be eighty-six and above. As a result, given the data in table 5.4, the one-year survival rate for the final cohort is computed as follows:

$$s_{85+,86+}^{2010,2011} = \frac{T_{86}}{T_{85}} = \frac{288,409}{334,927} = 0.8611. \qquad\qquad 5.13$$

Computing Multiple-Year Life Table Survival Rates

The preceding discussion computes one-year survival rates from a one-year life table. However, most demographic projection models require survival rates for five- or ten-year age cohorts to match the age cohorts of the fertility and migration data. In addition, life tables for sub-national areas are typically reported for multiple-year age cohorts because not enough data are available for one-year age cohorts. Multiple-year survival rates can be computed by combining the data from a one-year life table into the desired age groups and substituting them into modified versions of equations 5.11, 5.12, and 5.13.

Table 5.5 contains an abridged version of the one-year life table in table 5.4. The L_x values in column two are computed by summing the L_x values from the corresponding age groups in table 5.4. For example, the L_x value for the 0–4 age cohort in table 5.5, 497,062, is equal to the sum of the L_x column values for the 0–1 to 4–5 age cohorts in table 5.4 (99,514 + 99,424 + 99,393 + 99,373 + 99,358). The L_x value for the final (90+ years) includes only the sum of the L_x values for the 90–94 age intervals. The T_x values for the abridged life table in table 5.5 are identical to the equivalent values one-year life table in table 5.4. For example, the T_x value for the 5–9 age interval in both tables is 7,607,104.

The five-year survival rates for all but the first and last five-year cohorts are computed by applying equation 5.14.

$$s_{n,n+1}^{t,t+5} = \frac{L_{n+1}}{L_n}.$$

5.14

TABLE 5.5 **Computing Five-Year Survival Rates for Females: United States, 2010**

Cohort	Age Interval	Person-Years Lived between Ages x and x + 5	Total Number of Person-Years Lived above Age x	Survival Rate
n		L_x	T_x	$s_{n,n+1}^{2010,2015}$
(1)	(2)	(3)	(4)	(5)
0	Births	500,000		0.9941
1	0–4	497,062	8,104,166	0.9991
2	5–9	496,621	7,607,104	0.9995
3	10–14	496,377	7,110,483	0.9990
4	15–19	495,893	6,614,106	0.9982
5	20–24	494,981	6,118,213	0.9975
6	25–29	493,734	5,623,232	0.9968
7	30–34	492,172	5,129,498	0.9957
8	35–39	490,052	4,637,326	0.9938
9	40–44	486,997	4,147,274	0.9901
10	45–49	482,192	3,660,278	0.9846
11	50–54	474,776	3,178,086	0.9780
12	55–59	464,320	2,703,310	0.9682
13	60–64	449,566	2,238,990	0.9517
14	65–69	427,835	1,789,424	0.9248
15	70–74	395,676	1,361,589	0.8814
16	75–79	348,762	965,913	0.8092
17	80–84	282,224	617,151	0.6891
18	85–89	194,466	334,927	0.4194
—	90+	100,646	140,461	—

Source: Computed from data in table 5.4.

For example, the five-year survival rate from the first age cohort (ages 0–4) into the second age cohort (ages 5–9) is computed as follows:

$$s_{1,2}^{2010,2015} = \frac{L_2}{L_1} = \frac{496{,}621}{497{,}062} = 0.9991.$$

The five-year survival rate for newborns is computed by dividing the L_x value for the first age cohort by five hundred thousand, the total number of births over five years. That is,

$$s_{0,1}^{t,t+5} = \frac{L_1}{500{,}000}. \qquad 5.15$$

Given the data in table 5.5,

$$s_{0,1}^{2010,2015} = \frac{L_1}{500{,}000} = \frac{496{,}471}{500{,}000} = 0.9941.$$

The five-year survival rate for the population in the open-ended cohort eighteen made up of the population over the age of eighty-five is computed by applying equation 5.16, a modified version of equation 5.13:

$$s_{18,18}^{t,t+5} = \frac{T_{90}}{T_{85}}. \qquad 5.16$$

That is,

$$s_{18,18}^{2010,2015} = \frac{T_{90}}{T_{85}} = \frac{140{,}461}{334{,}927} = 0.4194.$$

Fertility Component

The fertility component of population change is incorporated in a cohort-component projection model as **age-specific fertility rates**, the probability that the women in an age cohort will give birth in each period. These rates are multiplied by the average of the observed and projected female populations to project future births.

Computing Age-Specific Fertility Rates

Fertility rates typically show more local variation than survival rates and thus contribute more to regional differences in growth or decline. However, they generally have less impact than migration. Since it takes several years for newborns to age through the population, different fertility rate assumptions also have a larger impact on forecasts for twenty, thirty, or more years in the future than for short-term forecasts.

Birth rates may vary considerably by race and sex, as well as by location. The fertility rates are very different for blacks in Mississippi and in the District of Columbia and for whites in Massachusetts and in Utah. Fertility rates can also be substantially different for different types of counties in the same state. Given this variability, it is best to calculate fertility rates specific to the study area when possible.

Computing local age-specific fertility rates requires information on the number of women in each age cohort as well as the number of live births by the age of the mother, preferably by the gender of the newborn. This information is usually available through the state or national agencies that provide vital statistics data online.[4] These data are typically reported at the county level for the mother's primary residence and not for the county in which the birth occurred. Birth counts for smaller areas are sometimes available through special requests to state vital statistics offices.

Caution should be used in computing age-specific fertility rates for small areas. The number of births to women in an age and race cohort is generally quite low and highly variable, making fertility rates for a single year unreliable in small areas. This is the case for DeKalb County, which had a total population of over 690,000 in 2010 and only 21,906 women between the ages of fifteen and nineteen. The county's fertility rate for women between the ages of fifteen and nineteen varied by 13.3 female births per thousand women between 1995 and 2012. Using one rate or the other would project 290 more, or fewer, annual female births between the ages of fifteen and nineteen.

Age-specific birth rates may also vary greatly as population bubbles like the millennial generation move through a community's age structure. As a result, it is advisable to use fertility data for at least three years centered on the launch year to estimate age-specific fertility rates. For example, the births in column four of table 5.6 are the average number of births for three years: 2009, 2010, and 2011.

One-year age-specific fertility rates for males and females are computed by applying equations 5.17 and 5.18:

$$ff_{n,0}^{t,t+1} = \frac{BF_n^t}{PF_n^t},$$ 5.17

$$fm_{n,0}^{t,t+1} = \frac{BM_n^t}{PF_n^t},$$ 5.18

where $ff_{n,0}^{t,t+1}$ = one-year age-specific birth rate for female births to women in cohort n between year t and year $t + 1$, $fm_{n,0}^{t,t+1}$ = one-year age-specific birth rate for male

TABLE 5.6   **Computing Five-Year Age-Specific Fertility Rates for Females: DeKalb County, 2010**

Cohort	Age of Mother	Female Population[a]	Female Births[b]	One-Year Fertility Rate	Five-Year Fertility Rate	Adjusted Five-Year Fertility Rate
n		PF_n^{2010}	BF_n^{2010}	$ff_{n,0}^{2010,2001}$	$ff_{n,0}^{2010,2015}$	$ff_{n,0}^{2010,2015}$
(1)	(2)	(3)	(4)	(5)	(6)	(7)
3	10–14	20,882	0	0.0000	0.0000	0.0516
4	15–19	21,906	452	0.0206	0.1032	0.1601
5	20–24	26,144	1,134	0.0434	0.2169	0.2181
6	25–29	31,514	1,382	0.0439	0.2193	0.2297
7	30–34	29,579	1,420	0.0480	0.2401	0.1881
8	35–39	28,245	769	0.0272	0.1361	0.0854
9	40–44	26,511	184	0.0069	0.0347	0.0182
10	45–49	26,424	9	0.0003	0.0017	0.0009
11	50–54	24,984	0	0.0000	0.0000	0.0000
Total	–	–	5,350	–	–	–

Source:
[a] US Bureau of the Census (2013b).
[b] National Center for Health Statistics (n.d.e).

births to women in cohort n between year t and year $t+1$, BF_n^t = live female births to women in cohort n in year t, BM_n^t = live male births to women in cohort n in year t, and PF_n^t = female population in cohort n in year t.

The one-year fertility rates for female births in column five of table 5.6 are computed by dividing the number of live female births in a cohort from column four by the female population in column three. For example, for the fourth cohort (ages 15–19),

$$ff_{4,0}^{2010,2011} = \frac{\text{BF}_4^{2010}}{\text{PF}_4^{2010}} = \frac{452}{21,906} = 0.0206.$$

The one-year age-specific fertility rates in column five of table 5.6 are multiplied by five to compute the five-year rates in column six. Unlike deaths, women are repeatedly "at risk" of giving birth every year during the five-year projection period.

The analysis must account for the time a women spends in each cohort as they age (Isserman 1993, 49–51). The adjusted five-year fertility rates in column seven are the average of the age-specific rate in a cohort and the next higher cohort. For example, the adjusted fertility rate for cohort four, ages 15–19 years (0.1601) in column seven is the average of the five-year rates for cohort four (0.1032) and cohort five (0.2169) in column six. It may seem odd to estimate age-specific rates for girls between the ages of ten and fourteen, but these rates are the probability that girls who are currently between ten and fourteen will give birth in the next five years, when they are older.

Equations 5.17 and 5.18 can be used if information is available on the number of births by the mother's age and child's gender. However, if the data on the child's sex is not available or is unreliable due to low birth counts, one-year age-specific fertility rates for female and male births can be estimated by applying equations 5.19 and 5.20 and an estimated **sex ratio** at birth for an appropriate reference area (e.g., a state or nation):

$$ff_{n,0}^{t,t+1} = \frac{B_n^t}{\text{PF}_n^t} \times \left(\frac{100}{100+\text{SR}}\right), \quad\quad 5.19$$

$$fm_{n,0}^{t,t+1} = \frac{B_n^t}{\text{PF}_n^t} \times \left(\frac{\text{SR}}{100+\text{SR}}\right), \quad\quad 5.20$$

where B_n^t = total live births to women in cohort n in year t, and SR = sex ratio at birth.

The sex ratio at birth is the number of male births divided by the number of female births, multiplied by 100. The average sex ratio at birth for the United States from 1915 to 1964 was 105.4, with a **standard deviation** of 0.3, indicating that it is quite stable. The sex ratio at birth varies by the age of the mother, the birth order, race, and ethnicity, and widely for different countries.[5]

Migration Component

Survival and fertility rates change gradually over time and are generally stable for short time periods. Migration is different. It is the most dynamic component of population change and often dominates population projections, especially for rapidly growing or declining areas. As a result, the accuracy of population projections is often contingent on the accuracy of the underlying migration projections.

Migration is also the most difficult component of change to measure and a common source of uncertainty and error in population projections. Migration rates vary greatly by age, sex, and race and are heavily influenced by factors such as business cycles, job prospects, housing costs, and how local economic conditions compare to other locations. There are also many types of migrants—in-migrants, out-migrants, domestic migrants, and international migrants—with their own characteristics and motivations for moving. Migration also includes **special populations** such as full-time college students, prisoners, and armed-forces personnel that tend to have their own migration patterns (Renski and Strate 2013).

The data needed to project migration are often difficult to obtain, particularly for small areas. Even when these data are available they are derived from population samples and not from population counts, introducing sampling error—again particularly so for small areas. These data limitations are so endemic that several cohort-component methods have been developed to overcome the practical limitations of migration data.

This section considers two approaches for dealing with the migration component of population change. The first approach, residual net-migration methods, do not directly estimate the number of people moving into or out of an area. Instead, they estimate the net effect of migration as the portion of population change that cannot be explained by births, deaths, and the aging of an area's residential population. The second approach, gross migration methods, estimate migration rates separately for in-migrants and out-migrants and can consider international migration.

Residual Net Migration Estimation Methods

Residual net migration estimation methods do not count the number of people who move into and out of an area. Instead, they estimate migration as the residual difference between a cohort's observed and expected populations, given natural processes of birth, death, and aging. As a result, these methods are particularly useful for projecting the population of smaller areas such as counties where reliable data on births and deaths are available but migration data are poor, incomplete, or not available.

The demographic balancing equation in equation 5.21 states that an area's population at time t (P^t) is equal to: (1) its population at an earlier time $t - 1$ (P^{t-1}), plus (2) the number of births between $t - 1$ and t ($B^{t-1,t}$); minus (3) the number of deaths between $t - 1$ and t ($D^{t-1,t}$), plus (4) the in-migrants who moved into the area between $t - 1$ and t ($\text{IM}^{t-1,t}$), minus (5) the out-migrants who moved out of the area between $t - 1$ and t ($\text{OM}^{-1,t}$). That is,

$$P^t = P^{t-1} + B^{t-1,t} - D^{t-1,t} + \text{IM}^{t-1,t} - \text{OM}^{t-1,t}. \qquad 5.21$$

In-migration and out-migration can be combined into a single net-migration term $\text{NM}^{t-1,t}$ that is equal to the number of in-migrants ($\text{IM}^{t-1,t}$) minus the number migrants ($\text{OM}^{t-1,t}$). Substituting $\text{NM}^{t-1,t}$ in equation 5.21 and rearranging terms yields equation 5.22:

$$\text{NM}^{t-1,t} = P^t - \left(P^{t-1} + B^{t-1,t} - D^{t-1,t} \right). \qquad 5.22$$

The bracketed portion of equation 5.22 is equal to PE^t, the expected population in year t that would exist if there were no migration between $t - 1$ and t. As a result, net migration is the population change that cannot be accounted for by **natural increase** (births and deaths).

Two residual net migration estimation methods can be used to estimate the births and deaths in equation 5.22. The **vital statistics net migration estimation method** uses the recorded number births and deaths in the study area to compute the expected population. The **survival rates net migration estimation method** uses fertility and survival rates to estimate the number of births and deaths that are used to compute the expected population.

It isn't clear which method provides better net migration estimates because no external benchmark can be used to compare them and they can be expected to yield similar results in most cases. The vital statistics method has some conceptual appeal because it uses observed birth and death counts that are specific to the study region. However, it requires complete and reliable information on births and deaths by age, sex, race, and ethnicity (if applicable), date of occurrence, and place of residence. These data are generally not available or too erratic for small areas so this method is generally used only for counties and larger areas.

The survival rates net migration estimation method has two advantages over the vital statistics net migration method. It averages survival and fertility rates for several years to minimize the effect of unusual demographic events, providing more stable long-term projections. It can also compute survival and fertility rates from national or state data or published life tables that may be useful if local vital statistics data are unavailable or unreliable. Only the survival rates method will be considered here.

Using the survival rates method to compute estimated net migration by age and sex (and, if desired, ethnicity) involves five steps: (1) estimating the number of births, (2) computing the expected population, (3) estimating the number of net migrants, (4) applying the at-risk principle, and (5) computing net migration rates. The following discussion will consider the process for estimating female net migration rates. The male population will be considered when its procedures differs from the ones for females.

Estimating Female Births The estimated number of female births, the fertility component of equation 5.22, is computed by applying equation 5.23:

$$\mathrm{BF}_n^{t,t+5} = \left(\frac{\mathrm{PF}_n^t + \mathrm{PF}_n^{t+5}}{2} \right) \times \left(\frac{ff_n^{t,t+5} + ff_n^{t+5,t+10}}{2} \right). \qquad 5.23$$

That is, the estimated number of female births to women in cohort n between year t and year $t + 5$ is equal to: (1) the average female population in cohort n between year t and year $t + 5$ multiplied by (2) the average fertility rate for female births to women in the cohort in the five-year period between year t and year $t + 5$ and the five-year period between year $t + 5$ and year $t + 10$.

Table 5.7 uses equation 5.23 to estimate the number of female births in DeKalb County between 2005 and 2010. The age-specific fertility rates in columns six and seven were computed using the methods in table 5.6. DeKalb County's female population in each cohort and their associated fertility rates for female births both changed between 2005 and 2010. To account for these changes, table 5.7 uses equation 5.22 to compute the estimated number of female births in each cohort. For example, the estimated number of births to women in cohort three (1,211) is equal to the average female population in column five (21,737) multiplied by the average fertility rate in column eight (0.0557). The estimated total number of female births between 2005 and 2010 ($BF^{2005,2010}$) at the bottom of column nine is 26,060.

Estimating Male Births The procedures for estimating the number of male births in is nearly identical to the procedure for estimating the number of female births in table 5.7. The only difference is in computing and applying the fertility rates. The one-year age-specific fertility rates for males are computed by applying equation 5.18. The adjusted five-year fertility rates for males are computed by substituting the one-year age-specific fertility rates for males into column five of table 5.6.

TABLE 5.7		Estimating Female Births: DeKalb County, 2010–2015						

Cohort	Age Interval	Female Population			Fertility Rate			Births
n		$\left(PF_n^{2010}\right)_a$	$\left(PF_n^{2040}\right)_b$	Average	$ff_n^{2005,2010}$	$\left(ff_n^{2010,2015}\right)_c$	Average	$BF_n^{2005,2010}$
(1)	(2)	(3)	(4)	(5)	(6)	(7)	(8)	(9)
1	0–4	23,916	24,551	24,234	—	—	—	—
2	5–9	21,499	22,180	21,840	—	—	—	—
3	10–14	22,591	20,882	21,737	0.0599	0.0516	0.0557	1,211
4	15–19	21,972	21,906	21,939	0.1822	0.1601	0.1711	3,754
5	20–24	25,213	26,144	25,679	0.2379	0.2181	0.2280	5,855
6	25–29	30,435	31,514	30,975	0.2218	0.2297	0.2257	6,991
7	30–34	30,093	29,579	29,836	0.1745	0.1881	0.1813	5,409
8	35–39	28,848	28,245	28,547	0.0811	0.0854	0.0832	2,375
9	40–44	28,252	26,511	27,382	0.0147	0.0182	0.0164	449
10	45–49	25,830	26,424	26,127	0.0003	0.0009	0.0006	16
11	50–54	23,625	24,984	24,305	—	—	—	—
12	55–59	19,652	22,198	20,925	—	—	—	—
13	60–64	12,875	18,134	15,505	—	—	—	—
14	65–69	9,260	11,793	10,527	—	—	—	—
15	70–74	7,676	8,254	7,965	—	—	—	—
16	75–79	6,592	6,552	6,572	—	—	—	—
17	80–84	5,303	5,287	5,295	—	—	—	—
18	85+	4,971	5,400	5,186	—	—	—	—
—	Total	348,603	360,538	354,576	—	—	—	26,060

Source:
[a] National Center for Health Statistics (n.d.e).
[b] US Bureau of the Census (2013b).
[c] Table 5.6.

Because only women give birth, equation 5.24 uses the observed female population of child-bearing age and the adjusted five-year fertility rates for males to project the number of male births. That is,

$$BM_n^{t,t+5} = \left(\frac{PF_n^t + PF_n^{t+5}}{2}\right) \times \left(\frac{fm_n^{t,t+5} + fm_n^{t+5,t+10}}{2}\right). \qquad 5.24$$

That is, the estimated number of male births to women in cohort n between year t and year $t+5$ is equal to: (1) the average female population in cohort n between year t and year $t+5$ multiplied by (2) the average fertility rate for male births to women in the cohort in the five-year period between year t and year $t+5$ and the five-year period between year $t+5$ and year $t+10$.

Computing the Expected Female Population Table 5.8 uses the life table survival rates computed in table 5.5 and the estimated number of births from table 5.7 to compute the expected female population and the number of female net migrants between 2005 and 2010.

TABLE 5.8 Estimating Female Net Migration: DeKalb County, 2005–2010

Cohort	Age Interval	Observed Population[a]	Survival Rate[b]	Expected Population	Observed Population[c]	Estimated Net Migration
n		PF_n^{2005}	$s_{n,n+1}^{2005,2010}$	PEF_n^{2010}	PF_n^{2010}	$NMF_n^{2005,2010}$
(1)	(2)	(3)	(4)	(5)	(6)	(7)
0	Births[c]	26,060	0.9941	—	—	—
1	0–4	23,916	0.9991	25,907	24,551	−1,356
2	5–9	21,499	0.9995	23,895	22,180	−1,715
3	10–14	22,591	0.9990	21,488	20,882	−606
4	15–19	21,972	0.9982	22,569	21,906	−663
5	20–24	25,213	0.9975	21,932	26,144	4,212
6	25–29	30,435	0.9968	25,149	31,514	6,365
7	30–34	30,093	0.9957	30,339	29,579	−760
8	35–39	28,848	0.9938	29,963	28,245	−1,718
9	40–44	28,252	0.9901	28,668	26,511	−2,157
10	45–49	25,830	0.9846	27,973	26,424	−1,549
11	50–54	23,625	0.9780	25,433	24,984	−449
12	55–59	19,652	0.9682	23,105	22,198	−907
13	60–64	12,875	0.9517	19,028	18,134	−894
14	65–69	9,260	0.9248	12,253	11,793	−460
15	70–74	7,676	0.8814	8,564	8,254	−310
16	75–79	6,592	0.8092	6,766	6,552	−214
17	80–84	5,303	0.6891	5,334	5,287	−47
18	85+	4,971	0.4194	5,739	5,400	−339
—	Total	348,603	—	364,105	360,538	−3,567

Source:
[a] National Center for Health Statistics (n.d.e).
[b] Table 5.5.
[c] US Bureau of the Census (2013b).

The expected female population in cohorts two (ages 5–9) through seventeen (ages 80–85) for a five-year cohort model is computed using equation 5.25

$$PEF_n^t = PF_{n-1}^{t-5} \times s_{n-1,n} \quad (2 \le n \le 17), \qquad 5.25$$

where PEF_n^t = expected female population in cohort n in year t, PF_{n-1}^{t-5} = observed female population in cohort $n-1$ in year $t-5$, and $s_{n-1,n}$ = survival rate from cohort $n-1$ to cohort n.

For example, the expected 2010 population in the second cohort (ages 5–9) is computed as follows:

$$PEF_2^{2010} = PF_1^{2005} \times s_{1,2} = 23,916 \times 0.9991 = 23,895.$$

The expected female population in the final, open-ended, cohort for a five-year cohort model is computed by applying equation 5.26:

$$PEF_{18}^t = (PF_{17}^{t-5} \times s_{17,18}) + (PF_{18}^{t-5} \times s_{18,18}). \qquad 5.26$$

For example, given the data in table 5.8, the expected population in the final cohort is computed as follows:

$$PF_{18}^{2010} = (PF_{17}^{2005} \times s_{17,18}) + (PF_{18}^{2005} \times s_{18,18})$$
$$= (5,303 \times 0.6891) + (4,971 \times 0.4194) = 5,739.$$

The expected population in the first cohort is computed by applying equation 5.27

$$PEF_1^t = BF^{t-5,t} \times s_{0,1}, \qquad\qquad 5.27$$

where $BF^{t-5,t}$ = projected female births between year $t - 5$ and year t, and $s_{0,1}$ = survival rate from birth to cohort one.

For example, given the data in table 5.8, the expected female population in the first cohort (ages 0–4) is computed as follows:

$$PEF_1^{2010} = BF^{2005,2010} \times s_{0,1} = 26,060 \times 0.9941 = 25,907. \qquad 5.28$$

Estimating the Number of Female Net Migrants The net migration estimates in column seven of table 5.8 are computed by subtracting the expected population values in column five from the observed population values in column six. For example, the estimated net migration for cohort one (−1,356) is equal to the observed population (24,551) minus the estimated population (25,907).

Table 5.8 and figure 5.8 reveal substantial net out-migration for DeKalb County's male and female population between 2010 and 2015. Only the population between twenty and thirty is estimated to have more in-migrants than out-migrants. The discussion at the end of this chapter will consider whether these estimates are reasonable in the light of the county's past population growth.

Applying the At-Risk Principle The next step computes the observed net migration rate for each cohort that will be used to project future net migration.[6] This process should recognize the **at-risk principle** of demography, which holds that the rates of population change should be defined for—and applied to—the people who will give birth, die, or migrate. Thus, for example, if the numerator of an age-specific birth rate is the number of births to women between the ages of twenty and

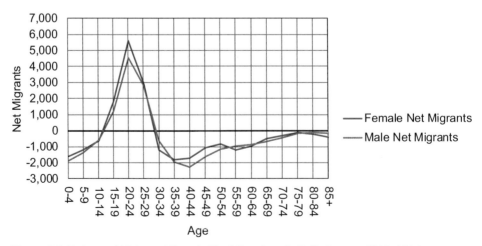

Figure 5.8 Estimated Male and Female Net Migration: DeKalb County, 2010–2015

twenty-four, the at-risk principle dictates that the denominator should be the number of women between the ages of twenty and twenty-four; they are the only people at risk of being mothers between the ages of twenty and twenty-four.

The at-risk principle suggests that migration rates should be computed by dividing the number of migrants by the population at risk for migration, which is different for in-migration and out-migration. The at-risk population for out-migration is the study area's current residents. The at-risk population for in-migration includes everyone who lives outside of the study area (both nationally and internationally) who could potentially move into the study area.

Net migration projection methods have traditionally computed net-migration rates by dividing the number of net migrants by study area's current population. This not only violates the at-risk principle for in-migration but also makes growing places grow faster and declining places decline faster, as a simple example illustrates.

Consider, for example, a hypothetical region that has a constant population of one hundred thousand, that is, no births, deaths, or in- and out-migration, and contains two subregions, A and B, each of which has an initial population of fifty thousand. Migration between the two subregions has a net effect of increasing the population of A by five thousand people per year and reducing the population of B by five thousand people per year. That is, the population of A is 50,000 in the year 2000, 55,000 in 2001, 60,000 in 2002, 65,000 in 2003, and so on. Correspondingly, the population of B is 50,000 in 2000, 45,000 in 2001, 40,000 in 2002, and so on.

If the 2000 and 2001 population values are known, the traditional net migration estimation method computes the observed net migration rates by dividing the number of net migrants by each subregion's initial population. That is, the net migration rate for subregion A is 5,000/50,000 or 0.10; the net migration rate for subregion B is −5,000/50,000 or −0.10. The net projected number of net migrants is equal to the estimated migration rate multiplied by the subregion's population in a year. The subregion's projected population is equal to its initial population plus the projected number of net migrants.

For example, the 2001 population of A (55,000) multiplied by the net migration rate (0.10) projects 5,500 net migrants between 2001 and 2002. Adding these net migrants to region's 2001 population (55,000) yields a projected 2002 population of 60,500. Applying the net migration rate to the 2002 population (60,500 × 0.10) projects an additional 6,050 net migrants and a total population of 66,550 in 2003. The compounding effect of applying the estimated net migration rate to a growing population yields projected population totals that increasingly outpace the actual population (60,000 in 2002, 65,000 in 2003, etc.). An identical compounding effect projects population declines for B that increasingly exceed the actual population declines. By 2008 all the region's population is projected to reside in county A, which is obviously impossible.

To offset this effect, Smith et al. (2013) suggest computing net migration rates by dividing the estimated number of net migrants by the nation's population. Although this approach violates the at-risk principle for out-migration, it satisfies the at-risk principle for in-migration and reduces the exaggerated migration trends for growing and declining areas.

Applying this approach to the preceding example yields a net migration rate of 0.05 (5,000/100,000) for A and a net migration rate of −0.05 (−5,000/100,000) for B. The projected number of net migrants is computed by multiplying the net migration rate by the region's projected population. That is, the projected number of net migrants for subregion A in each projection period is equal to the net migration rate (0.05) multiplied by the projected regional population (100,000) or 5,000. Adding the projected number of net migrants to the projected population yields projections that correspond exactly to the actual population growth, that is, 55,000 in 2001, 60,000 in 2002, and so on. The projected declines for the population of subregion B are also exactly equal to the actual declines.

Computing Female Net Migration Rates Table 5.9 uses the at-risk principle to estimate the net migration rates for DeKalb County's female population between 2005 and 2010. The nation's female population in 2005 is reported in column three. The net migration estimates in column four are taken from column seven of table 5.8. The net migration rates in column five are computed by dividing the estimated number of net migrants in a cohort by the nation's female population in that cohort. For example, the net migration rate for cohort one (-1.392×10^{-4}) is equal to the estimated number of net migrants ($-1,356$) divided by the nation's female population (9,741,805).

The adjusted net migration rates in column six are the averages for the successive net migration rates in column five. For example, the adjusted net migration rate for cohort one (-1.601×10^{-4}) is the average of the net migration rate for cohort one (-1.392×10^{-4}) and the net migration rate for cohort two (-1.810×10^{-4}). The adjusted net migration rates will be used to project the county's future net migration.

TABLE 5.9   **Estimating Female Net Migration Rates: DeKalb County, 2005–2010**

Cohort	Age Interval	National Female Population[a]	Estimated Net Migration[b]	Net Migration Rate	Adjusted Net Migration Rate
n		PNF_n^{2005}	$NMF_n^{2005,2010}$	$nm_n^{2005,2010}$	$nm_n^{2005,2010}$
(1)	(2)	(3)	(4)	(5)	(6)
1	0–4	9,741,805	−1,356	-1.392×10^{-4}	-1.601×10^{-4}
2	5–9	9,473,900	−1,715	-1.810×10^{-4}	-1.198×10^{-4}
3	10–14	10,348,382	−606	-5.856×10^{-5}	-6.105×10^{-5}
4	15–19	10,434,789	−663	-6.354×10^{-5}	1.742×10^{-4}
5	20–24	10,222,806	4,212	4.120×10^{-4}	5.337×10^{-4}
6	25–29	9,712,544	6,365	6.553×10^{-4}	2.890×10^{-4}
7	30–34	9,835,154	−760	-7.727×10^{-5}	-1.211×10^{-4}
8	35–39	10,411,620	−1,718	-1.650×10^{-4}	-1.767×10^{-4}
9	40–44	11,453,723	−2,157	-1.883×10^{-4}	-1.624×10^{-4}
10	45–49	11,357,088	−1,549	-1.364×10^{-4}	-9.008×10^{-5}
11	50–54	10,257,441	−449	-4.377×10^{-5}	-7.223×10^{-5}
12	55–59	9,007,859	−907	-1.007×10^{-4}	-1.155×10^{-4}
13	60–64	6,860,551	−894	-1.303×10^{-4}	-1.072×10^{-4}
14	65–69	5,469,500	−460	-8.410×10^{-5}	-7.483×10^{-5}
15	70–74	4,729,069	−310	-6.555×10^{-5}	-5.767×10^{-5}
16	75–79	4,299,256	−214	-4.978×10^{-5}	-3.169×10^{-5}
17	80–84	3,454,579	−47	-1.361×10^{-5}	-5.897×10^{-5}
18	85+	3,249,455	−339	-1.043×10^{-4}	-1.043×10^{-4}
—	Total	150,319,521	−3,567	—	—

Source:
[a] National Center for Health Statistics (n.d.e).
[b] Table 5.8.

Gross Migration Methods

Net migration estimation methods, while intuitively appealing and easily applied, are conceptually weak in computing a single net-migration rate that includes both in- and out-migrants. Migration rates should ideally measure the probability that a person will move into or out of the study area and not the net effect of their movement.

Gross migration methods consider in- and out-migration separately and use the study area's current residents to compute out-migration rates and the rest of the nation (or possibly the world) to compute in-migration rates. Gross migration methods have several advantages. They are conceptually more appealing than net migration methods because they measure true migration probabilities. They also connect an area's population to broader national trends and provide additional information on the dynamics of population change that may be useful for public policy making. For example, gross migration methods often distinguish between domestic and international migrants, which net migration methods cannot do. This information is useful for identifying recent immigrants with special service needs such as English as a Second Language programs in local schools.

Nevertheless, net migration methods are generally preferred for most planning applications. Detailed in- and out-migration estimates by age, sex, and race are not available for many counties and smaller areas. Even when they are available, they often have large errors, making the projected migration rates unreliable. In these cases simple net migration methods with modest data requirements may provide better forecasts than more complicated methods using poor data (Smith and Swanson 1998). There may also be only minor differences between projections developed with gross and net migration methods, particularly when population change is stable and only short-term forecasts are required.

The information needed to calculate in- and out-migration rates is available in the United States from the **American Community Survey (ACS)** described in appendix B. The ACS collects detailed information on where respondents lived in the previous year that can be used to estimate the number of domestic in- and out-migrants. Unfortunately, the standard ACS reports do not provide enough information to compute migration rates for detailed age-sex cohorts. This information can only be tabulated from the ACS Public Use Microsample (PUMS) files, a random sample of person- and household-level records from the ACS survey data.

Single-year PUMS migration estimates are available only for relatively large areas such as states and major metropolitan areas. Estimates for smaller places are included in the three- or five-year rolling ACS releases that pool data for multiple years to get larger samples. But because they are based on random subsets of ACS records, even the five-year estimates are often too unreliable for cohort-specific migration models, particularly for smaller regions.

Working with ACS PUMS files is also computationally demanding. The ACS PUMS files are very large, requiring specialized software and substantial analytic skills for their application. For example, estimating domestic in- and out-migration rates requires analyzing the PUMS files for the entire nation, over fifteen million records. Files this large cannot be analyzed with standard spreadsheet software such as Excel and require specialized software such as SAS or STATA.

In addition, the PUMS data are reported by Public Use Microdata Areas (PUMAs) and not for more common geographic units such as metropolitan areas, counties, or places. The PUMAs are nonoverlapping, substate areas containing at least one hundred thousand residents that often transcend city and county boundaries, particularly in sparsely populated areas. PUMAs also change over time making it difficult to construct consistent study areas for extended time periods.

Gross migration methods will not be considered in the book because of the substantial practical difficulties of applying them to cities and counties. Extensive discussions of these methods are provided in Isserman (1993), Smith (1986a), Renski and Strate (2013), and Smith et al. (2013, 134–138).

DeKalb County Cohort-Component Projections

Cohort-component methods project an area's population by dividing it into uniform age, sex, racial, or ethnic cohorts and applying the estimated rates for the three components of population change to each cohort. They project an area's population at intervals equal to the length of the age cohorts; this allows the population in each cohort to move completely into the next higher cohort in a projection period. For example, the Hamilton-Perry method described in the previous section projects an area's population at ten-year intervals corresponding to the time between the decennial population censuses. The mortality, fertility, and net migration methods described in the preceding sections use five-year age cohorts and project the population in five-year intervals. Cohort-component methods can also be used to project a population for other projection periods and other age intervals if the required age-specific data are available.

The following discussion uses the estimated DeKalb County mortality, fertility, and net migration rates computed in the preceding parts of this chapter to project the county's 2040 population for five-year age cohorts.

Projecting with Constant Rates

The first example assumes that the following rates remain constant throughout the projection period: (1) the life table survival rates computed in table 5.5, (2) the adjusted five-year age-specific fertility rates in table 5.6, and (3) the adjusted net migration rates in table 5.9. Later examples will consider the implications of modifying these rates.

The following steps are required to project the female population by age: (1) projecting female net migration, (2) projecting the surviving female population, (3) projecting the female population by age, (4) projecting the number of female births, and (5) projecting the female population in the first cohort. The procedure for projecting the male population will be considered after discussing the procedure for projecting the female population.

Projecting Female Net Migration Table 5.10 projects DeKalb County's female net migration by age between 2010 and 2015.

The population data in column three are from the decennial census of population and housing. The net migration rates in column four were computed in table 5.9. Given this information, the projected female net migration in each cohort is computed by applying equation 5.29:

$$\text{NMF}_n^{t,t+5} = \text{PNF}_n^t \times nm_n^{t,t+5}, \qquad 5.29$$

where $\text{NMF}_n^{t,t+5}$ = projected five-year female net migration in cohort n between year t and $t+5$, PNF_n^t = national female population in cohort n in year t, and $nm_n^{t,t+5}$ = net migration rate for cohort n for period t and $t+5$.

For example, the projected net migration for cohort one is computed as follows:

$$\text{NMF}_1^{2000,2015} = \text{PNF}_1^{2010} \times nm_1^{2010,2015} = 9{,}881{,}935 \times -1.601 \times 10^{-4} = -1{,}582.$$

| TABLE 5.10 | Projecting Female Net Migration: DeKalb County, 2010–2015 |

Cohort	Age Group	2010 Population[a]	Net Migration Rate[b]	Net Migrants
n		PNF_n^{2010}	$nm_n^{2010,2015}$	$nm_n^{2010,2015}$
(1)	(2)	(3)	(4)	(5)
1	0–4	9,881,935	-1.601×10^{-4}	−1,582
2	5–9	9,959,019	-1.198×10^{-4}	−1,193
3	10–14	10,097,332	-6.105×10^{-5}	−616
4	15–19	10,736,677	1.742×10^{-4}	1,871
5	20–24	10,571,823	5.337×10^{-4}	5,642
6	25–29	10,466,258	2.890×10^{-4}	3,025
7	30–34	9,965,599	-1.211×10^{-4}	−1,207
8	35–39	10,137,620	-1.767×10^{-4}	−1,791
9	40–44	10,496,987	-1.624×10^{-4}	−1,704
10	45–49	11,499,506	-9.008×10^{-5}	−1,036
11	50–54	11,364,851	-7.223×10^{-5}	−821
12	55–59	10,141,157	-1.155×10^{-4}	−1,171
13	60–64	8,740,424	-1.072×10^{-4}	−937
14	65–69	6,582,716	-7.483×10^{-5}	−493
15	70–74	5,034,194	-5.767×10^{-5}	−290
16	75–79	4,135,407	-3.169×10^{-5}	−131
17	80–84	3,448,953	-5.897×10^{-5}	−203
18	85+	3,703,754	-1.043×10^{-4}	−386
—	Total	156,964,212	—	−3,025

Source:
[a] US Bureau of the Census (2013c).
[b] Table 5.9.

Projecting the Surviving Female Population Table 5.11 projects the female population in each cohort in 2010 who survive to make up the population of the next higher cohort in 2015. The surviving female population in cohorts two (ages 5–9) through seventeen (ages 80–85) for a five-year cohort model is computed by applying equation 5.30:

$$PF_n^{t+5} = PF_{n-1}^t \times s_{n-1,n} \quad (2 \leq n \leq 17).$$ 5.30

For example, the surviving female population in cohort two (ages 5–9) is computed as follows:

$$PF_2^{2015} = PF_1^{2010} \times s_{1,2} = 24,551 \times 0.9991 = 24,529.$$

The surviving female population in the final, open-ended cohort (ages 85+) for a five-year cohort model is computed by applying equation 5.31:

$$PF_{18}^{t+5} = (PF_{17}^t \times s_{17,18}) + (PF_{18}^t \times s_{18,18}).$$ 5.31

Given the data in table 5.11, the surviving female population in the final cohort is computed as follows:

$$PF_{18}^{2015} = (PF_{17}^{2010} \times s_{17,18}) + (PF_{18}^{2010} \times s_{18,18})$$
$$= (5,287 \times 0.6891) + (5.400 \times 0.4194) = 5,908.$$

TABLE 5.11	Projecting the Female Population by Age with Constant Rates: DeKalb County, 2015

Cohort	Age Interval	Initial Population[a]	Survival Rate[b]	Surviving Population	Net Migration[c]	Age-Specific Fertility Rate[d]	Births by Age of Mother	Projected Population
n		PF_n^{2010}	$s_{n,n+1}$	PF_n^{2015}	$\text{NMF}_n^{2010,2015}$	$ff_n^{2010,2015}$	$\text{BF}_n^{2010,2015}$	PF_n^{2015}
(1)	(2)	(3)	(4)	(5)	(6)	(7)	(8)	(9)
0	Births	–	0.9941	–	–	–	–	–
1	0–4	24,551	0.9991	25,840	–1,582	–	–	24,258
2	5–9	22,180	0.9995	24,529	–1,193	–	–	23,336
3	10–14	20,882	0.9990	22,169	–616	0.0516	1,095	21,553
4	15–19	21,906	0.9982	20,862	1,871	0.1601	3,572	22,733
5	20–24	26,144	0.9975	21,866	5,642	0.2181	5,851	27,508
6	25–29	31,514	0.9968	26,078	3,025	0.2297	6,962	29,103
7	30–34	29,579	0.9957	31,414	–1,207	0.1881	5,623	30,207
8	35–39	28,245	0.9938	29,452	–1,791	0.0854	2,387	27,661
9	40–44	26,511	0.9901	28,069	–1,704	0.0182	481	26,365
10	45–49	26,424	0.9846	26,249	–1,036	0.0009	22	25,213
11	50–54	24,984	0.9780	26,018	–821	–	–	25,197
12	55–59	22,198	0.9682	24,434	–1,171	–	–	23,263
13	60–64	18,134	0.9517	21,493	–937	–	–	20,556
14	65–69	11,793	0.9248	17,257	–493	–	–	16,764
15	70–74	8,254	0.8814	10,907	–290	–	–	10,617
16	75–79	6,552	0.8092	7,275	–131	–	–	7,144
17	80–84	5,287	0.6891	5,302	–203	–	–	5,099
18	85+	5,400	0.4194	5,908	–386	–	–	5,522
Total	–	360,538	–	375,122	–3,025	–	25,993	372,097

Source:
[a] US Bureau of the Census (2013b).
[b] Table 5.5.
[c] Table 5.10.
[d] Table 5.6.

Projecting the Female Population by Age The 2015 female population for all but the first cohort is computed by adding the surviving population in column five and the number of net migrants in column six. For example, the 2015 female population in cohort two (ages 5–9) in column nine of table 5.11 (23,336) is equal to the surviving population (24,529) plus the number of net migrants (–1,193).

Projecting the Number of Female Births The number of female births between 2010 and 2015 is computed by applying equation 5.23. That is,

$$\text{BF}_n^{2010,2015} = \left(\frac{\text{PF}_n^{2010} + \text{PF}_n^{2015}}{2} \right) \times \left(\frac{ff_n^{2010,2015} + ff_n^{2015,2020}}{2} \right).$$

Assuming constant fertility rates, the projected 2015–2020 age-specific fertility rate for cohort n ($ff_n^{2015,2020}$) is identical to cohort's observed fertility rates for 2010–2015 ($ff_n^{2010,2015}$) computed in table 5.6. Given this, the projected number of female births to women in cohort three (ages 10–14) is computed as follows:

$$\text{BF}_3^{2010,2015} = \left(\frac{\text{PF}_3^{2010} + \text{PF}_3^{2015}}{2} \right) \times \left(\frac{f\!f_3^{2010,2015} + f\!f_3^{2010,2015}}{2} \right)$$

$$= \left(\frac{20{,}882 + 21{,}553}{2} \right) \times \left(\frac{0.0516 + 0.0516}{2} \right) = 1{,}095.$$

The projected total number of births between 2000 and 2015 ($\text{BF}^{2010,2015}$) is 25,993.

Projecting the Population in the First Cohort The surviving female population in the first cohort (ages 0–4) in column five of table 5.11 is computed by multiplying the projected number of female births ($\text{BF}^{t,t+5}$) by the survival rate for births ($s_{0,1}$). That is,

$$\text{PF}_1^{2015} = \text{BF}^{2010,2015} \times s_{0,1} = 25{,}993 \times 0.9941 = 25{,}840.$$

The projected population in the first cohort (23,258) in column nine is equal to the cohort's surviving population (25,840) plus its projected net migration (−1,582).

Projecting the Male Population by Age The procedures for projecting DeKalb County's 2015 male population in five-year age cohorts in table 5.12 is nearly identical to the procedure for projecting the female population in table 5.11. The only difference is the procedure for computing and applying the fertility rates. The one-year

	TABLE 5.12			Projecting the Male Population by Age with Constant Rates: DeKalb County, 2015				

Cohort	Age Interval	Initial Population[a]	Survival Rate[b]	Surviving Population	Net Migration	Age-Specific Fertility Rate	Births by Age of Mother	Projected Population
n		PM_n^{2010}	$s_{n,n+1}$	PM_n^{2015}	$\text{NMM}_n^{2010,2015}$	$fm_n^{2010,2015}$	$\text{BM}_n^{2010,2015}$	PM_n^{2015}
(1)	(2)	(3)	(4)	(5)	(6)	(7)	(8)	(9)
0	Births	—	0.9929	—	—	—	—	—
1	0–4	25,856	0.9988	27,364	−1,877	—	—	25,487
2	5–9	23,110	0.9994	25,826	−1,378	—	—	24,448
3	10–14	21,915	0.9980	23,097	−567	0.0542	1,149	22,530
4	15–19	22,737	0.9950	21,871	1,219	0.1696	3,785	23,090
5	20–24	25,555	0.9933	22,623	4,282	0.2320	6,224	26,905
6	25–29	29,422	0.9930	25,385	2,615	0.2425	7,350	28,000
7	30–34	28,134	0.9921	29,216	−661	0.1999	5,974	28,555
8	35–39	27,346	0.9898	27,911	−2,061	0.0907	2,535	25,850
9	40–44	25,043	0.9842	27,066	−2,365	0.0192	506	24,701
10	45–49	23,810	0.9752	24,647	−1,546	0.0014	36	23,101
11	50–54	21,027	0.9628	23,220	−1,037	—	—	22,183
12	55–59	17,928	0.9471	20,245	−800	—	—	19,445
13	60–64	14,530	0.9252	16,980	−668	—	—	16,312
14	65–69	9,311	0.8897	13,444	−584	—	—	12,860
15	70–74	6,107	0.8333	8,284	−394	—	—	7,890
16	75–79	4,182	0.7428	5,089	−134	—	—	4,955
17	80–84	3,046	0.6047	3,106	−106	—	—	3,000
18	85+	2,296	0.3563	2,660	−145	—	—	2,515
Total	—	331,355	—	347,277	−6,207	—	27,559	341,827

Source:
[a] US Bureau of the Census (2013b).
[b] Computed by the authors from data in Arias (2014).

age-specific fertility rates for males are computed by applying equation 5.18. The adjusted five-year rates for males are computed by substituting the one-year age-specific fertility rates for males into column five of table 5.6.

The projected number of male births is computed by apply equation 5.24. For example, for cohort three (ages 10–14):

$$BM_3^{2010,2015} = \left(\frac{PF_3^{2010} + PF_3^{2010}}{2} \right) \times \left(\frac{fm_3^{2010,2015} + fm_3^{2015,2020}}{2} \right).$$

Assuming constant fertility rates, the projected number of male births to women in cohort three (ages 10–14) is computed by multiplying the average of the cohort's initial and projected female population from table 5.11 by the fertility rates in column seven of table 5.12. That is,

$$BM_3^{2010,2015} = (20,882 + 21,533) \times \left(\frac{0.0542 + 0.0542}{2} \right) = 1,149.$$

Projecting Future Populations Projecting the male and female population for additional years follows a similar process. The procedure for projecting DeKalb County's 2020 female population is illustrated in table 5.13.

TABLE 5.13	Projecting the Female Population by Age with Constant Rates: DeKalb County, 2020

Cohort	Age Interval	Initial Population[a]	Survival Rate[b]	Surviving Population	Net Migrants	Age-Specific Fertility Rate[c]	Births by Age of Mother	Projected Population
n		PF_n^{2015}	$s_{n,n+1}$	PF_n^{2020}	$NMF_n^{2015,2020}$	$ff_n^{2015,2020}$	$BF_n^{2015,2020}$	PF_n^{2020}
(1)	(2)	(3)	(5)	(6)	(7)	(8)	(9)	(10)
0	Births	—	0.9941	—	—	—	—	—
1	0–4	23,377	0.9991	25,999	−2,437	—	—	23,563
2	5–9	23,391	0.9995	23,356	−1,144	—	—	22,211
3	10–14	21,556	0.9990	23,379	−611	0.0516	1,144	22,768
4	15–19	22,740	0.9982	21,535	1,801	0.1601	3,687	23,336
5	20–24	27,520	0.9975	22,698	5,916	0.2181	6,121	28,614
6	25–29	29,115	0.9968	27,450	3,199	0.2297	6,864	30,650
7	30–34	30,218	0.9957	29,023	−1,291	0.1881	5,450	27,732
8	35–39	27,676	0.9938	30,087	−1,781	0.0854	2,390	28,307
9	40–44	26,383	0.9901	27,504	−1,631	0.0182	476	25,873
10	45–49	25,236	0.9846	26,123	−925	0.0009	21	25,198
11	50–54	25,225	0.9780	24,847	−792	—	—	24,055
12	55–59	23,264	0.9682	24,669	−1,294	—	—	23,376
13	60–64	20,563	0.9517	22,524	−1,059	—	—	21,465
14	65–69	16,807	0.9248	19,570	−579	—	—	18,990
15	70–74	10,644	0.8814	15,544	−323	—	—	15,221
16	75–79	7,129	0.8092	9,382	−160	—	—	9,222
17	80–84	4,776	0.6891	5,769	−517	—	—	5,252
18	85+	4,864	0.4194	5,331	−1,162	—	—	4,168
Total	—	370,483	—	384,791	−4,791	—	26,153	380,000

Source:
[a] National Center for Health Statistics (n.d.e).
[b] Table 5.5.
[c] Table 5.6.

The initial female population in column three of table 5.13 is equal to the projected 2015 female population in column nine of table 5.11. This example assumes constant rates so the survival rates in column five and the fertility rates in column eight of table 5.13 are identical to the ones used to project the 2015 population in table 5.11. The procedure for projecting the number of net migrants in column seven is like the procedure in table 5.10 in applying equation 5.29:

$$\text{NMF}_n^{t,t+5} = \text{PNF}_n^t \times nm_n^{t,t+5}.$$

That is, projecting the number of female net migrants between 2015 and 2020 ($\text{NMF}_n^{2015,2020}$) requires a projection for the nation's female population in 2015 (PNF_n^{2015}). Projections for the male and female population of the United States by age, sex, race, Hispanic origin, and nativity are available from the US Bureau of the Census (2016h).

The procedures for projecting the surviving female population in column six, the number of births in column nine, and the projected 2020 population in column ten of table 5.13 are identical to the procedures in table 5.11. DeKalb County's 2020 male population can be similarly projected by applying the procedures in table 5.12. The county's population by age and sex in 2025 and beyond can be computed in an identical way.

Projecting Survival and Fertility Rates

The preceding analysis assumed that DeKalb County's current survival and fertility rates will continue. This assumption is questionable in the face of the declining mortality rates displayed in figures 5.1 and 5.2 and the declining fertility rates for younger women and increasing fertility rates for older women. As a result, it seems reasonable to assume that future survival and fertility rates will be different from today's rates. However, it is not always clear that using projected rates are better than assuming current rates continue. Local rates are highly variable and provide an uncertain basis for projecting future rates. National trends and forecasts may not apply to a specific area. It is also difficult to predict the changes in medical technology and lifestyles that affect births and deaths.

As a result, it is advisable to prepare a series of projections that use constant rates as a baseline and alternate projections that incorporate projected fertility, survival, and net migration rates. In any case, it is always wise to solicit the judgment of outside experts such as public health experts and medical professionals who can provide an important "reality check" on the assumptions underlying projected demographic rates.

Two methods can be used to project future survival and fertility rates. The trend projection methods discussed in chapter 3 can be used to project long-term trends in survival and fertility rates.[7] However, this approach only works well if sufficient historical data for the study area are available, past trends are generally stable, and these trends can be assumed to continue. Unfortunately, these conditions are rarely satisfied for sub-state areas where data are limited, demographic rates are highly variable for small age and sex cohorts, and past trends cannot be assumed to continue.

The second approach relates the study area's future mortality and fertility rates to the rates incorporated in the population projections prepared the US Bureau of the Census and other agencies.[8] The following analysis uses this approach. It assumes that the future trends in DeKalb County's survival and fertility rates will parallel the trends for the mortality and fertility inputs to the Census Bureau's national population projections. The county's future population by age and sex is computed by substituting the projected survival rates for each projection period into column five of tables 5.11 and 5.12 and substituting the projected fertility rates for each projection period into column eight of the two tables.[9]

Table 5.14 records DeKalb County's estimated five-year female survival rates for 2010–2015 and the county's projected female survival rates for 2035–2040. The survival rates are projected to go down slightly for the population under nine and increase gradually for the population over twenty. The survival rate for the population aged eighty-five and above is projected to increase by more than ten percent over the twenty-five-year projection period. The projected growth trends for the male survival rates are similar.

Table 5.15 records DeKalb County's estimated five-year fertility rates for female births for 2010–2015 and the county's projected fertility rates for 2035–2040. The county's fertility rates are projected to decline for all but the ten- to fourteen-year age interval. The projected fertility rates for male births decline in the same way.

Table 5.16 records DeKalb County's 2010 female population in five-year age cohorts and three projections for the county's 2040 female population. The projections in column four assume that the county's 2010–2015 survival and fertility

TABLE 5.14		Observed and Projected Five-Year Survival Rates: DeKalb County Females, 2010–2015 and 2035–2040		
Cohort	**Age Interval**	**Female Survival Rate**		
		2010–2015	**2035–2040**	**Percent Change**
n		$\left(s_{n,n+1}^{2010-2015}\right)^a$	$\left(s_{n,n+1}^{2035-2040}\right)^b$	
(1)	(2)	(3)	(4)	(5)
0	Births	0.9941	0.9923	−0.2
1	0–4	0.9991	0.9957	−0.3
2	5–9	0.9995	0.9956	−0.4
3	10–14	0.9990	0.9996	0.1
4	15–19	0.9982	0.9993	0.1
5	20–24	0.9975	0.9989	0.1
6	25–29	0.9968	0.9985	0.2
7	30–34	0.9957	0.9979	0.2
8	35–39	0.9938	0.9974	0.4
9	40–44	0.9901	0.9960	0.6
10	45–49	0.9846	0.9917	0.7
11	50–54	0.9780	0.9859	0.8
12	55–59	0.9682	0.9790	1.1
13	60–64	0.9517	0.9666	1.6
14	65–69	0.9248	0.9449	2.2
15	70–74	0.8814	0.9089	3.1
16	75–79	0.8092	0.8453	4.5
17	80–84	0.6891	0.7408	7.5
18	85+	0.4194	0.4650	10.9

Source:
[a] Table 5.5.
[b] Computed by the authors.

TABLE 5.15	Observed and Projected Five-Year Fertility Rates: DeKalb County Female Births, 2010–2015 and 2035–2040

Cohort	Age Interval	Fertility Rate		Percent Change
		2010–2015	2035–2040	
n		$\left(ff_{n,0}^{2010-2015}\right)^{a}$	$ff_{n,0}^{2035-2040}$	
(1)	(2)	(3)	(4)	(5)
3	10–14	0.0516	0.0516	0.0
4	15–19	0.1601	0.1597	−0.2
5	20–24	0.2181	0.2087	−4.3
6	25–29	0.2297	0.2233	−2.8
7	30–34	0.1881	0.1766	−6.1
8	35–39	0.0854	0.0819	−4.1
9	40–44	0.0182	0.0174	−4.6
10	45–49	0.0017	0.0016	−4.7

Source:
[a] Table 5.6.

rates will remain constant for the entire projection period. This assumption projects a 13.6 percent growth in the county's female population. The population below thirty-five is projected to increase slightly; the population between the ages of forty and fifty-five is projected to decline slightly; and the population over the age of seventy is projected to more than double.

The projected female population values in column six of table 5.16 assume that the county's survival rates grow at the same rate as the Census Bureau's population projections; the fertility rates are assumed to be constant. The total female population is projected to grow by 15.7 percent compared to a 13.6 percent increase for the constant rate projection. The projected population between the ages of forty and fifty-four is projected to decline but not as much as the projected decline for the constant rates projection. The population below thirty grows more slowly for the projected survival rates projection and the population over fifty-five grows more rapidly. The dramatic increase in the population over the age of seventy reflects the impact of applying the larger projected survival rates reported in table 5.14 for six projection periods (2015 through 2040).

The projected female population values in column eight of table 5.16 assume that DeKalb County's fertility rates will grow at the same rate as the Census Bureau's projections; the county's survival rates are assumed to be constant. The total female population is projected to grow by 12.1 percent. The projected population below twenty-nine is smaller than the other two projections, reflecting the smaller projected fertility rates reported in table 5.15. The projected population values for ages thirty and above are identical for the constant rate and projected fertility rate projections because the reduced fertility rates are only applied for six projection periods that only affect the first six age cohorts.

TABLE 5.16	Observed and Projected Female Population for Different Projection Methods: DeKalb County, 2010 and 2040

Cohort	Age Interval	2010 Female Population[a]	Constant Rates		Projected Survival Rates		Projected Fertility Rates	
			2040 Female Population	Percent Change	2040 Female Population	Percent Change	2040 Female Population	Percent Change
n		PF_n^{2010}	PF_n^{2040}		PF_n^{2040}		PF_n^{2040}	
(1)	(2)	(3)	(4)	(5)	(6)	(7)	(8)	(9)
1	0–4	24,551	26,276	7.0	26,117	6.4	24,993	1.8
2	5–9	22,180	24,602	10.9	24,400	10.0	23,481	5.9
3	10–14	20,882	23,456	12.3	23,192	11.1	22,415	7.3
4	15–19	21,906	24,769	13.1	24,511	11.9	23,741	8.4
5	20–24	26,144	30,095	15.1	29,817	14.1	29,050	11.1
6	25–29	31,514	33,202	5.4	33,027	4.8	33,221	5.4
7	30–34	29,579	31,865	7.7	31,867	7.7	31,865	7.7
8	35–39	28,245	28,456	0.7	28,641	1.4	28,456	0.7
9	40–44	26,511	25,859	−2.5	26,119	−1.5	25,859	−2.5
10	45–49	26,424	23,586	−10.7	23,935	−9.4	23,586	−10.7
11	50–54	24,984	21,301	−14.7	21,727	−13.0	21,301	−14.7
12	55–59	22,198	22,385	0.8	22,960	3.4	22,385	0.8
13	60–64	18,134	20,101	10.8	20,780	14.6	20,101	10.8
14	65–69	11,793	18,751	59.0	19,552	65.8	18,751	59.0
15	70–74	8,254	16,530	100.3	17,403	110.8	16,530	100.3
16	75–79	6,552	15,169	131.5	16,236	147.8	15,169	131.5
17	80–84	5,287	11,996	126.9	13,252	150.7	11,996	126.9
18	85+	5,400	11,207	107.5	13,536	150.7	11,207	107.5
Total	—	360,538	409,607	13.6	417,071	15.7	404,109	12.1

Source:
[a] US Bureau of the Census (2013b).

The three projections in table 5.16 all project declines in DeKalb County' female population between the ages of forty and fifty-five and dramatic increases in the county's population over the age of seventy. The projections for the county's male population are similar. These projections suggest that DeKalb County and its communities should consider improving the services they provide to their senior citizens and implementing policies that will increase the county's employment base that will attract more middle-aged people to the county.

Projecting Net Migration Rates

Projections can also be prepared that represent different assumptions about future net migration rates. Migration is important because it is typically the most dynamic component of population change and is often the primary reason why some places grow faster or slower than others. Migration is also the most difficult component of population change to project. It is determined by a complex interaction of local, regional, and national economic conditions opportunities, demographic trends, and lifestyle preferences—some of which may change quickly while others change gradually over years or decades. The data for estimating migration is also less accurate than the data for analyzing the other components of population change, requiring the application of indirect estimation methods that add increased uncertainty.

The academic literature suggests a range of sophisticated methods for projecting future migration, none of which has been widely used in practice.[10] Given the extreme difficulty of projecting net migration rates, the best strategy is preparing projections incorporating different migration assumptions that suggest different demographic futures. If possible, the projections should be vetted by experts and other stakeholders who understand the driving economic and demographic forces of the region.

Adjusting Net Migration Values Table 5.17 uses the *plus-minus adjustment method* described in chapter 4 to project a 10 percent increase in the female net out-migration estimates in table 5.8. That is, it assumes that the total number of female net out-migrants will increase from −3,567 to −3,924 (1.1 × −3,567).

The plus-minus method employs the two adjustment factors. The f_1 adjustment in equation 5.32 is applied to positive net migration values. The f_2 adjustment in equation 5.33 is applied to negative net migration estimates.

TABLE 5.17	Plus-Minus Adjustment for Projected Ten Percent Net Migration Increase: DeKalb County Females, 2010–2015			
Cohort	Age Interval	Estimated Net Migration[a]	Absolute Net Migration	Adjusted Net Migration
n		$NMF_n^{2005,2010}$		$NMF_n^{2010,2015}$
(1)	(2)	(3)	(4)	(5)
1	0–4	−1,356	1,356	−1,376
2	5–9	−1,715	1,715	−1,740
3	10–14	−606	606	−615
4	15–19	−663	663	−673
5	20–24	4,212	4,212	4,151
6	25–29	6,365	6,365	6,273
7	30–34	−760	760	−771
8	35–39	−1,718	1,718	−1,743
9	40–44	−2,157	2,157	−2,188
10	45–49	−1,549	1,549	−1,571
11	50–54	−449	449	−455
12	55–59	−907	907	−920
13	60–64	−894	894	−907
14	65–69	−460	460	−467
15	70–74	−310	310	−314
16	75–79	−214	214	−217
17	80–84	−47	47	−48
18	85+	−339	339	−344
Total	–	−3,567	24,721	−3,924

Source:
[a] Table 5.8.

$$f_1 = \frac{\sum |n_i| + (N - n)}{\sum |n_i|} \quad (n_i > 0) \qquad\qquad 5.32$$

and

$$f_2 = \frac{\sum |n_i| - (N - n)}{\sum |n_i|} \quad (n_i < 0), \qquad\qquad 5.33$$

where f_1 = adjustment for positive net migration estimates, f_2 = adjustment for negative net migration estimates, n_i = estimated net migration value, n = sum of the net migration estimates recognizing signs, N = sum of the adjusted net migration estimates recognizing signs, and $\sum |n_i|$ = sum of the absolute net migration estimates.

The sum of the net migration estimates recognizing signs, n, in column three is −3,567. The sum of the adjusted net migration estimates recognizing signs, N, is −3,924. The sum of absolute net migration estimates, $\sum |n_i|$, in column four is 24,721. Substituting these values into equations 5.32 and 5.33 yields the following values for the f_1 and f_2 adjustment factors:

$$f_1 = \frac{24,721 + [-3,924 - (-3,567)]}{24,721} = 0.9856$$

and

$$f_2 = \frac{24,721 - [-3,924 - (-3,567)]}{24,721} = 1.0144.$$

The sum of the plus-minus adjustments is 2.0, as it always is.

The estimated net migration for cohort five (ages 20–24) is 4,212. Applying the f_1 adjustment factor for positive net migration estimates, 4,212 × 0.9856 = 4,151, the adjusted net migration estimate in column five. The estimated net migration for cohort four (ages 15–19) is −663. Applying the f_2 adjustment factor for negative net migration estimates, −663 × 1.0144 = −673, the adjusted net migration value in column five.

Alternate Net Migration Projections Table 5.18 reports three different projections for DeKalb County's 2040 female population. The projections in column four assume that the total number of female net out-migrants is equal to the 2005–2010 estimate in table 5.8 for the entire projection period. The projections in column six assume that the projected number of female net out-migrants is 10 percent less than the 2005–2010 estimate. The projections in column eight assume that the projected number of female net out-migrants is 10 percent higher than the 2005–2010 estimate. All three of the projections assume constant mortality and fertility rates.

The projected female population for a 10 percent net migration decline in column six is larger than the projected population for constant net migration rates in column four because the projected total net migration is negative (net out-migration) and less net out-migration increases the projected population. Similarly, the projected population for a 10 percent net migration increase in column eight is smaller than the constant rate projection because the increased net out-migration reduces the projected population.

The difference between the projections for a 10 percent decrease and a 10 percent increase in table 5.18 is only 1.4 percent, less than the 1.8 percent difference between the constant rate and projected survival rate projections in table 5.16. This is contrary to the usual situation where migration has a bigger effect on population change than the mortality and fertility components of population change.

Table 5.19 suggests why changes in the projected number of net migrants has a relatively small impact on DeKalb County's projected population. It decomposes

| TABLE 5.18 | | Observed and Projected Female Population for Different Net Migration Projections: DeKalb County, 2010 and 2040 | | | | | | |

Cohort	Age Interval	2010 Female Population	Constant Rates		Ten Percent Decrease		Ten Percent Increase	
			2040 Female Population	Percent Change	2040 Female Population	Percent Change	2040 Female Population	Percent Change
n		PF_n^{2010}	PF_n^{2040}		PF_n^{2040}		PF_n^{2040}	
(1)	(2)	(3)	(4)	(5)	(6)	(7)	(8)	(9)
1	0–4	24,551	26,276	7.0	26,497	7.9	26,053	6.1
2	5–9	22,180	24,602	10.9	24,811	11.9	24,393	10.0
3	10–14	20,882	23,456	12.3	23,644	13.2	23,268	11.4
4	15–19	21,906	24,769	13.1	24,959	13.9	24,580	12.2
5	20–24	26,144	30,095	15.1	30,325	16.0	29,862	14.2
6	25–29	31,514	33,202	5.4	33,451	6.1	32,953	4.6
7	30–34	29,579	31,865	7.7	32,089	8.5	31,641	7.0
8	35–39	28,245	28,456	0.7	28,692	1.6	28,221	−0.1
9	40–44	26,511	25,859	−2.5	26,120	−1.5	25,598	−3.4
10	45–49	26,424	23,586	−10.7	23,813	−9.9	23,359	−11.6
11	50–54	24,984	21,301	−14.7	21,451	−14.1	21,151	−15.3
12	55–59	22,198	22,385	0.8	22,490	1.3	22,281	0.4
13	60–64	18,134	20,101	10.8	20,201	11.4	20,001	10.3
14	65–69	11,793	18,751	59.0	18,836	59.7	18,666	58.3
15	70–74	8,254	16,530	100.3	16,602	101.1	16,458	99.4
16	75–79	6,552	15,169	131.5	15,226	132.4	15,112	130.7
17	80–84	5,287	11,996	126.9	12,037	127.7	11,955	126.1
18	85+	5,400	11,207	107.5	11,248	108.3	11,167	106.8
Total	—	360,538	409,607	13.6	412,491	14.4	406,717	12.8

the observed population change between 2005 and 2010 used to estimate DeKalb County's net migration in table 5.8 into two components. The first part, recorded in column six, is the portion of the population change caused by natural increase: mortality and fertility. The second part, recorded in column seven, is the portion of the population change that is assumed to be caused by net migration. As the totals at the bottom of the table indicate, natural increase accounts for more than four times of the observed population change than net migration. As a result, changes in the projected net migration have less effect on the projected population than changes in the survival and fertility rate assumptions. This unusual situation reflects the fact that net migration estimates were derived from data for 2005–2010 that included the global "Great Recession," as will be described in evaluating the DeKalb County cohort-component projections.

Evaluating the DeKalb County Cohort-Component Projections

Planners typically produce a range of cohort-component projections that emphasize the uncertainty of the future and reveal the implications that different assumptions play in projecting what the future may be. The projections can help a community understand what its future may be and help guide choices and policies in the present that can help bring about a more desired future.

| TABLE 5.19 | | Components of Population Change: DeKalb County Female Population, 2005–2010 | | | | |

Cohort	Age Interval	Female Population		Change		
n		$\left(PF_n^{2005}\right)^a$	$\left(PF_n^{2015}\right)^b$	Total	Natural Increase[c]	Net Migration[c]
(1)	(2)	(3)	(4)	(5)	(6)	(7)
1	0–4	23,916	24,551	635	1,991	−1,356
2	5–9	21,499	22,180	681	2,396	−1,715
3	10–14	22,591	20,882	−1,709	−1,103	−606
4	15–19	21,972	21,906	−66	597	−663
5	20–24	25,213	26,144	931	−3,281	4,212
6	25–29	30,435	31,514	1,079	−5,286	6,365
7	30–34	30,093	29,579	−514	246	−760
8	35–39	28,848	28,245	−603	1,115	−1,718
9	40–44	28,252	26,511	−1,741	416	−2,157
10	45–49	25,830	26,424	594	2,143	−1,549
11	50–54	23,625	24,984	1,359	1,808	−449
12	55–59	19,652	22,198	2,546	3,453	−907
13	60–64	12,875	18,134	5,259	6,153	−894
14	65–69	9,260	11,793	2,533	2,993	−460
15	70–74	7,676	8,254	578	888	−310
16	75–79	6,592	6,552	−40	174	−214
17	80–84	5,303	5,287	−16	31	−47
18	85+	4,971	5,400	429	768	−339
—	**Total**	**348,603**	**360,538**	**11,935**	**15,502**	**−3,567**

Source:
[a] National Center for Health Statistics (n.d.e).
[b] US Bureau of the Census (2013b).
[c] Table 5.8.

Population projections are often developed for the public and for advisory committees and review panels comprised of community leaders, outside experts, and other interested stakeholders. These people may not be familiar with the technical aspects of the projection methods, but they often have a deep intuitive sense of their community and its long-term development trends. As a result, they may object if the projections do not match their perceptions of a place or their visions of what the community will become. Some people may question the projected values, for example, whether the projected values are too high or too low. Others may question the projection's underlying assumptions, for example, whether the projected mortality, fertility, and migration rates are reasonable in the face of past trends.

This means that planners must often justify their projections and provide a range of projections based on different assumptions about what the future will be. The future in unwritten, and it is impossible to know whether a projection will be accurate or if its underlying assumptions will be correct. However, planners can, and should, carefully scrutinize their projections to ensure that they present a reasonable portrait of a likely future.

Three procedures can be used to consider different cohort-component projections: (1) evaluating the projected population trends, (2) evaluating the projected population by age and sex, and (3) evaluating the components of population change.

Evaluating the Projected Population Trends

Planners should first insure that the projected population values are consistent with historic trends. Projections for extreme growth or decline are often warnings that something is wrong with the analysis. Even the most careful analyst makes mistakes, especially when working with complicated methods that have many interconnected pieces. Even if nothing is technically and the high and wrong with the analysis, extreme breaks in historic patterns are often questioned by public officials and stakeholders.

Table 5.20 records DeKalb County's 2010 population and the county's projected 2040 population for the five projection assumptions considered in the previous section. The projections range from a low of 770,835 for the projected fertility rates projection to a high of 798,884 for the projected survival rates projection, a difference of 3.5 percent.

Figure 5.9 plots the projected fertility and projected survival rate projections for DeKalb County's 2040 population. The other projections lie between these two extremes. It reveals that the county's population grew less between 2000 and 2010 than it had been for any decade in the previous fifty years. The cohort-component projections are based on the data for 2005–2010 so they continue the reduced population growth for this period, which does not appear to be a reasonable continuation of the county's long-term growth trend.

The slowdown in DeKalb County's population growth between 2000 and 2010 is due in part to the global "Great Recession" of 2007–2009. The recession in the United States was the longest and most severe in sixty years. From the recession's peak in 2007 to its trough in 2009, the United States' gross domestic product fell by 4.3 percent; the country's unemployment rate doubled from 5 to 10 percent; home prices fell by approximately 30 percent on average, and the net worth of US households fell from approximately $69 trillion in 2007 to roughly $55 trillion in 2009 (Rich 2013). The dramatic decline in the national economy was reflected in substantial declines in birth rates and migration throughout the country and in DeKalb County's reduced population growth for the decade.

Figure 5.9 reveals that the high and low cohort-component projections provide a reasonable continuation of the 2010–2020 population trend on which they are based. However, it also reveals that the 2010–2020 period was an unusual period in the county's past, which questions all the cohort-component projections.

TABLE 5.20 **Observed and Projected Total Population for Different Growth Rates: DeKalb County, 2010 and 2040**

Projection	2010 Population[a] P^{2010}	2040 Population P^{2040}	Percent Change
(1)	(2)	(3)	(4)
Constant rates	691,893	783,455	13.2
Projected fertility rates	691,893	770,835	11.4
Projected survival rates	691,893	798,884	15.5
Ten percent net migration decrease	691,893	791,159	14.3
Ten percent net migration increase	691,893	775,751	12.1

Source:
[a] US Bureau of the Census (2013b).

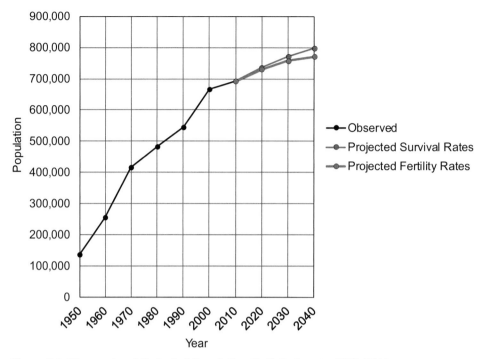

Figure 5.9 Observed and Projected Population: DeKalb County, 1950–2040

Evaluating the Projected Population Change by Age and Sex

DeKalb County's observed 2010 population and 2040 projected population by age and sex are reported in table 5.21 and figure 5.10. The projected values assume constant mortality, fertility, and net migration rates, but the population structures for the other projections are similar. The county's male and female population are projected to grow for the population under forty, decline for the population between forty and fifty-five, and more than double for the population over seventy. The dramatic growth in the county's older population suggests that DeKalb County and its communities should plan on improving the services it provides its senior citizens.

Evaluating the Components of Population Change

Analysts must also consider the factors that shape the projected population trends. This requires examining the three components of population change—births, deaths, and migration—to determine whether they correspond to recent trends and the expected changes in the future.

   Figure 5.11 plots the observed number of deaths in DeKalb County between 1995 and 2010 and the projected number of deaths from 2010 to 2040 for constant survival rates and the projected survival rates. The projected deaths are substantially lower for the projected survival rates reflecting the survival rate increases in table 5.14. The constant survival rates seem to provide a better continuation of the county's past trends.

   Figure 5.12 plots the projected number of live births from 2010 to 2040 for constant fertility rates and the projected fertility rates in table 5.15. Births for the constant rates projection grow steadily as the county's population grows, which

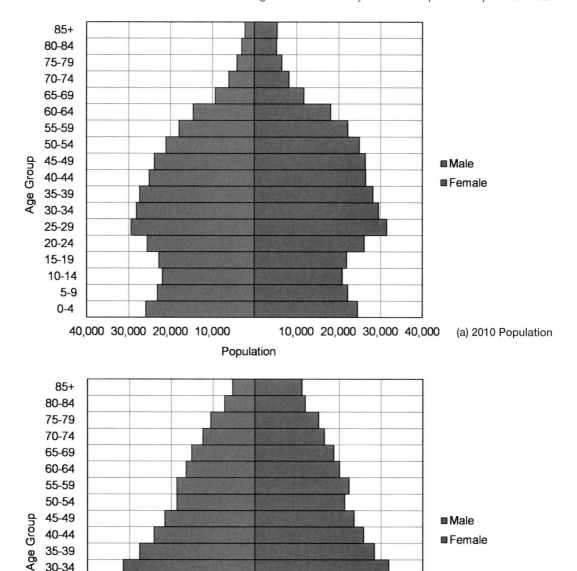

Figure 5.10 Observed and Projected Population by Age and Sex: Constant Rates Projection, DeKalb County, 2010 and 2040.

seems reasonable. Births for the projected fertility rates decline by more than 1,500 between the 2005–2010 projection period and the 2015–2020 projection period and do not match the earlier rates for twenty years. This is unlikely in a county with a steadily growing population.

| TABLE 5.21 | | Observed and Projected Population by Age and Sex, Constant Rates Projection: DeKalb County, 2010 and 2040 | | | | | |

Cohort	Age Interval	Male Population			Female Population		
		2010	2040	Percent Change	2010	2040	Percent Change
n		$\left(PM_n^{2010}\right)^a$	PM_n^{2040}		$\left(PF_n^{2010}\right)^a$	PF_n^{2040}	
(1)	(2)	(3)	(4)	(5)	(6)	(7)	(8)
1	0–4	25,856	27,606	6.8	24,551	26,276	7.0
2	5–9	23,110	25,710	11.3	22,180	24,602	10.9
3	10–14	21,915	24,587	12.2	20,882	23,456	12.3
4	15–19	22,737	25,230	11.0	21,906	24,769	13.1
5	20–24	25,555	29,390	15.0	26,144	30,095	15.1
6	25–29	29,422	32,227	9.5	31,514	33,202	5.4
7	30–34	28,134	31,497	12.0	29,579	31,865	7.7
8	35–39	27,346	27,553	0.8	28,245	28,456	0.7
9	40–44	25,043	24,101	–3.8	26,511	25,859	–2.5
10	45–49	23,810	21,506	–9.7	26,424	23,586	–10.7
11	50–54	21,027	18,622	–11.4	24,984	21,301	–14.7
12	55–59	17,928	18,633	3.9	22,198	22,385	0.8
13	60–64	14,530	16,459	13.3	18,134	20,101	10.8
14	65–69	9,311	15,108	62.3	11,793	18,751	59.0
15	70–74	6,107	12,415	103.3	8,254	16,530	100.3
16	75–79	4,182	10,532	151.8	6,552	15,169	131.5
17	80–84	3,046	7,293	139.4	5,287	11,996	126.9
18	85+	2,296	5,380	134.3	5,400	11,207	107.5
Total	–	331,355	373,848	–	360,538	409,607	–

Source:
[a] US Bureau of the Census (2013b).

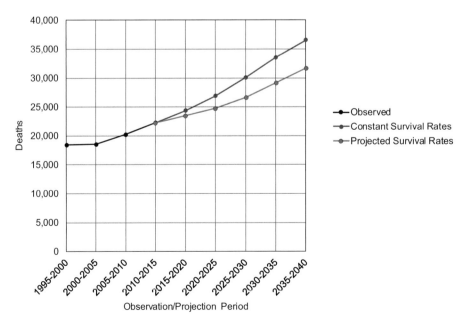

Figure 5.11 Observed and Projected Deaths: DeKalb County, 1995–2040

Figure 5.13 displays DeKalb County's projected net migration for constant net migration projection and 10 percent increases and decreases in the projected net migration rates. The projected number of net out-migrants is consistently smaller for the reduced net migration rates and consistently larger for the increased net migration rates, as reported in table 5.18.

Table 5.22 examines the effect that the three components of population change have on DeKalb County's projected population change between 2010 and 2040. The projected population for the projected survival rates is the largest, reflecting the smaller number of deaths reported in figure 5.11. The projected population for

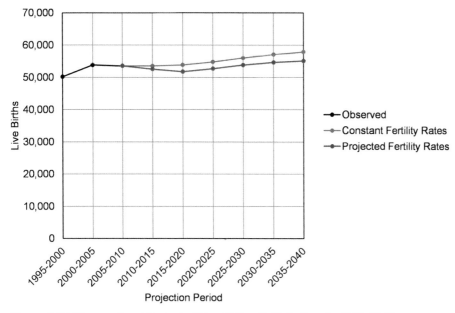

Figure 5.12 Observed and Projected Live Births: DeKalb County, 2005–2040

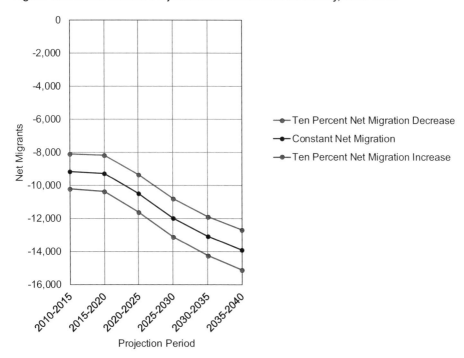

Figure 5.13 Projected Net Migration: DeKalb County, 2010–2040

TABLE 5.22   **Components of Population Change: DeKalb County, 2010–2040**

Projection	2010 P^{2010}	Births $B^{2010,2040}$	Deaths $D^{2010,2040}$	Net Migrants $NM^{2010,2040}$	2040 P^{2040}
(1)	(2)	(3)	(4)	(5)	(6)
Constant rates	691,893	333,196	173,758	−67,876	783,455
Projected fertility rates	691,893	320,456	173,638	−67,876	770,835
Projected survival rates	691,893	332,865	157,998	−67,876	798,884
Ten percent net migration decrease	691,893	334,626	174,335	−61,025	791,159
Ten percent net migration increase	691,893	331,759	173,181	−74,720	775,751

the projected fertility rates is the smallest, reflecting the lower number of projected births reported in figure 5.12. The other population projections are between these extremes. Together, figures 5.11, 5.12, and 5.13 and table 5.22 suggest that the constant rate projection of roughly 783,000 provides the best cohort-component projection for DeKalb County's 2040 population.

Notes

[1] The discussion in this section draws on Molly et al. (2011) and Ihrke (2014).
[2] The discussion of life tables is based on Klosterman (1990, 65–77).
[3] See, e.g., Centers for Disease Control and Prevention (2016b, 2016c).
[4] E.g., Centers for Disease Control and Prevention (2016d) provides live birth counts to residents and non-residents of the United States by state, county, the child's gender, the mother's age and race, and many other fertility-related variables.
[5] Shryock et al. (1975, 195–196); Mathews and Hamilton (2005).
[6] The following discussion draws on Isserman (1993).
[7] Other survival and fertility rate projection methods are described in Smith et al. (2001).
[8] See, e.g., Colby and Ottman (2015) and US Bureau of the Census (2014a).
[9] Several adjustments are required to use the Census Bureau's projected mortality and fertility data. First, the Census Bureau's projected mortality data are only reported by race and ethnicity, and not for the population as a whole. To account for this, the projected national rates were weighted by DeKalb County's 2010 population shares by race and ethnicity. Second, the published age-specific fertility rates are not reported for male and female births. The fifty-year average sex ratio at birth, 105.4, was used to convert the national rates for the total population to the equivalent rates for male and female births. And, third, the Census Bureau's fertility and mortality rate projections are for one-year age cohorts, which had to be converted into adjusted five-year rates.
[10] See, e.g., Isserman (1985).

Economic Analysis Methods 6

A **local economy** consists of producers and consumers exchanging goods and services, some of which occurs in a city, town, or region. The local economy includes residents and in-commuters who work in the place; locally based businesses, public or not-for-profit organizations producing goods or services for local consumption or export; and resident and nonresident consumers buying goods and services from local firms. It also includes residents receiving income from outside the place such as workers commuting to jobs in nearby localities, retirees drawing pensions from nonlocal sources, and low-income families receiving transfer payments from the government. Nearly all local economies are part of larger and more complex regional economies which, in turn, operate within even larger and more complex national and global economies.

Understanding the local economy's linkages to larger economies and its prospects for growth or decline is essential for sound planning. The local economy influences—and is influenced by—an area's physical, demographic, and social characteristics. Local economies drive the property, sales, and other tax and fee receipts that fund public services and investments that are essential for implementing a community's plans. An underperforming local economy that fails to provide adequate jobs and incomes for its residents exerts demands on local public services and resources that could be used to meet other community needs. Conversely, a rapidly growing economy can drive substantial in-migration that outpaces available public goods such as schools, roads, and parks, leading to congestion, environmental degradation, and an overall reduction in the local quality of life.

This chapter describes several simple techniques for understanding the core features of a local economy. The techniques are applied to the City of Decatur and DeKalb County, Georgia. It also offers guidelines for informative local economic analysis. The general lesson is that local economic analysis that is useful for informing city planning does not depend on vast quantities of data and highly sophisticated statistical models and methods. It does, however, require assembling data from a diversity of sources, paying careful attention to the sources and their limitations, and then thoughtfully interpreting the information that has been collected.

Concepts and Definitions
Establishments, Firms, Sectors, and Industries

Information on businesses and jobs can be organized in different ways. The information can be reported for establishments or for firms. An **establishment** (or **plant**) is a place where business is conducted, services are provided, or industrial operations are performed. Establishments include factories, stores, hotels, movie theaters, mines, farms, airline terminals, sales offices, warehouses, and central administrative offices. An establishment is typically a single physical location, though administratively distinct operations at a single location may be treated as separate establishments. Each establishment is classified within an industry by the primary good or service being developed or produced there.

An **firm** (or **enterprise**) consists of one or more than one location performing the same or different types of economic activities. A firm may operate a single establishment or multiple establishments; enterprises operating more than one

establishment are called multisite firms. Most readily available sources of local industry economic data are reported for establishments and not for firms. Firm-level data
generally are available only from proprietary sources.

An **industry** is made up of business units (usually establishments but sometimes
firms) that are grouped together because they produce similar products or use similar
production methods or technologies. A **sector** is a group of industries. There is no
technical difference between industries and sectors; indeed, the terms are often used
interchangeably. However, sectors generally tend to be larger, broader, and more het-
erogeneous than industries. As a practical consideration, data by industry are among
the most readily available kinds of disaggregated information on local economies.

It is important to determine how the data obtained from different organizations
are collected and defined. For example, information on an area's employment may be
available from different sources that cover different populations and time intervals.
Analysts using data from secondary sources implicitly assume that the data properly
measure the concept being studied and are consistent with the data obtained from
other sources.

North American Industrial Classification System

The **North American Industry Classification System** (**NAICS**, pronounced
"nakes") is the standard for classifying information on establishments and firms in
the United States, Canada, and Mexico. NAICS groups establishments and firms
into industries according to similarities in the processes they use to produce goods or
services.

The NAICS codes are a two- through six-digit hierarchical classification system
that offers five levels of detail. The digits in the code define a series of progressively
narrower categories; more digits in the code signify a greater classification detail.
The first two digits identify an economic sector; the third digit identifies a subsector; the fourth digit identifies an industry group; the fifth digit identifies a NAICS
industry; and the sixth digit identifies a national industry. For example, NAICS 11,
Agriculture, Forestry, Fishing, and Hunting, includes NAICS 111, Crop Production;
which includes NAICS 1113, Fruit and Tree Nut Farming; which includes NAICS
11133, Noncitrus Fruit and Tree Nut Farming; which includes NAICS 111331,
Apple Orchards. The five-digit NAICS codes are comparable for most of the NAICS
sectors in the three countries participating in NAICS. The six-digit level allows each
country to add country-specific detail. A complete NAICS code contains six digits.

The North American Industry Classification System replaced the **Standard
Industrial Classification** (**SIC**) system in 1997. Economic data generally are
reported by SIC code prior to 1997 and by NAICS code after that date. The NAICS
system is revised every five years to keep pace with the changing economy. For example, there were 1,175 industries in the 2007 NAICS for the United States and 1,065
industries in the 2012 version. The changes included content revisions for some
industries, several title changes, and the clarification of a few industry definitions. The
changes most often are made for detailed industries (four-, five-, or six-digit NAICS
codes) but analysts must be alert to changes in the NAICS categories because the
industry codes and labels may not be consistent over time.

Estimating Missing Values

Many public data providers are required by law or regulation to withhold data that
could violate the privacy of individual respondents or compromise mandated con-
fidentiality protection. The withheld data may be presented in an abridged form

(e.g., providing a range of values and not a precise value) or may be unavailable entirely. The portion of suppressed data can be high for smaller geographic areas or populations and for more disaggregated classifications. Private data vendors provide estimates for some information that is suppressed in government sources, typically for a fee.

Title 13 of the US Code prohibits the disclosure of responses, addresses, or personal information on individuals or establishments by the Bureau of the Census and its component agencies. As a result, the Bureau of the Census does not report the number of employees or establishments when this information would reveal, or permit a reader to deduce, information pertaining to an individual establishment or firm. In these cases, it uses an alphabetic flag to identify the employment range in which the industry lies. Thus, for example, **County Business Patterns (CBP)** provides an "a" when an industry's employment is between 0 and 9, a "b" when the employment is between 20 and 99, a "c" when it is between 100 and 249, and so on. As a result, analysts must do their best to estimate the employment or income for industries that only record an alphabetic flag.

It may seem that the employment estimate should be simply the **arithmetic mean** (or **average**) of the endpoints of an employment range. Thus, for example, the estimated employment for industries with an "a" flag would be the average of 0 and 9, or 4.5; the employment estimate for industries with a "b" flag would be an average of 20 and 99, or 54.5; and so on. However, the arithmetic mean does not provide the best estimate because the establishments in a range typically are more frequent at the bottom of a range so that the arithmetic mean overestimates the number of employees in a range. The employment estimates for missing values should be computed with the **geometric mean**, not the arithmetic mean.[1]

A similar situation occurs in estimating the average income of individuals and households which are reported in broad income categories. For example, table 6.1 uses the arithmetic mean and the geometric mean to estimate the average income of Decatur's households for the three-year period between 2011 and 2013. The observed average household income is $99,554. Column two records the number of households in the four income ranges in column one. Columns three and four record the low and high values for each range. The high value for the $150,000 or more range is an estimate; this value is not provided in the census data. Column five records the arithmetic mean of the low and high-income ranges. For example, the

TABLE 6.1   **Estimating Average Household Income: Decatur, 2011–2013**

Income	Households[a]	Income Range		Arithmetic Mean		Geometric Mean	
		Low	High	Mean Income	Total Income	Mean Income	Total Income
(1)	(2)	(3)	(4)	(5)	(6)	(7)	(8)
Less than $25,000	2,238	$1	$24,999	$12,500	$27,975,000	$158	$353,852
$25,000 to $74,999	2,090	$25,000	$74,999	$50,000	$104,498,955	$43,301	$90,499,051
$75,000 to $149,999	2,447	$75,000	$149,999	$112,500	$275,286,277	$106,066	$259,542,679
$150,000 or more	1,810	$150,000	$500,000	$325,000	$588,250,000	$273,861	$495,688,915
Total	8,585	—	—	—	$996,010,232	—	$846,084,497
Average	$99,594	—	—	—	$116,017	—	$98,554

[a]*Source:* US Bureau of the Census (2015c).

arithmetic mean for the first income range is ($1 + $24,999)/2 or $12,500. The total income values in column six are computed by multiplying the arithmetic mean by the number of households in each range. The total income for these estimates is computed at the bottom of the column and, when divided by the total number of households to compute the estimated average income, $116,017, is substantially larger than the true average value reported in column two.

The geometric mean income values in column seven are the square roots of the products of the low- and high-income values. For example, the geometric mean for the lowest income range is $\sqrt{\$1 \times \$24,999}$ or $158. The geometric mean for the second income range is $\sqrt{\$25,000 \times \$74,999}$ or $43,301. The estimated average household income for the geometric mean, $96,554, is much closer to the true value than the arithmetic mean. The geometric mean estimates are also less sensitive to the estimated maximum household income in the top income range. Increasing or reducing this estimate by $100,000 changes the arithmetic estimate by $10,000 and the geometric estimate by $6,000. Adjusting the maximum household income to $525,000 and the geometric mean yields an average household income of $99,980, very close to the observed value ($99,598).

The geometric mean can be used in a similar way to estimate the employment in industries with suppressed data.[2] For example, CBP report a suppressed data flag of "b" (20–99 employees) for DeKalb County's 2013 employment in NAICS 21, Mining, Quarrying, Fishing, and Hunting. The employment in this industry can be estimated by computing the geometric mean of the upper and lower range values, $\sqrt{20 \times 99}$ or 44.

Fortunately, CBP also report the number of establishments in several employment-size categories, 1–4 employees, 5–9 employees, and so on, which can be used to improve employment estimates for suppressed values. Table 6.2 uses this information to estimate DeKalb County's 2013 employment in Mining, Quarrying, and Oil and Gas Extraction (NAICS 21). Column two records the number of establishments in the employment-size categories in column one. Column three reports the geometric mean of the upper and lower values in each employment-size category. The total employment in each employment size category is equal to the number of establishments in column two multiplied by the geometric mean employment in column three. The estimated total employment, 97, is within the county's reported employment range (20–99) for NAICS 21.

TABLE 6.2	**Estimating Employment in Mining, Quarrying, and Oil and Gas Extraction: DeKalb County, 2013**

| Employment-Size Class | Number of Establishments | Employees | |
		Geometric Mean	Total
(1)	(2)	(3)	(4)
1–4	3	2.00	6
5–9	1	6.71	7
10–19	1	13.78	14
20–49	0	31.30	0
50–99	1	70.36	70
Total	**6**	—	**97**

Source: US Bureau of the Census (2015a).

Similar procedures can be used to estimate the employment for all industries with suppressed values. For example, Decatur's 2013 employment estimates in column three of table 6.3 were computed with geometric means as described above. However, the sum of the employment estimates, 12,686, does not agree with the CBP total, 12,257. The adjusted employment estimates in column four are computed by multiplying the employment estimates in column three by the ratio of the CBP total and the estimated total. Thus, for example, the adjusted estimate for NAICS 23, Construction, is equal to $(12,257/12,686) \times 260$ or 251. The sum of the adjusted values equals the CBP total, as it should.[3]

Analysis Guidelines

An effective economic analysis requires the selection and application of appropriate methods and the critical and insightful assessment of key relationships in the data. Experienced analysts not only present important and useful information to their clients, they also develop and present the information within a narrative that ties together otherwise disparate or discordant parts of the analysis and places the key pieces of evidence in their most informative context.

Creating an effective economic analysis demands sound judgment based on considerable expertise. This judgment and expertise can be obtained in three ways: observation, exposure, and practice. Local economic analysts must be keen *observers* of their local economies and how they are situated within the regional, national, and global economies. Developing familiarity with trade and industry publications, becoming knowledgeable of the characteristics and concerns of major local employers, engaging

TABLE 6.3 — **Estimating Employment by Major Sector: Decatur, 2013**

NAICS Code	NAICS Sector	Employment	
		Estimated	Adjusted
(1)	(2)	(3)	(4)
11	Agriculture, Forestry, Fishing, and Hunting	0	0
21	Mining, Quarrying, and Oil and Gas Extraction	0	0
22	Utilities	7	7
23	Construction	260	251
31	Manufacturing	381	368
42	Wholesale Trade	870	841
44	Retail Trade	864	835
48	Transportation and Warehousing	86	83
51	Information	377	364
52	Finance and Insurance	229	221
53	Real Estate and Rental and Leasing	223	215
54	Professional, Scientific, and Technical Services	1,364	1,318
55	Management of Companies and Enterprises	2	2
56	Administrative and Support Services	729	704
61	Educational Services	1,716	1,658
62	Health Care and Social Assistance	2,669	2,579
71	Arts, Entertainment, and Recreation	122	118
72	Accommodation and Food Services	1,699	1,642
81	Other Services (except Public Administration)	1,072	1,036
99	Industries not classified	16	15
—	**Total**	**12,686**	**12,257**

Source: US Bureau of the Census (2015a).

other local economic experts in the public and private sectors, and tracking local and regional economic trends on a continuous basis are essential. *Exposure* to economic analyses prepared by both novice and experienced analysts also is important. Closely examining the methods, interpretations, and explanations of other analysts, and considering how they might be improved helps develop the depth of understanding necessary for good analysis. And, lastly, *practice*: there is no substitute for assembling the raw materials of data and geography, devising and undertaking analyses, and then presenting the results to audiences, informal and formal, untrained and expert, to gather feedback, gain experience, and develop mature judgment.

Selecting an Appropriate Geographic Scale

The geographic scale best suited for investigating a local economy often is not obvious and, therefore, demands explicit consideration. Economies do not follow political boundaries within a nation's borders: they operate simultaneously across jurisdictions and at multiple scales. Therefore, there is a modicum of truth in the contention that "there is no such thing as a local economy." People commute from their residences to workplaces both near and far; firms purchase inputs from suppliers and sell products to purchasers both within and outside the region; and legislation, regulations, and customs at the state, national, and international scales may be as influential as policies and plans enacted at the regional and local levels. As a result, many, perhaps most, of the factors that influence an area's economy may be outside the control of a local planner.

Conceptually, the perfect spatial unit for analyzing an economy is a **functional economic area**—a region that encompasses all the activities, flows, and influences relevant to the economy in question. This idealized unit, of course, does not exist. It would be impossible to place a spatial boundary that unambiguously separates a functional economic area from its neighbors; some degree of overlap and exchange is present at any spatial scale. Moreover, the choice of borders is constrained in practice by the availability of data conforming to political jurisdictions. In the United States, the multicounty **metropolitan statistical areas (MSAs)** described in appendix A are commonly adopted as approximations of functional economic areas for urban and suburban regions. When working with smaller areas, such as counties or municipalities, it is important to understand how the activities taking place within its borders depend on—and interact with—the functioning of a larger economic area. For example, this chapter examines the economy of Decatur along with, and in comparison to, relevant features of the surrounding areas of DeKalb County and the Atlanta metropolitan region.

Selecting an Analysis Period

It is generally best to use the latest information to analyze an area's economy. For example, 2013 was used in this analysis because it is the most recent year for which data were available when this chapter was written. However, it may be preferable to exclude the most recent year, or years, to avoid unusual circumstances or data oddities, if the decision can be explained easily to the intended audience. Thus, for example, 2008, the first whole year of the "Great Recession" of 2007–2009 might wisely be avoided.

The comparison years for the past should be selected to include enough time to include appreciable changes in the economy but not so long that fundamental economic changes make them irrelevant for current or emerging planning concerns. Three to ten years is generally a reasonable observation period. Longer time periods can be divided into several shorter time spans that correspond to major structural changes in the local, regional, or national economy.

National macroeconomic conditions should also be considered in selecting the comparison period. If data limitations make this impossible, the data limitations should be considered in interpreting the data. Comparison years should be selected that are at similar points in the national business cycle to facilitate direct comparisons and avoid misleading temporary upswings and downswings. For instance, analyzing data for a period spanning from the trough of an economic cycle to its peak overstates the rate of local growth.

Figure 6.1 reveals that 2013 continues America's slow employment recovery from a dramatic economic downturn between 2007 and 2011. 2003 corresponds to the end of a mild nationwide recession that interrupted a general growth trend from 1999 to 2007. It also reveals that Decatur's employment growth does not parallel the nation's. The city's employment declined between 2001 and 2007 when the national economy grew, and it grew between 2007 and 2011, while the nation's employment declined substantially. The section Analyzing Decatur's Local Economy below attempts to understand these counter-cyclical trends.

Ensuring Consistency

Consistency is an important objective of economic analysis and projection methods. Two types of consistency are important. The projected values should first be consistent across the relevant levels of sectoral or geographic aggregation. The projected employment for an industry sector should equal the sum of projected employment for all its component industries. Similarly, the projected employment for all the industrial sectors in an economy should match the projection for the entire economy.

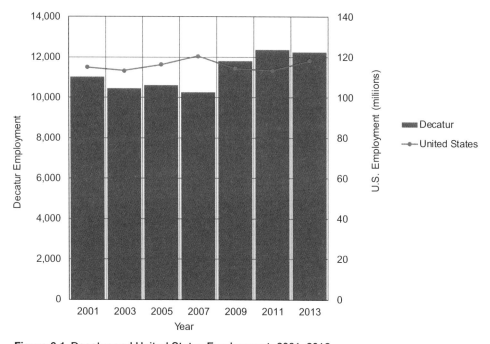

Figure 6.1 Decatur and United States Employment, 2001–2013

Source: US Bureau of the Census (n.d.b, n.d.c).

In addition, a city's projected employment should correspond to the corresponding projections for the county, metropolitan area, or other encompassing region in which it is located. This **aggregate consistency** is particularly valuable when the forecasts serve as inputs to other planning actions and decisions.

Aggregate consistency can be achieved by privileging selected projections or levels of aggregation and adjusting the remaining projections to match the privileged projections. Table 6.4 illustrates three ways this can be done. The projected employment values for the six NAICS subsectors in column three sum to 120; the projected total employment for the sector is 100. Columns four through six illustrate three adjustment methods: a top-down adjustment, a bottom-up adjustment, and a selective top-down adjustment.

Top-Down Adjustment The **top-down adjustment** retains the top-level value and adjusts the bottom-level values to match the top-level value. The top-level adjustment factor is equal to the ratio of the top-level value to the computed total, that is, 100/120 or 0.833 for this example. The bottom-level values are multiplied by the adjustment factor. For example, the adjusted value for NAICS 511 in column four is equal to 60.00 × 0.833 or 50.00. The sum of the adjusted values is equal to the projected total employment, as this adjustment requires.

Bottom-Up Adjustment The **bottom-up adjustment** preserves the projections for the lower level of aggregation and replaces the top-level projection by the sum of the lower-level projections. Thus, all the lower-level values for the bottom-up adjustment in column five are retained and the sector total is replaced by their total, 120.

Selective Top-Down Adjustment The **selective top-down adjustment** preserves the projections for the top level of aggregation and for one or more lower-level projections. The values in column six assume that the projected employment for Sector 51 and in Subsector 511 should be retained and the other subsector values should be adjusted. The adjustment factor is equal to the difference between the top-level value (100) and the value to be retained (60), divided by the sum of the remaining bottom-level values (60), that is, $(100-60)/60$ or 0.667. The bottom-level values to be modified are multiplied by this ratio. For example, the adjusted value

TABLE 6.4 **Employment Adjustment Methods**

NAICS Code	NAICS Sector/Subsector	Projected Employment	Top-down Adjustment	Bottom-up Adjustment	Selective Adjustment
(1)	(2)	(3)	(4)	(5)	(6)
511	Publishing Industries (except Internet)	60.00	50.00	60.00	60.00
512	Motion Picture and Sound Recording Industries	20.00	16.67	20.00	13.33
513	Broadcasting (except Internet)	10.00	8.33	10.00	6.67
517	Telecommunications	10.00	8.33	10.00	6.67
518	Data Processing, Hosting, and Related Services	10.00	8.33	10.00	6.67
519	Other Information Services	10.00	8.33	10.00	6.67
51	**Information**	**100.00**	**100.00**	**120.00**	**100.00**

for NAICS 512 is equal to 20×0.667 or 13.33. The adjusted values in column six retain the total employment in Sector 51 and Subsector 511, as the method ensures.

Other approaches are possible, such as adjusting the projections for the smallest, largest, or most volatile industries while preserving the remaining projections. In each case, the essential principle is retaining the most reliable or important projections and ensuring consistency by modifying the remaining projections.

A second type of consistency reflects the linkages between the projected variable and other characteristics of the study area. Employment change in a closed economy is determined in conjunction with demographic and social features such as population change, government activity, and economic policy. For instance, an area's employment often is related to its population because everything else being equal, larger populations are associated with more employment. This association reflects the fact that an area's population provides a reasonably accurate indicator of the demand for local products and services and also of the labor available for filling job openings. However, the relationship between population and employment can be modified by local characteristics such as the population's age distribution, labor-force participation (i.e., how many people are able and willing to be employed), and the degree to which the skills of the populace match the needs of employers.

Substantial effort in developing employment projections often goes into taking an area's other characteristics into account.[4] However, this consistency is not relevant for all local projections. Population, employment, and other regional features can be assumed to be highly interrelated in an independent or nearly independent region, such as a metropolitan area or a geographically isolated municipality. However, consistency among local features is not as relevant for a locality such as Decatur that is socially embedded and economically intermingled within a larger region. In this case, the large share of employees that commute into or out of Decatur overwhelms the relationship between employment and population characteristics at the municipal scale.

In some cases, consistency between an area's employment projections and other local features is central. Industry sectors such as education and social services are functionally dependent on the local population. Other sectors such as municipal government employment may be regulated at the local level. And sectors such as natural resource extraction and public services may be strongly influenced by local assets or attitudes. Methods for ensuring consistency in these situations are dependent on the relationships in a location that will not be considered here.

Making Comparisons

Comparisons give economic analyses much of their potency. Both quantitative metrics and qualitative assessments convey useful information about a local economy and inform planning and policy making by invoking comparisons to larger encompassing regions, competing jurisdictions, and exemplar or archetypal locations. The most useful comparisons depend on the purpose of the analysis, the character of the study area, and the available information.

Another type of comparison explores changes over time, an approach that is particularly helpful when benchmark locations are difficult to identify or comparable information for those locations is not available. Comparing the current situation of an area to its past conditions or within ongoing trends emphasizes what is different—and the same—now. Both comparison strategies—across locations and over time—are illustrated in this chapter.

Analyzing Decatur's Local Economy

The residents and businesses in Decatur are part of the large and dynamic Atlanta metropolitan region, which is part of a growing state within the rapidly changing southeastern part of the United States. Data on the economic linkages between municipalities and their immediate region and beyond are extremely limited. Nevertheless, we can draw an initial picture of Decatur's economy and its economic role in the Atlanta region with a few basic pieces of information gleaned from local government websites, other mapping sources, and the Census Bureau's QuickFacts website (http://www.census.gov/quickfacts).

As pointed out in chapter 2, Decatur is in DeKalb County, about six miles east of Atlanta's city center. The city is tied to the rest of the region by three stops on the Metropolitan Atlanta Regional Transit Authority (MARTA) blue line and ten MARTA bus routes. On average the city's residents are wealthier and better educated than the remainder of the county, the region, and the state.

Decatur is the county seat, and DeKalb County government is its largest employer. Other major employers include Agnes Scott College, DeVry University, the US Postal Service, the City of Decatur, DeKalb Medical Hospital, the Columbia Theological Seminary, and satellite offices and clinics of nearby Emory University. Most of the employers are concentrated in Decatur's town center. Because there is very little vacant land for accommodating extensive industrial or commercial development, the city's economic development strategy prioritizes selective mixed-use development.

This initial picture—formed quickly from readily accessible online sources—is of a prosperous local economy concentrated in professional, health, education, and government services that provides desirable neighborhoods and homes with easy access to Atlanta's core and Emory University's main campus. Even without digging into the data further, the juxtaposition of a services-based economy, that relies on both high- and low-skilled labor (and thus both high and low wages), and comparatively high housing costs suggests that Decatur has considerable cross-commuting, with lower-skilled workers commuting in from lower cost neighborhoods and higher skilled workers commuting out to job centers elsewhere in the region.

Decatur's Aggregate Economy

A detailed analysis of a local economy begins by examining it as a single unit. Ideally, this analysis would include both private sector and public sector activities. However, this may not be possible in many cases. Public sector employment data in the United States are rarely measured or made available in the same way as private sector employment data, particularly for small areas. This means that reliable public sector employment estimates can be prepared for counties and states but not for subcounty areas. As a result, municipal-scale economic analyses in the United States generally focus on the private sector.

Figure 6.1 highlights three years that are important for understanding Decatur's recent economic history: 2003, the beginning of a major local recession; 2007, the recession's low point; and 2013, the most recent year for which employment data are available when this was written. As table 6.5 reports, employment in Decatur's private sector businesses was 12,257 in 2013.[5] That is about 8 percent more jobs than in 2007, representing a substantial gain over a period that included a national recession (2007–2011) and a slow recovery (2011–2013). The rise in Decatur's private sector employment over this period contrasts with the performance of DeKalb County, the Atlanta region, the state of Georgia, and the United States, which all had declining employments between 2007 and 2013. In contrast, Decatur's employment declined between 2011 and 2013 while the employment grew in all the other areas. The differences between Decatur and the larger regions of which it is a part indicate that the

TABLE 6.5	Employment and Establishment Comparisons, Decatur 2003, 2007, and 2011						

Area	Variable	Year			Percent Change		
		2007[a]	2011[b]	2013[c]	2007–2011	2011–2013	2007–2013
(1)	(2)	(3)	(4)	(5)	(6)	(7)	(8)
Decatur	Employees	11,303	12,824	12,257	13.5	−4.4	8.4
	Establishments	1,079	1,121	1,159	3.9	3.4	7.4
	Employees per Establishment	10.5	11.4	10.6	9.2	−7.6	1.0
DeKalb County	Employees	282,022	252,171	260,194	−10.6	3.2	−7.7
	Establishments	17,233	16,003	16,052	−7.1	0.3	−6.9
	Employees per Establishment	16.4	15.8	16.2	−3.7	2.9	−1.0
Atlanta Region	Employees	2,214,273	2,003,708	2,119,397	−9.5	5.8	−4.3
	Establishments	137,758	128,050	131,783	−7.0	2.9	−4.3
	Employees per Establishment	16.1	15.6	16.1	−2.6	2.8	0.1
Georgia	Employees	3,648,670	3,328,033	3,458,050	−8.8	3.9	−5.2
	Establishments	231,810	214,635	217,559	−7.4	1.4	−6.1
	Employees per Establishment	15.7	15.5	15.9	−1.5	2.5	1.0
United States	Employees	120,604,265	113,427,241	118,266,253	−6.0	4.3	−1.9
	Establishments	7,705,018	7,354,043	7,488,353	−4.6	1.8	−2.8
	Employees per Establishment	15.7	15.4	15.8	−1.5	2.4	0.9

Source:
[a] US Bureau of the Census (2009b).
[b] US Bureau of the Census (2013d).
[c] US Bureau of the Census (2013b).

Decatur economy does not simply mirror the economy of the county, region, and nation in which it is located.

As table 6.5 indicates, there were 1,159 private sector establishments in Decatur in 2013, 3.4 percent more than in 2011 and 7.4 percent more than in 2007. The average establishment in Decatur employed fewer than eleven people in 2013; the county, region, state, and nation averaged more than fifteen employees per establishment. This suggests that Decatur's businesses may simply be smaller on average or the city's industrial mix may be tilted toward business sectors with smaller establishments.

The employment data for the City of Decatur in table 6.5 are for zip code 30030, which includes Decatur and adjacent parts of DeKalb County; its boundaries changed very little between 2003 and 2013. As a result, the estimates for the city's employment and establishments are slightly overstated. The employment values for industries with suppressed values were estimated by using the procedures described earlier in this chapter.

Decatur's Industry Composition

The economy of any place is not a monolith, and this is true for Decatur. A local economy is made up of large and small businesses that provide many kinds of goods and services and employ workers with a wide range of skills. Examining the local industrial composition—its major economic sectors and industries and how they combine to form the local economy—can be useful in understanding this variation and its implications for local planning.

As table 6.6 shows, Decatur has substantial employment in four industries with more than one thousand workers in 2011 and 2013: Professional, Scientific, and Technical Services (NAICS 54); Educational Services (NAICS 61); Health Care and Social Assistance (NAICS 62); and Accommodation and Food Services (NAICS 72). The Health Care and Social Assistance sector is considerably larger than the others, accounting for more than one fifth of the city's private sector jobs in 2013.

TABLE 6.6   **Employment by Major Employment Sector: Decatur 2011 and 2013**

NAICS Code	NAICS Sector	Decatur Employment		Change (2011–2013)		Employees per Establishment (2013)	
		2011[a]	2013[b]	Number	Percent	Decatur	United States[c]
(1)	(2)	(3)	(4)	(5)	(6)	(7)	(8)
11	Agriculture, Forestry, Fishing, and Hunting	0	0	0	n.a.	n.a.	7.0
21	Mining, Quarrying, and Oil and Gas Extraction	0	0	0	n.a.	n.a.	25.5
22	Utilities	7	7	0	1.0	6.8	35.5
23	Construction	259	251	−8	−2.9	4.6	8.3
31	Manufacturing	289	368	79	27.5	12.3	38.6
42	Wholesale Trade	143	841	698	488.3	25.5	14.1
44	Retail Trade	1,432	835	−597	−41.7	8.6	14.1
48	Transportation and Warehousing	150	83	−67	−44.5	10.4	19.9
51	Information	135	364	229	169.5	11.8	24.1
52	Finance and Insurance	319	221	−97	−30.6	5.4	12.8
53	Real Estate and Rental and Leasing	182	215	33	18.1	3.7	5.5
54	Professional, Scientific, and Technical Services	1,601	1,318	−283	−17.7	4.1	9.5
55	Management of Companies and Enterprises	2	2	0	0.1	1.9	58.6
56	Administrative and Support Services	1,057	704	−353	−33.4	19.6	25.9
61	Educational Services	1,545	1,658	113	7.3	40.4	35.8
62	Health Care and Social Assistance	2,806	2,579	−227	−8.1	14.9	22.0
71	Arts, Entertainment, and Recreation	92	118	26	28.5	4.4	16.6
72	Accommodation and Food Services	1,440	1,642	201	14.0	18.2	18.4
81	Other Services (except Public Administration)	914	1,036	121	13.3	10.0	7.2
99	Industries not classified	10	15	6	60.1	1.9	0.9
—	**Total**	**12,381**	**12,257**	**−124**	**−1.0**	**10.6**	**15.8**

Note: n.a = not applicable

Source:
[a] US Bureau of the Census (2013d).
[b] US Bureau of the Census (2013b).
[c] US Bureau of the Census (2015a).

Decatur's Industrial Specialization

It is hard to identify the features of a local economy that are interesting, unusual, or important by considering the economy in isolation. Relevant comparisons—to larger regions, states, similar local economies elsewhere, or the nation—help uncover what is unique about a place. Table 6.7 compares Decatur's industry employment mix to the nation's. Decatur's economy is clearly different from the nation's. The four sectors that dominate Decatur's employment (highlighted in table 6.7) do not dominate the national economy. Decatur's employment share also is much smaller for two industries that are important for the nation: Manufacturing (NAICS 31) and Retail Trade (NAICS 44).

| TABLE 6.7 | Employment by Major Employment Sector: Decatur and United States, 2013 |

NAICS Code	NAICS Sector	Decatur		United States	
		Employment[a]	Percent	Employment[b]	Percent
(1)	(2)	(3)	(4)	(5)	(6)
11	Agriculture, Forestry, Fishing, and Hunting	0	0.0	154,496	0.1
21	Mining, Quarrying, and Oil and Gas Extraction	0	0.0	732,186	0.6
22	Utilities	7	0.1	638,575	0.5
23	Construction	251	2.0	5,470,181	4.6
31	Manufacturing	368	3.0	11,276,438	9.5
42	Wholesale Trade	841	6.9	5,908,763	5.0
44	Retail Trade	835	6.8	15,023,362	12.7
48	Transportation and Warehousing	83	0.7	4,287,236	3.6
51	Information	364	3.0	3,266,084	2.8
52	Finance and Insurance	221	1.8	6,063,761	5.1
53	Real Estate and Rental and Leasing	215	1.8	1,972,105	1.7
54	Professional, Scientific, and Technical Services	1,318	10.8	8,275,350	7.0
55	Management of Companies and Enterprises	2	0.0	3,098,762	2.6
56	Administrative and Support Services	704	5.7	10,185,297	8.6
61	Educational Services	1,658	13.5	3,513,469	3.0
62	Health Care and Social Assistance	2,579	21.0	18,598,711	15.7
71	Arts, Entertainment, and Recreation	118	1.0	2,112,000	1.8
72	Accommodation and Food Services	1,642	13.4	12,395,387	10.5
81	Other Services (except Public Administration)	1,036	8.5	5,282,688	4.5
99	Industries not classified	15	0.1	11,402	0.0
—	**Total**	**12,257**	**100.0**	**118,266,253**	**100.0**

Source:
[a] US Bureau of the Census (2015b).
[b] US Bureau of the Census (2015a).

The information in tables 6.6 and 6.7 answers a question raised earlier. The average size of Decatur's business establishments (reported in table 6.6) is smaller than the national average in all but four industries and substantially lower than the national average for three industries: Manufacturing (NAICS 31); Management of Companies and Enterprises (NAICS 55); and Arts, Entertainment, and Recreation (NAICS 71). Decatur's employment shares in Manufacturing and Retail Trade (reported in table 6.7) also are smaller than the nation's. As a result, Decatur's small average establishment size is due partially to its economic composition favoring industries that tend to have smaller establishments and partially to a smaller average establishment size in most of the city's industries than in the nation.

Location Quotients

Location quotients provide a useful statistic for comparing the industrial composition of different economies. A location quotient (LQ) measures a region's specialization in an attribute relative to a reference area's specialization in the attribute. An employment LQ is the local share of the employment in each industry divided by the industry's employment share in the reference area. Although LQs are most commonly used to characterize the industrial composition of economies (usually using employment data) as they compare to the nation, they can be used to compare shares of other variables such as occupations, patents, and land uses across any spatial units.

As shown in equation 6.1, an employment LQ is computed by dividing an industry's share of the local employment by the industry's share of the employment in a reference area;

$$\mathrm{LQ}_i^t = \frac{e_i^t / e_T^t}{E_i^t / E_T^t},$$

6.1

where LQ_i^t = location quotient for industry i in year t, e_i^t = local employment in industry i in year t, e_T^t = total local employment in year t, E_i^t = reference area employment in industry i in year t, and E_T^t = total reference area employment in year t.

Employment LQs range from 0, when there is no local employment in an industry, to very large values when the local share of an industry is very large, and/or the reference area's employment share is very small. A LQ of 1.0 indicates that an industry's employment share is identical in the local and the reference region. A LQ below 1.0 indicates that the local employment share is smaller than its share in the reference region. A LQ greater than 1.0 indicates that the local employment share is larger than the reference area's share. Location quotients larger than 1.2 are often interpreted as identifying local concentrations or specializations. Location quotients less than 0.8 generally are taken to imply the relative absence of an economic activity.

Location quotients help identify industries that stand out in comparison to a typical or average economy. Therefore, it generally makes sense to select a reference region that is relatively large and diverse, such as the metropolitan area, state, or nation that encompasses the local area. In some cases, however, it may be valuable to select a different reference region that exemplifies an important characteristic of the local economy. For instance, it may be more useful to compare the composition of a rural county to the nonurban portion of the state or to a rural multistate region than to a state or region which is substantially urban.

The LQs in table 6.8 were computed by applying equation 6.1. For example, the LQ for NAICS 61, Educational Services in 2013, using the United States as the reference area in column seven, is computed as follows:

$$LQ_{61}^{2013} = \frac{e_{61}^{2013} / e_T^{2013}}{E_{61}^{2013} / E_T^{2013}} = \frac{1,658 / 12,257}{3,513,469 / 118,266,253} = \frac{0.1353}{0.0297} = 4.55.$$

Table 6.8 reveals that private sector employment in Decatur is concentrated in several business and consumer service sectors. Educational Services (NAICS 61) has the largest LQ (ignoring the Industries Not Classified sector), more than four times its national share. The city's educational services employment is driven by Agnes State College and the Columbia Theological Seminary, which are in Decatur, and Emory University, which is located near to Decatur. Emory University employs more than twenty-seven thousand people across the Atlanta metropolitan area, and while the university's employment is not included in Decatur's total, many businesses in Decatur serve the university and its students, faculty, and staff. Five other employment sectors also are concentrated in Decatur, although to a much lesser degree.

| TABLE 6.8 | **Employment and Employment Location Quotients: Decatur, Atlanta Region, and United States, 2013** |

NAICS Code	NAICS Sector	Employment			Location Quotient	
		Decatur[a]	Atlanta Region[b]	United States[b]	Atlanta Region	United States
(1)	(2)	(3)	(4)	(5)	(6)	(7)
11	Agriculture, Forestry, Fishing, and Hunting	0	474	154,496	0.00	0.00
21	Mining, Quarrying, and Oil and zGas Extraction	0	1,231	732,186	0.00	0.00
22	Utilities	7	12,883	638,575	0.09	0.10
23	Construction	251	93,085	5,470,181	0.47	0.44
31	Manufacturing	368	133,107	11,276,438	0.48	0.31
42	Wholesale Trade	841	140,033	5,908,763	1.04	1.37
44	Retail Trade	835	251,210	15,023,362	0.57	0.54
48	Transportation and Warehousing	83	113,940	4,287,236	0.13	0.19
51	Information	364	108,594	3,266,084	0.58	1.08
52	Finance and Insurance	221	107,971	6,063,761	0.35	0.35
53	Real Estate and Rental and Leasing	215	40,884	1,972,105	0.91	1.05
54	Professional, Scientific, and Technical Services	1,318	181,918	8,275,350	1.25	1.54
55	Management of Companies and Enterprises	2	95,044	3,098,762	0.00	0.01
56	Administrative and Support Services	704	190,962	10,185,297	0.64	0.67
61	Educational Services	1,658	57,420	3,513,469	4.99	4.55
62	Health Care and Social Assistance	2,579	237,948	18,598,711	1.87	1.34
71	Arts, Entertainment, and Recreation	118	30,089	2,112,000	0.68	0.54
72	Accommodation and Food Services	1,642	224,403	12,395,387	1.26	1.28
81	Other Services (except Public Administration)	1,036	98,008	5,282,688	1.83	1.89
99	Industries not classified	15	193	11,402	13.85	13.08
—	**Total**	**12,257**	**2,119,397**	**118,266,253**	—	—

Source:
[a] US Bureau of the Census (2015b).
[b] US Bureau of the Census (2015a).

Another anchor institution that is not directly included in the private sector employment figures—DeKalb County government—probably explains the concentrations in Health Care and Social Assistance and other services in Decatur's employment mix. Two of the industries that are underrepresented in Decatur, Agriculture, Forestry, Fishing, and Hunting (NAICS 11), and Mining and Quarrying (NAICS 21), are no surprise: these industries are rare in suburban cities in the center of metropolitan regions for obvious reasons. Decatur also has relatively few jobs in traditional "blue collar" sectors—Manufacturing (NAICS 31), Construction (NAICS 23), Utilities (NAICS 22), Wholesale Trade (NAICS 32), and Transportation and Warehousing (NAICS 48)—and in consumer and business service sectors such as Information (NAICS 51), Finance and Insurance (NAICS 52), Management of Companies and Enterprises (NAICS 55), and Arts, Entertainment, and Recreation (NAICS 71). Overall, the profile of Decatur summarized in table 6.8 is consistent with a relatively high income, mostly residential community.

Because it is larger, DeKalb County's economy more closely parallels the national economy. As a result, DeKalb County's LQs in column seven of table 6.9 are generally less extreme (closer to one) than the LQs for Decatur in column seven of table 6.8. DeKalb County's employment also is concentrated in Educational Services (NAICS 61) with smaller concentrations in Transportation and Warehousing (NAICS 48), Information (NAICS 51), and Management of Companies and Businesses (NAICS 55). Like Decatur, DeKalb's employment shares in Agriculture, Forestry, Fishing, and Hunting; Mining and Quarrying; Manufacturing; Finance and Insurance; and Arts, Entertainment, and Recreation are smaller than the nation's shares, reflecting Decatur's location and function within the Atlanta region.

In general, larger regions tend to be more diverse than smaller areas. As a result, comparisons to smaller and less diverse reference regions tend to result in more extreme LQs. The LQs calculated for Decatur and DeKalb County in tables 6.8 and 6.9 that use the Atlanta region and the United States as reference areas exhibit this pattern. However, there are some exceptions reflecting the Atlanta region's smaller employment shares in these sectors than the industries' employment shares in the United States.

Location quotients also can help one understand the significance of changes in an area's industrial composition over time. For example, table 6.10 reveals that Decatur's employment in the Health Care and Social Assistance sector (NAICS 62) increased between 2007 and 2011 and declined slightly between 2011 and 2013. Location quotients help determine whether this change reflected a particular specialization or advantage for Decatur in this industry.

The LQs for Health Care and Social Assistance increased from 1.17 to 1.42 between 2007 and 2011, indicating that the increases in Decatur's employment in this sector exceeded the nation's growth in this sector. This suggests an increased local concentration in the Health Care and Social Assistance sector between 2007 and 2011. The sector's LQ declined slightly (from 1.42 to 1.34) between 2011 and 2013, indicating that the change in this sector's share of the local employment was matched by an increased share in the national economy. In contrast, Decatur's employment in Educational Services (NAICS 61) grew between 2007 and 2011, but its LQ declined from 5.07 to 4.18 over this period, indicating that the industry's local growth was less than its national growth. In both cases, trends in the LQs over time help reveal the importance of changes in the local economy by placing them in a wider context.

| TABLE 6.9 | Employment and Employment Location Quotients: DeKalb County, Atlanta Region, and United States, 2013 |

NAICS Code	NAICS Sector	Employment[a]			Location Quotient	
		DeKalb County	Atlanta Region	United States	Atlanta Region	United States
(1)	(2)	(3)	(4)	(5)	(6)	(7)
11	Agriculture, Forestry, Fishing, and Hunting	4	398	156,520	0.08	0.01
21	Mining, Quarrying, and Oil and Gas Extraction	111	1,084	651,204	0.79	0.07
22	Utilities	1,230	12,759	639,795	0.74	0.84
23	Construction	10,035	93,292	5,190,921	0.83	0.84
31	Manufacturing	10,725	125,613	10,984,361	0.66	0.43
42	Wholesale Trade	14,434	133,387	5,626,328	0.83	1.12
44	Retail Trade	30,861	245,002	14,698,563	0.97	0.92
48	Transportation and Warehousing	12,788	105,617	4,106,359	0.93	1.36
51	Information	14,795	96,589	3,121,317	1.18	2.07
52	Finance and Insurance	8,984	105,714	5,886,602	0.65	0.67
53	Real Estate and Rental and Leasing	6,293	38,486	1,917,640	1.26	1.43
54	Professional, Scientific, and Technical Services	18,941	166,568	7,929,910	0.88	1.04
55	Management of Companies and Enterprises	10,658	99,000	2,921,669	0.83	1.59
56	Administrative and Support Services	19,302	171,232	9,389,950	0.87	0.90
61	Educational Services	22,254	56,148	3,386,047	3.05	2.87
62	Health Care and Social Assistance	38,691	228,951	18,059,112	1.30	0.93
71	Arts, Entertainment, and Recreation	2,491	27,907	2,003,129	0.69	0.54
72	Accommodation and Food Services	24,821	202,370	11,556,285	0.94	0.94
81	Other Services (except Public Administration)	12,758	93,394	5,181,801	1.05	1.07
99	Industries not classified	18	197	19,728	0.70	0.40
—	**Total**	**260,194**	**2,003,708**	**113,427,241**	—	—

Source:
[a] US Bureau of the Census (2015a).

Location quotients also can be useful for analyzing other features of a local economy, such as its establishment counts by industry, occupational employment, tax revenues by source, and so on. They can also be used at any level of industrial disaggregation for which data are available. Analysis at greater levels of detail and for smaller areas generally reveals substantial differences within broader employment categories with correspondingly more extreme LQs.

Although Decatur's manufacturing sector is relatively small for the city's size, table 6.11 reveals that it has substantial concentrations in Printing and Related Support Activities (NAICS 323) and Nonmetallic Mineral Products Manufacturing (NAICS 327). However, LQs can be misleading at this level of aggregation, due to small denominators. While Decatur's LQ for Furniture and Related Products Manufacturing (NAICS 337) is a healthy 1.87 in 2013, the industry employs only twenty people, a decline of nearly 90 percent from its employment in 2007 when it had a robust LQ of 11.12. As this example suggests, analysts should consider the number of employees in an industry when using LQs to focus on local employment concentrations or underconcentrations that are of substantial size and local importance.

TABLE 6.10	Employment and Employment Location Quotients: Decatur 2007, 2011, and 2013

NAICS Code	NAICS Sector	Employment			Location Quotient		
		2007[a]	2011[b]	2013[c]	2007	2011	2013
(1)	(2)	(3)	(4)	(5)	(6)	(7)	(8)
11	Agriculture, Forestry, Fishing and Hunting	0	0	0	0.00	0.00	0.00
21	Mining, Quarrying, and Oil and Gas Extraction	0	0	0	0.00	0.00	0.00
22	Utilities	2	7	7	0.03	0.10	0.10
23	Construction	311	259	251	0.50	0.46	0.44
31	Manufacturing	373	289	368	0.33	0.24	0.31
42	Wholesale Trade	141	143	841	0.28	0.23	1.37
44	Retail Trade	1,448	1,432	835	1.08	0.89	0.54
48	Transportation and Warehousing	37	150	83	0.10	0.33	0.19
51	Information	135	135	364	0.47	0.40	1.08
52	Finance and Insurance	234	319	221	0.42	0.50	0.35
53	Real Estate and Rental and Leasing	204	182	215	1.07	0.87	1.05
54	Professional, Scientific, and Technical Services	1,456	1,601	1,318	2.09	1.85	1.54
55	Management of Companies and Enterprises	6	2	2	0.02	0.01	0.01
56	Administrative and Support Services	573	1,057	704	0.67	1.03	0.67
61	Educational Services	1,311	1,545	1,658	5.07	4.18	4.55
62	Health Care and Social Assistance	1,677	2,806	2,579	1.17	1.42	1.34
71	Arts, Entertainment, and Recreation	254	92	118	1.49	0.42	0.54
72	Accommodation and Food Services	1,351	1,440	1,642	1.37	1.14	1.28
81	Other Services (except Public Administration)	756	914	1,036	1.61	1.62	1.89
99	Industries not classified	2	10	15	1.69	4.48	13.08
—	**Total**	**10,271**	**12,381**	**12,257**	—	—	—

Source:
[a] US Bureau of the Census (2009).
[b] US Bureau of the Census (2011
[c] US Bureau of the Census (2015b).

| TABLE 6.11 | Manufacturing Sector Employment and Employment Location Quotients: Decatur, 2007, 2011, and 2013 |

NAICS Code	NAICS Sector	Employment			Location Quotient		
		2007[a]	2011[b]	2013[c]	2007	2011	2013
(1)	(2)	(3)	(4)	(5)	(6)	(7)	(8)
311	Food Manufacturing	28	41	20	0.63	1.05	0.45
312	Beverage and Tobacco Product Manufacturing	0	0	2	0.00	0.00	0.40
313	Textile Mills	0	0	0	0.00	0.00	0.00
314	Textile Product Mills	0	2	2	0.00	0.66	0.56
315	Apparel Manufacturing	0	0	0	0.00	0.00	0.00
316	Leather and Allied Product Manufacturing	0	0	0	0.00	0.00	0.00
321	Wood Product Manufacturing	7	7	7	0.43	0.74	0.64
322	Paper Manufacturing	0	0	0	0.00	0.00	0.00
323	Printing and Related Support Activities	42	44	49	2.15	3.35	3.38
324	Petroleum and Coal Products Manufacturing	0	0	0	0.00	0.00	0.00
325	Chemical Manufacturing	2	2	2	0.08	0.10	0.09
326	Plastics and Rubber Products Manufacturing	34	20	33	1.28	1.07	1.49
327	Nonmetallic Mineral Product Manufacturing	74	106	191	5.07	11.45	17.44
331	Primary Metal Manufacturing	2	2	0	0.15	0.20	0.00
332	Fabricated Metal Product Manufacturing	0	0	2	0.00	0.00	0.05
333	Machinery Manufacturing	16	14	14	0.45	0.53	0.42
334	Computer and Electronic Product Manufacturing	11	9	7	0.34	0.38	0.27
335	Electrical Equipment and Appliance Manufacturing	0	0	0	0.00	0.00	0.00
336	Transportation Equipment Manufacturing	14	0	0	0.29	0.00	0.00
337	Furniture and Related Product Manufacturing	178	18	20	11.12	1.97	1.87
339	Miscellaneous Manufacturing	4	35	4	0.19	2.28	0.23
—	**Total**	**412**	**300**	**353**	—	—	—

Source:
[a] US Bureau of the Census (2009b).
[b] US Bureau of the Census (2013d).
[c] US Bureau of the Census (2015b).

Decatur's Occupations

The preceding analysis used industries as the primary analysis units to learn about the people who work in Decatur and DeKalb County. These workers may live in Decatur, DeKalb County, the Atlanta region, or elsewhere, but they all work in the city or county. In contrast, **occupational data** provide information on the people who live in Decatur or DeKalb County; these people may work in the city, the county, in other parts of the Atlanta region, or elsewhere but they all live in the city or county. That is, the occupational data are place of residence information on the people who reside in an area; the **employment data** are "place of work" information on the people who work in an area.

Table 6.12 reveals that more than half of Decatur residents worked in three occupational categories in the period between 2011 and 2013: (1) Management, Business, and Finance; (2) Computer, Engineering, and Science; and (3) Education, Training, and Library. Many Decatur residents worked in three other occupations: (1) Healthcare Practitioners and Technicians, (2) Sales and Related, and (3) Office and Administrative Support. The occupation LQs for the 2011 to 2013 period indicate that Decatur residents are highly specialized in Legal Services, with an occupation LQ greater than 5.0. The occupation LQs also reveal significant concentrations in three other occupations: (1) Computer, Engineering and Science; (2) Education, Training, and Library; and (3) Arts, Design, Entertainment, Sports, and Media.

Decatur's concentrations in occupations such as Education, Training, and Library, and Computer, Engineering, and Science match the city's employment concentrations described in table 6.7. The concentrations in other occupations such as Legal and Arts, Design, Entertainment, Sports, and Media are not reflected in employment data. This suggests that a substantial number of commuters live in Decatur but work elsewhere and are not included in Decatur's employment information.

The fact that more than twelve thousand people were working in Decatur in 2013 (table 6.7) and ten thousand employed people lived in the city (table 6.12) suggests that commuting provides a net inflow of workers into the city. However, this difference may also reflect the possibility that Decatur's residents hold multiple jobs in the same occupational categories, some of which may be part-time. Many of the occupations in which Decatur is well-represented require higher levels of education and pay better wages, which fits with Decatur's character as a relatively affluent suburban community.

DeKalb County is a much larger and more diverse place than the City of Decatur and one of the two most central counties in the Atlanta metropolitan area. As a result, its occupational LQs reported in table 6.12 are much closer to 1.0 than they are for Decatur. They suggest that only three occupations are concentrated in the county: (1) Computer, Engineering, and Science; (2) Legal; and (3) Arts, Design, Entertainment, Sports, and Media. The Farming, Fishing, and Forestry and Production (primarily manufacturing) occupations are substantially underrepresented among DeKalb County residents, reflecting the county's suburban status.

Table 6.12 indicates that Decatur's occupational employment grew slightly between 2000 and the 2011–2013 period, in contrast to more than a 10 percent growth across the nation. This fact and the city's employment growth since 2007 suggest an ongoing evolution of Decatur from a bedroom community for residents who work elsewhere to a local employment center. DeKalb County exhibits a similar pattern on a larger scale. Both trends are consistent with a gradual decentralization of employment in the Atlanta metropolitan area.

TABLE 6.12	Occupations and Occupation Location Quotients: Decatur and DeKalb County, 2000 and 2011–2013

Occupational Category	Decatur				DeKalb County			
	2000		2011–2013		2000		2011–2013	
	Employed People[a]	LQ	Employed People[b]	LQ	Employed People[a]	LQ	Employed People[b]	LQ
(1)	(2)	(3)	(4)	(5)	(6)	(7)	(8)	(9)
Management, Business, and Finance	1,890	1.48	2,620	1.80	56,141	1.26	56,347	1.17
Computer, Engineering, and Science	1,115	2.08	1,113	2.08	24,764	1.32	21,869	1.24
Community and Social Service	298	2.00	222	1.33	5,734	1.10	5,902	1.08
Legal	454	4.21	596	5.06	5,956	1.57	6,326	1.63
Education, Training, and Library	832	1.48	1,434	2.34	19,945	1.02	22,791	1.13
Arts, Design, Entertainment, Sports, and Media	698	3.68	489	2.54	9,008	1.35	8,185	1.29
Healthcare Practitioners and Technicians	612	1.34	766	1.35	15,845	0.99	18,049	0.96
Healthcare Support	101	0.51	19	0.07	4,907	0.71	6,745	0.79
Protective Services	131	0.67	128	0.57	6,054	0.89	6,872	0.93
Food Preparation and Serving	394	0.83	352	0.60	14,799	0.88	20,583	1.07
Building and Grounds Services	246	0.76	14	0.03	11,129	0.98	13,704	1.02
Personal Care	286	1.03	262	0.71	8,689	0.89	9,189	0.75
Sales and Related	815	0.73	834	0.76	36,632	0.94	33,354	0.92
Office and Administrative Support	1,335	0.87	951	0.70	60,547	1.13	47,116	1.05
Farming, Fishing, and Forestry	0	0.00	0	0.00	450	0.10	182	0.08
Construction and Extraction	208	0.38	117	0.23	18,549	0.97	12,965	0.78
Installation, Maintenance, and Repair	146	0.37	0	0.00	10,272	0.75	6,629	0.61
Production	116	0.14	59	0.10	17,984	0.61	14,293	0.72
Transportation and Material Moving	231	0.38	124	0.20	20,005	0.94	21,637	1.06
Total	**9,908**	—	**10,100**	—	**347,410**	—	**332,738**	—

Source:
[a] US Bureau of the Census (2003a).
[b] US Bureau of the Census (2015d).

Shift-Share Analysis

Table 6.13 reports that Decatur's private sector economy declined slightly between 2011 and 2013; table 6.14 reports that the nation's economy grew by 4.3 percent over this period. Decatur's employment in Wholesale Trade (NAICS 42) and Information (NAICS 51) increased dramatically over the two-year period. These gains were offset by significant declines in four sectors: (1) Retail Trade (NAICS 44), (2) Transportation and Warehousing (NAICS 48), (3) Finance and Insurance (NAICS 52), and (4) Administration and Support Services (NAICS 56). The nation's private sector employment increased in all but three sectors: (1) Agriculture, Forestry, Fishing, and Hunting (NAICS 11); (2) Utilities (NAICS 22); and (3) Industries not classified (NAICS 99). Shift-share analysis is helpful for understanding the factors that caused Decatur's economy to decline while the nation's economy grew.

Shift-share analysis methods examine the changes in an area's economic composition that occurred during a specified period, relative to a reference region, which is very often the nation. Sometimes called classical shift-share to distinguish it from numerous variants,[6] shift-share analysis decomposes the changes (or shifts) in a local economy and a reference area into three components: reference shift, industry mix,

| TABLE 6.13 | Employment and Employment Change: Decatur, 2011 and 2013 | | | | |

NAICS Code	NAICS Sector	Employment		Change	
		2011[a]	2013[b]	Number	Percent
(1)	(2)	(3)	(4)	(5)	(6)
11	Agriculture, Forestry, Fishing, and Hunting	0	0	0	n.a.
21	Mining, Quarrying, and Oil and Gas Extraction	0	0	0	n.a.
22	Utilities	7	7	0	0.1
23	Construction	259	251	−8	−2.9
31	Manufacturing	289	368	79	27.5
42	Wholesale Trade	143	841	698	488.3
44	Retail Trade	1,432	835	−597	−41.7
48	Transportation and Warehousing	150	83	−67	−44.5
51	Information	135	364	229	169.5
52	Finance and Insurance	319	221	−97	−30.6
53	Real Estate and Rental and Leasing	182	215	33	18.1
54	Professional, Scientific, and Technical Services	1,601	1,318	−283	−17.7
55	Management of Companies and Enterprises	2	2	0	0.1
56	Administrative and Support Services	1,057	704	−353	−33.4
61	Educational Services	1,545	1,658	113	7.3
62	Health Care and Social Assistance	2,806	2,579	−227	−8.1
71	Arts, Entertainment, and Recreation	92	118	26	28.5
72	Accommodation and Food Services	1,440	1,642	201	14.0
81	Other Services (except Public Administration)	914	1,036	121	13.3
99	Industries not classified	10	15	6	60.1
—	**Total**	**12,381**	**12,257**	**−124**	**−1.0**

Source:
[a] US Bureau of the Census (2013d).
[b] US Bureau of the Census (2015b).

| TABLE 6.14 | Employment and Employment Change: United States, 2011 and 2013 |

NAICS Code	NAICS Sector	Employment		Change	
		2011[a]	2013[b]	Number	Percent
(1)	(2)	(3)	(4)	(5)	(6)
11	Agriculture, Forestry, Fishing, and Hunting	156,520	154,496	–2,024	–1.3
21	Mining, Quarrying, and Oil and Gas Extraction	651,204	732,186	80,982	12.4
22	Utilities	639,795	638,575	–1,220	–0.2
23	Construction	5,190,921	5,470,181	279,260	5.4
31	Manufacturing	10,984,361	11,276,438	292,077	2.7
42	Wholesale Trade	5,626,328	5,908,763	282,435	5.0
44	Retail Trade	14,698,563	15,023,362	324,799	2.2
48	Transportation and Warehousing	4,106,359	4,287,236	180,877	4.4
51	Information	3,121,317	3,266,084	144,767	4.6
52	Finance and Insurance	5,886,602	6,063,761	177,159	3.0
53	Real Estate and Rental and Leasing	1,917,640	1,972,105	54,465	2.8
54	Professional, Scientific, and Technical Services	7,929,910	8,275,350	345,440	4.4
55	Management of Companies and Enterprises	2,921,669	3,098,762	177,093	6.1
56	Administrative and Support Services	9,389,950	10,185,297	795,347	8.5
61	Educational Services	3,386,047	3,513,469	127,422	3.8
62	Health Care and Social Assistance	18,059,112	18,598,711	539,599	3.0
71	Arts, Entertainment, and Recreation	2,003,129	2,112,000	108,871	5.4
72	Accommodation and Food Services	11,556,285	12,395,387	839,102	7.3
81	Other Services (except Public Administration)	5,181,801	5,282,688	100,887	1.9
99	Industries not classified	19,728	11,402	–8,326	–42.2
—	**Total**	**113,427,241**	**118,266,253**	**4,839,012**	**4.3**

Source:
[a] US Bureau of the Census (2011).
[b] US Bureau of the Census (2015a).

and local shift. The three components can be interpreted as the outcomes of different process which have their own characteristics. As was true for LQs, the reference region is generally a large and diverse area that encompasses the study area and provides an appropriate benchmark for the local economy. Also, like LQs, shift-share analysis generally is used to analyze changes in the local industrial composition but can also be used to analyze any situation that involves changes over time in the composition of a study area and an appropriate reference area.

The classical shift-share analysis method decomposes the changes in an area's total employment change into the three components in equation 6.2:

$$\Delta e_i^{t,t'} = \mathrm{RS}_i^{t,t'} + \mathrm{IM}_i^{t,t'} + \mathrm{LS}_i^{t,t'}, \qquad\qquad 6.2$$

where $\Delta e_i^{t,t'}$ = observed employment change in industry i between time t and t', $\mathrm{RS}_i^{t,t'}$ = reference shift for industry i between time t and t', $\mathrm{IM}_i^{t,t'}$ = industrial mix for industry i between time t and t', and $\mathrm{LS}_i^{t,t'}$ = local shift for industry i between time t and t'.

The three components are computed as follows:

$$\mathrm{RS}_i^{t,t'} = e_i^t \left(\frac{\Delta E_T^{t,t'}}{E_T^t} \right),$$
6.3

$$\mathrm{IM}_i^{t,t'} = e_i^t \left(\frac{\Delta E_i^{t,t'}}{E_i^t} - \frac{\Delta E_T^{t,t'}}{E_T^t} \right),$$
6.4

$$\mathrm{LS}_i^{t,t'} = e_i^t \left(\frac{\Delta e_i^{t,t'}}{e_i^t} - \frac{\Delta E_i^{t,t'}}{E_i^t} \right),$$
6.5

where e_i^t = study area employment in industry i in year t, E_T^t = total reference area employment in year t, E_i^t = reference area employment in industry i in year t, $\Delta E_T^{t,t'}$ = change in the total reference area employment between year t and t', $\Delta E_i^{t,t'}$ = change in the reference area employment in industry i between year t and t', and $\Delta e_i^{t,t'}$ = change in the study area's employment in industry i between year t and t'. The three shift components for the entire local economy are calculated by adding the shift components for the individual industries.

Table 6.15 uses the data in tables 6.13 and 6.14 to compute the three shift-share components for the observed changes in Decatur's employment between 2011 and 2013. For example, the reference shift for Educational Services (NAICS 61) is computed by applying equation 6.3:

$$\mathrm{RS}_{61}^{2011,2013} = e_{61}^{2011} \left(\frac{\Delta E_T^{2011,2013}}{E_T^{2011}} \right) = 1{,}545 \left(\frac{4{,}839{,}012}{113{,}427{,}241} \right) = 1{,}545(0.0427) = 66.$$

The industrial mix is computed by applying equation 6.4:

$$\mathrm{IM}_{61}^{2011,2013} = e_{61}^{2011} \left(\frac{\Delta E_{61}^{2011,2013}}{E_{61}^{2011}} - \frac{\Delta E_T^{2011,2013}}{E_T^{2011}} \right)$$
$$= 1{,}545 \left(\frac{127{,}422}{3{,}386{,}047} - \frac{4{,}839{,}012}{113{,}427{,}241} \right)$$
$$= 1{,}545(0.0376 - 0.0427) = -8.$$

The local shift is computed by applying equation 6.5:

$$\mathrm{LS}_{61}^{2011,2013} = e_{61}^{2011} \left(\frac{\Delta e_{61}^{2011,2013}}{e_{61}^t} - \frac{\Delta E_{61}^{2011,2013}}{E_{61}^{2011}} \right)$$
$$= 1{,}545 \left(\frac{113}{1{,}545} - \frac{127{,}422}{3{,}386{,}047} \right)$$
$$= 1{,}545(0.0732 - 0.0376) = 55.$$

The sums of the three shift components for the total economy and for all the employment sectors in table 6.15 are equal to the employment changes in table 6.13, as they should be.

Reference Shift The **Reference Shift** (often called the **National Shift** because the national economy is commonly the reference region) applies the reference region's rate of change to a local industry. It measures the change in the industry's local employment that would have occurred if its growth or decline matched the reference

TABLE 6.15	Shift-Share Analysis: Employment Change, Decatur, 2011–2013				
NAICS Code	NAICS Sector	Reference Shift	Industry Mix	Local Shift	Total
(1)	(2)	(3)	(4)	(5)	(6)
11	Agriculture, Forestry, Fishing, and Hunting	0	0	0	0
21	Mining, Quarrying, and Oil and Gas Extraction	0	0	0	0
22	Utilities	0	0	0	0
23	Construction	11	3	−21	−8
31	Manufacturing	12	−5	72	79
42	Wholesale Trade	6	1	691	698
44	Retail Trade	61	−29	−629	−597
48	Transportation and Warehousing	6	0	−73	−67
51	Information	6	1	223	229
52	Finance and Insurance	14	−4	−107	−97
53	Real Estate and Rental and Leasing	8	−3	28	33
54	Professional, Scientific, and Technical Services	68	1	−353	−283
55	Management of Companies and Enterprises	0	0	0	0
56	Administrative and Support Services	45	44	−442	−353
61	Educational Services	66	−8	55	113
62	Health Care and Social Assistance	120	−36	−311	−227
71	Arts, Entertainment, and Recreation	4	1	21	26
72	Accommodation and Food Services	61	43	96	201
81	Other Services (except Public Administration)	39	−21	104	121
99	Industries not classified	0	−4	10	6
—	**Total**	**528**	**−16**	**−637**	**−124**

area's rate of change. For example, table 6.13 indicates that 1,545 people in Decatur were employed in Educational Services (NAICS 61) in 2011. Table 6.14 indicates that the national economy increased by 4.3 percent over this period. Applying the national rate of 4.3 percent to the 1,545 employees in Educational Services yields a reference shift of 66 jobs highlighted in table 6.15. This value is a counterfactual; it indicates that Decatur's employment in Educational Services would have increased by sixty-six people *if* its performance matched the overall national economy, that is, increased by 4.3 percent.

Industry Mix The **Industry Mix** is the growth or decline that would occur if the local industry's employment change matched the rate of change for the industry in the reference region *relative* to the entire reference region's economy. Stated another way, the Industry Mix term assumes that the employment in a local industry grew or declined by the difference between the industry's rate of increase or decline in the reference region and the reference region's overall change rate. For example, table 6.14 indicates that the nation's employment in Educational Services increased by 3.8 percent between 2011 and 2013 and the national economy increased by 4.3 percent. Applying the difference between these two rates (3.8 percent minus 4.3 percent) to the local employment of 1,545 employees in Educational Services yields an Industry Mix for this industry of −8. Like the Reference Shift, the Industry Mix is a counterfactual. It indicates that Decatur's employment in the Educational Services sector would have declined by eight more people between 2011 and 2013 *if* its growth matched the industry's national growth rate rather than the nation's overall growth rate.

Summed across all the industries, the Industry Mix term accounts for the differences in the composition of the local and the reference economies (the nation in this example). An area with its employment concentrated in industries that are growing nationally can be expected to fare better than an area with its employment concentrated in industries that are stable or declining nationally. Put differently, part of an area's growth or decline over a period is determined by its portfolio of industries, the area's "industry mix."

Local Shift The **Local Shift** term captures the effect of differences between an industry's growth or decline in the study area and its growth or decline in the reference area. Table 6.13 indicates that Decatur's employment in the Educational Services sector increased by 7.3 percent between 2011 and 2013. The sector's national employment grew by 3.8 percent, reported in table 6.14. Applying the difference between these rates (7.3 percent minus 3.8 percent) to the 1,545 Education Services employees in 2011 yields a Local Shift of 55.

The Local Shift term is sometimes called the **Residual Shift** or the **Differential Shift** because it incorporates all the factors that remain after accounting for the reference area's growth and the differences in the composition of the local and reference area's economies. It is also called the **Competitive Shift** because it incorporates the factors that determine an area's competitiveness in an industry.

The Local Shift is the most important piece of information for policy making and planning because it measures the effect of an area's distinctive features, some of which are shaped by local efforts to improve the area's economy. In contrast, it is difficult to change the Reference Shift and Industry Mix effects through public action, though changes may be possible over very long time periods.

Summing the three shift-share components of all of Decatur's employment sectors provides a general picture of the relationship between its economy and the nation's. Decatur lost 124 private sector jobs between 2011 and 2013. Most of this decline can be attributed to large declines in some of its local industries, reflected in the Local Shift of –637. The city also had a slightly unfavorable Industry Mix in 2011, that yielded a total Industry Mix effect of –16. The Local Shift and Industry Mix effects were enough to offset the positive influence of the nation's employment growth, reflected in the positive Reference Shift term of 528.

The values for three shift-share components reflect the fact that Decatur's economy is substantially different from the nation's. The LQ analysis in table 6.8 indicated that Decatur's economy was concentrated in three sectors: Professional, Scientific, and Technical Services (NAICS 54); Educational Services (NAICS 61); and Educational Services (NAICS 61). The city's economy is underconcentrated in Manufacturing (NAICS 31), Retail Trade (NAICS 44), and other sectors that are important nationally. The Industry Mix in table 6.15 reflects the extent to which Decatur's economic performance reflects the growth or decline of different employment sectors nationally, in contrast to the nation's 4.3 percent overall growth rate. Decatur's Industry Mix identifies an employment loss of sixteen jobs due to the local industrial composition.[7]

The Local Shift in table 6.15 identifies sectors in Decatur that performed better or worse than their national counterparts. The Local Shift reflects both the differences between the local and national growth rates in different industries and the size of Decatur's industries in 2011. Health Care and Social Assistance is the largest contributor to Decatur's 2011 employment; this sector and five others comprise most of Decatur's employment. The Local Shift indicates that 311 more jobs were lost in the

Health Care and Social Assistance sector than would be expected given the industry's national growth and Decatur's local concentration in the sector. Local Shifts with losses of more than three hundred people are reported for three other sectors: Retail Trade (NAICS 44); Professional, Scientific, and Technical Services (NAICS 54); and Administrative and Support Services (NAICS 56). The declines in these sectors offset a positive Local Shift of 691 in Wholesale Trade (NAICS 42). The local conditions in these industries include factors that contribute to their competitiveness, some of which may reflect supportive public policies.

The preceding analysis considers the Decatur economy and its major employment sectors, but shift-share analysis also can be applied at a greater level of detail, for example to examine an employment sector and its component industries. In general, shift-share is applicable to any changes in a local area and an appropriate reference area that can be disaggregated into the three mutually exclusive and exhaustive shift-share components.

Projecting Decatur's Economic Future

This chapter has examined Decatur's current economy and suggested how recent trends have shaped the city's economic composition. The discussion here uses the trend and share method projection methods described in chapters 3 and 4 to project Decatur's private sector employment. Employment is a popular economic measure because it is comparable over time, readily available at different levels of spatial and industrial classification, comprehensible, and of concern to policy makers and local stakeholders. Nevertheless, the projection methods considered here also can be used to project other economic variables such as the number of establishments, earnings levels, and tax revenues. The projection methods are the same but particular considerations are important for projecting local employment.

The base and launch years are particularly important in applying the share projection methods because they are used to project future economic conditions from only one or two points of time. As a result, unstable or unusual observation years in recessions or short-lived economic increases or declines can produce misleading results and should be avoided. Share projection methods should not be used to project employment in volatile or highly cyclical industries. Trend projections are less sensitive to the base and launch years because they use several observations to establish a pattern. Nevertheless, periods of instability or abnormal economic circumstances should be avoided because they can generate unrealistic projections.

Trend Employment Projections

Trend projections often are used to project an area's total employment and rarely are used to project the employment in particular industries that can vary widely in a short time period. However, they can be used to project the employment in particular industries for large areas such as counties or multicounty regions or for clear patterns in large employment sectors in smaller areas.

For example, figure 6.2 plots the employment in Decatur's most important sectors from 1998 (the first year for which ZIP code employment data were available) through 2013 (the last year for which these data were available when this was written). It reveals the following employment trends: (1) employment in Educational Services declined from 2000 to 2007 and grew slowly from 2007 to 2013, (2) employment in Health Care and Social Assistance declined since 2009, (3) employment in Professional Services has remained stable from 2003 to 2013, (4) employment in

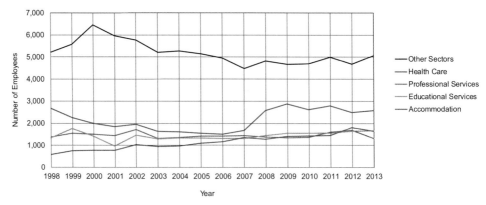

Figure 6.2 Employment by Major Employment Sector: Decatur, 1998–2013

Source: US Bureau of the Census (n.d.c).

TABLE 6.16   Trend Employment Projections: Decatur, 2022 and 2040

NAICS Code	NAICS Sector	Employment			Percent Change	
		2013[a]	2022	2040	2013–2022	2013–2040
(1)	(2)	(3)	(4)	(5)	(6)	(7)
54	Professional Services	1,318	1,556	1,561	18.1	18.4
61	Educational Services	1,658	1,915	2,487	15.5	50.0
62	Health Care	2,579	1,858	1,532	−27.9	−40.6
72	Accommodation	1,642	2,452	2,633	49.4	60.4
—	Other Sectors	5,061	5,372	8,038	6.1	58.8
—	**Total**	**12,257**	**13,153**	**16,251**	**7.3**	**32.6**

Source:
[a] US Bureau of the Census (n.d.c).

Accommodation and Food Services increased steadily since 2009; and (5) employment in the residual Other Sectors increased slightly since 2008.

Table 6.16 extends the trends for each employment sector described in the previous section to project the employment for Decatur's total economy and its major employment sectors in 2022 and 2040.[8] The table suggests that Decatur's private sector employment will grow by 7.3 percent by 2022 and by 32.6 percent by 2040. It projects dramatic employment declines in Health Care and Social Assistance that are offset by increases in all the other employment sectors.

Constant-Share Employment Projections

The constant-share and other share projection methods project local economic conditions by relating them to the projected growth for a larger and more diverse reference area that generally encompasses the study area. The following discussion uses employment projections for DeKalb County's 2022 employment prepared by the Georgia Department of Labor to project Decatur's 2022 employment in its largest employment sectors.

The **constant-share projection method** assumes that the growth or decline in all sectors of a local economy will equal the sectors' projected growth or decline in the reference area. The constant-share employment projections are computed by applying equation 6.6:

$$e_i^{ta} = s_i^l \times E_i^{ta},$$

6.6

where ta = target year for projecting the study area employment, l = launch year when employment data are available for the study area and reference area, e_i^{ta} = projected study area employment in industry i in target year ta, s_i^l = study area's employment share in industry i in launch year l, and E_i^{ta} = projected reference area employment in industry i in target year ta.

The study area's employment share is computed by dividing the local employment in industry i by the reference area's employment in that industry. That is,

$$s_i^l = \frac{e_i^l}{E_i^l},$$

where e_i^l = study area employment in industry i in launch year l, and E_i^l = reference area employment in industry i in launch year l.

Table 6.17 uses the constant-share method to project the 2022 employment in Decatur's largest employment sectors. For example, Decatur's employment share and projected employment in Professional Services (NAICS 54) are computed as follows:

$$s_{54}^{2013} = \frac{e_{54}^{2013}}{E_{54}^{2013}} = \frac{1{,}318}{18{,}941} = 0.0696,$$

$$e_{54}^{2022} = s_{54}^{2013} \times E_{54}^{2022} = 0.0696 \times 21{,}670 = 1{,}508.$$

The constant-share method projects a 28 percent growth in Decatur's total employment, substantial employment growth in Educational Services, and a decline in Accommodation employment. The projected change rates for all the sectors are equal to DeKalb County's rates, as the constant-share projection method assumes.

TABLE 6.17   **Constant-Share Employment Projections: Decatur, 2022**

NAICS Code	NAICS Sector	Decatur			DeKalb County		
		2013[a]	2022	Percent Growth	2013[b]	2022[c]	Percent Growth
(1)	(2)	(3)	(4)	(5)	(6)	(7)	(8)
54	Professional Services	1,318	1,508	14.4	18,941	21,670	14.4
61	Educational Services	1,658	3,036	83.1	22,254	40,750	83.1
62	Health Care	2,579	3,029	17.4	38,691	45,440	17.4
72	Accommodation	1,642	1,400	−14.7	24,821	21,170	−14.7
—	Other Sectors	5,061	6,748	33.3	155,487	207,310	33.3
—	**Total**	**12,257**	**15,720**	**28.3**	**260,194**	**336,340**	**28.3**

Source:
[a] US Bureau of the Census (2015b).
[b] US Bureau of the Census (2015a).
[c] Georgia Department of Labor (n.d.).

Shift-Share Employment Projections

The **shift-share projection method** recognizes that the changes in a local area's employment sectors rarely will match their changes in the reference area. It modifies the constant-share projection formula by adding a "shift" term that accounts for the observed differences between the employment changes in the local sectors and their changes in the reference area. "Shift-share" is used here as it was in chapter 4 to identify a projection method. The projection method should not be confused with the shift-share analysis method for analyzing observed changes in an area's employment or other characteristics described earlier in this chapter.

The shift-share employment projection equation is:

$$e_i^{ta} = \left(s_i^l + \Delta s_i^{l,ta}\right) E_i^{ta}, \qquad 6.7$$

where e_i^{ta} = projected study area employment in industry i in target year ta, s_i^l = observed study area employment share in launch year l, $\Delta s_i^{l,ta}$ = projected employment shift for industry i between launch year l and target year ta, and E_i^{ta} = projected reference area employment in industry i in target year ta.

The projected shift term, $\Delta s_i^{l,ta}$ is computed as follows:

$$\Delta s_i^{l,ta} = \left(\frac{ta - l}{l - b}\right)\left(s_i^l - s_i^b\right), \qquad 6.8$$

where ta = target year for projecting the study area employment, l = launch year when employment data are available for the study area and reference area, b = base year when employment data are available for the study area and reference area, s_i^l = employment share for in industry i in launch year l, and s_i^b = employment share for industry i in base year b.

As pointed out previously, it is important to select base and launch years that provide an appropriate foundation for projecting an area's future economic growth or decline. A graph of the local area's past employment shares can be helpful in this regard. For example, figure 6.3 plots Decatur's employment share of DeKalb County's employment for the city's major employment sectors from 1998 to 2013. It reveals that the employment shares for all but the Professional Services sector had relatively consistent trends from 2009 to 2013. It also suggests that 2009 is an appropriate base year for projecting Decatur's future employment growth.

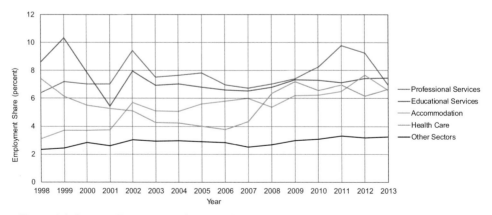

Figure 6.3 Decatur Employment Share, 1998–2013

Table 6.18 uses equation 6.8 to compute the projected employment shift for Decatur's major employment sectors. For example, the projected shift for Professional Services (NAICS 54) is computed as follows:

$$\Delta s_{54}^{2013,2022} = \left(\frac{2022-2013}{2013-2009}\right)\left(s_{54}^{2013} - s_{54}^{2009}\right) = \left(\frac{9}{4}\right)(0.0696 - 0.0739) = -0.0097.$$

Table 6.19 projects the 2022 employment in Decatur's largest employment sectors by applying equation 6.7. For example, the 2022 employment in Professional Services (NAICS 54) is computed as follows:

$$e_{54}^{2022} = \left(s_{54}^{2022} + \Delta s_{54}^{2013,2022}\right)E_{54}^{2022} = (0.0696 + [-0.0097])21,670 = 1,270.$$

TABLE 6.18   **Projected Employment Shift: Decatur, 2013–2022**

NAICS Code	NAICS Sector	Decatur Employment[a]		DeKalb County Employment[a]		Employment Share (Percent)		Projected Shift (Percent)
		2009[a]	2013[b]	2009[c]	2013[d]	2009	2013	2013–2022
(1)	(2)	(3)	(4)	(5)	(6)	(7)	(8)	(9)
54	Professional Services	1,331	1,318	18,012	18,941	7.39	6.96	−0.97
61	Educational Services	1,552	1,658	21,135	22,254	7.34	7.45	0.24
62	Health Care	2,874	2,579	39,988	38,691	7.19	6.66	−1.18
72	Accommodation	1,395	1,642	22,463	24,821	6.21	6.61	0.91
—	Other Sectors	4,672	5,061	157,930	155,487	2.96	3.25	0.67
—	Total	11,824	12,257	259,528	260,194	4.56	4.71	0.35

Source:
[a] US Bureau of the Census (2009b).
[b] US Bureau of the Census (2015b).
[c] US Bureau of the Census (2011).
[d] US Bureau of the Census (2015a).

TABLE 6.19   **Shift-Share Employment Projections: Decatur, 2022**

NAICS Code	NAICS Sector	2013 Employment Share (Percent)[a]	2013–2022 Projected Shift (Percent)[a]	Projected 2022 Employment	
				DeKalb County[b]	Decatur
(1)	(2)	(3)	(4)	(5)	(6)
54	Professional Services	6.96	−0.97	21,670	1,297
61	Educational Services	7.45	0.24	40,750	3,136
62	Health Care	6.66	−1.18	45,440	2,494
72	Accommodation	6.61	0.91	21,170	1,592
—	Other Sectors	3.25	0.67	207,310	8,131
—	Total	4.71	0.35	336,340	16,650

Source:
[a] Table 6.18.
[b] Georgia Department of Labor (n.d.).

| TABLE 6.20 | Projected Shift-Share Employment Changes: Decatur, 2022 | | | | |

NAICS Code	NAICS Sector	Employment		Change	
		2013[a]	2022	Number	Percent
(1)	(2)	(3)	(4)	(5)	(6)
54	Professional Services	1,318	1,297	−21	−1.6
61	Educational Services	1,658	3,136	1,478	89.1
62	Health Care	2,579	2,494	−85	−3.3
72	Accommodation	1,642	1,592	−50	−3.0
–	Other Sectors	5,061	8,131	3,070	60.7
–	**Total**	**12,257**	**16,650**	**4,393**	**35.8**

Source:
[a] US Bureau of the Census (2015b).

As table 6.20 reports, the shift-share method projects more than a 35 percent growth in Decatur's total employment, substantial growth in the Educational Services and Other Sectors, and slight declines in the remaining sectors.

Share-of-Change Employment Projections

The constant-share and shift-share methods assume that an area's future growth or decline is a function of its *size*, that is, that larger areas will grow or decline more rapidly than smaller areas. The **share-of-change projection method** assumes that an area's future growth or decline is proportional to its *change* over the base period, that is, that areas which grew (or declined) more rapidly than the reference area in the past will grow (or decline) more rapidly in the future. This assumption is expressed in the share-of-change employment projection equation:

$$e_i^{ta} = e_i^l + \left(c_i^{b,l} \times \Delta E_i^{l,ta}\right),\qquad 6.9$$

where e_i^{ta} = projected study area employment in industry i in target year ta, e_i^l = observed study area employment in industry i in launch year l, $c_i^{b,l}$ = observed share of employment change for industry i between base year b and launch year l, and $\Delta E_i^{l,ta}$ = projected reference area employment change in industry i between launch year l and target year ta.

The observed share of employment change share, $c_i^{b,l}$, is computed as follows:

$$c_i^{b,l} = \frac{\Delta e_i^{b,l}}{\Delta E_i^{b,l}},\qquad 6.10$$

where $\Delta e_i^{b,l}$ = observed study area employment change in industry i between base year b and launch year l, and $\Delta E_i^{b,l}$ = observed reference area employment change in industry i between base year b and launch year l.

Table 6.21 computes Decatur's share of the changes in DeKalb County's employment between 2009 and 2013. The share-of-change values in column nine are computed by applying equation 6.10. For example, Decatur's share of the changes in DeKalb County's employment in Professional Services (NAICS 54) is computed as follows:

$$c_{54}^{2009,2013} = \frac{\Delta e_{54}^{2009,2013}}{\Delta E_{54}^{2009,2013}} = \frac{1,318-1,331}{18,941-18,012} = -1.4\%.$$

TABLE 6.21 — Share-of-Change: Decatur, 2009–2013

NAICS Code	NAICS Sector	Decatur Employment		DeKalb County Employment		Share-of-Change (Percent)
		2009[a]	2013[b]	2009[c]	2013[d]	2009–2013
(1)	(2)	(3)	(4)	(6)	(7)	(8)
54	Professional Services	1,331	1,318	18,012	18,941	−1.4
61	Educational Services	1,552	1,658	21,135	22,254	9.5
62	Health Care	2,874	2,579	39,988	38,691	22.8
72	Accommodation	1,395	1,642	22,463	24,821	10.4
—	Other Sectors	4,672	5,061	157,930	155,487	−15.9
—	**Total**	**11,824**	**12,257**	**259,528**	**260,194**	**65.0**

Source:
[a] US Bureau of the Census (2009b).
[b] US Bureau of the Census (2015b).
[c] US Bureau of the Census (2011).
[d] US Bureau of the Census (2015a).

The share-of-change projections for the employment in Decatur's major sectors are computed in table 6.22 by applying equation 6.9. For example, Decatur's 2022 employment in Professional Services (NAICS 54) is computed as follows:

$$e_{54}^{2022} = e_{54}^{2013} + \left(c_{54}^{2009,2013} \times \Delta E_{54}^{2013,2022} \right) = 1{,}318 + (-0.014 \times 2{,}729) = 1{,}279.$$

The share-of-change projections are obviously unsatisfactory because they project negative employment in the residual Other Sectors sector. This value reflects the combined effect of a large projected growth for DeKalb County's employment in this sector and a negative share of Decatur's share of the county's past growth. This does not mean that the share-of-change method should never be used to project an area's employment; it only means that it is not appropriate for these data.

Determining when particular methods are appropriate is largely a matter of experience and practice, as pointed out in the section "Using Economic Projection Methods" below. A defensible balance must be attained that favors more reasonable results without rejecting approaches that produce unanticipated results. However, at a minimum, logically impossible or unexplainable results should be discarded—after rechecking the calculations!

Share-Trend Employment Projections

The **share-trend projection method** projects the study area's employment in different sectors by multiplying the projected reference area employment in each employment sector by the study area's projected employment share. That is,

$$e_i^{ta} = s_i^{ta} \times E_i^{ta},\qquad\qquad 6.11$$

where e_i^{ta} = projected study area employment in industry i in target year ta, s_i^{ta} = projected study area's employment share in industry i in target year ta, and E_i^{ta} = projected reference area employment in industry i in target year ta.

Table 6.23 uses the share-trend method to project Decatur's employment in 2022. The projected share values in column five were computed by applying the trend projection method described in chapter 3 to Decatur's share of the DeKalb County's employment from 2009 to 2013. The projected employment values in

TABLE 6.22	Share-of-Change Employment Projections: Decatur, 2022						
NAICS Code	NAICS Sector	Decatur Employment	DeKalb County Employment			Share-of-Change (Percent)	Decatur Employment
		2013[a]	2013[b]	2022[c]	Change	(2009–2013)	2022
(1)	(2)	(3)	(4)	(5)	(6)	(7)	(8)
54	Professional Services	1,318	18,941	21,670	2,729	−1.4	1,279
61	Educational Services	1,658	22,254	40,750	18,496	9.5	3,415
62	Health Care	2,579	38,691	45,440	6,749	22.8	4,116
72	Accommodation	1,642	24,821	21,170	−3,651	10.4	1,260
—	Other Sectors	5,061	155,487	207,310	51,823	−15.9	−3,190
—	**Total**	**12,257**	**260,194**	**336,340**	**76,146**	—	**6,881**

Source:
[a] US Bureau of the Census (2015b).
[b] US Bureau of the Census (2015a).
[c] Georgia Department of Labor (n.d.).

column six were computed by applying equation 6.11. For example, Decatur's 2022 employment in Professional Services (NAICS 54) is computed as follows:

$$e_{54}^{2022} = s_{54}^{2022} \times E_{54}^{2022} = 0.088 \times 21,670 = 1,907.$$

Table 6.23 suggests that Decatur's economy will grow by more than 45 percent by 2022, with substantial growth in the Educational Services and Other Sectors and a slight decline in the Health Care sector's employment.

Adjusted-Share Employment Projections

Adjusted-share employment projection methods recognize that Decatur's place in the Atlanta region and its economy may provide advantages or disadvantages for the city's economy that are not reflected in the standard share projection methods. For example, Decatur's employment in the Health Care and Social Assistance sector may be increased by its residents growing incomes that allow them to purchase more services such as health care.

As one example, the constant-share projection method can be modified as follows to account for the projected difference between the study area and reference area's per capita income:

$$e_i^{ta} = \left(\frac{\text{pci}^{ta} / \text{PCI}^{ta}}{\text{pci}^{l} / \text{PCI}^{l}} \right) \times \frac{e_i^{l}}{E_i^{l}} \times E_i^{ta}, \qquad 6.12$$

where e_i^{ta} = projected study area employment in industry i in target year ta, pci^{ta} = projected study area per capita income in target year ta, PCI^{ta} = projected reference area per capita income in target year ta, pci^{l} = observed study area per capita income in launch year l, PCI^{l} = observed reference area per capita income in launch year l, e_i^{l} = study area's employment in industry i in launch year l, E_i^{l} = reference area employment in industry i in launch year l, and E_i^{ta} = projected reference area employment in industry i in target year ta.

For example, table 6.24 indicates that the ratio between Decatur's per capita income and DeKalb County's per capita income has increased from 1.42 in 2008 to 1.49 in 2013. Trend projections for the two area's per capita incomes indicate that this ratio will increase to 1.62 by 2022. Inserting the projected per capita income values into equation 6.12 yields the adjusted constant-share projections for 2022.

TABLE 6.23 Share-Trend Employment Projection: Decatur, 2022

NAICS Code	NAICS Sector	Employment				
		Decatur[a]	DeKalb County[b]	Projected Share (Percent)	Decatur Employment	Percent Change
		2013	2022	2022	2022	2013–2022
(1)	(2)	(3)	(4)	(5)	(6)	(7)
54	Professional Services	1,318	21,670	8.8	1,907	44.7
61	Educational Services	1,658	40,750	7.9	3,219	94.2
62	Health Care	2,579	45,440	5.3	2,408	−6.6
72	Accommodation	1,642	21,170	9.4	1,990	21.2
—	Other Sectors	5,061	207,310	4.0	8,292	63.9
—	**Total**	**12,257**	**336,340**	**—**	**17,817**	**45.4**

Source:
[a] US Bureau of the Census (2015b).
[b] Georgia Department of Labor (n.d.).

For example, Decatur's 2022 employment in Professional Services (NAICS 54) is projected as follows:

$$e_{54}^{2022} = \left(\frac{(\text{pci}^{2022} / \text{PCI}^{2022})}{(\text{pci}^{2013} / \text{PCI}^{2013})} \right) \times \frac{e_{54}^{2013}}{E_{54}^{2013}} \times E_{54}^{2022}$$

$$= \frac{(53,542 / 33,060)}{(43,032 / 28,971)} \times \frac{1,318}{18,941} \times 21,670 = 1,383.$$

Table 6.25 projects a 40 percent growth in Decatur's employment by 2022, in contrast to a 29 percent increase in DeKalb County' employment. It projects a doubling of Decatur's employment in Educational Services, a 45 percent growth in Other Sectors and a slight decline in Accommodations.

The per capita income adjustment can also be applied to the other share projection methods. In addition, per capita income is only one adjustment factor that can be used with the share projection methods. Other adjustments include local industry productivity, key sector costs (e.g., energy or resource prices and wage rates), and industry size. These adjustments require data for the study and reference area over time which can be projected into the future by applying the trend projection methods described in chapter 3.

Evaluating the Decatur Employment Projections

Table 6.26 reports the projected employment for Decatur's major employment sectors that were prepared using four projection methods: trend projection, constant-share projection, shift-share projection, and share-trend projection. The projections for Decatur's total employment in 2022 vary from 13,153 for the trend projection to 17,817 for the share-trend projection. The projections for individual sectors vary even more dramatically, as is generally the case due to the greater employment volatility over time and poorer data quality for individual sectors and the effect of using smaller denominators to compute the projected values.

It is extremely difficult to choose between the projections in table 6.26. There are no hard and fast rules for identifying the most likely projection. However, analysts can select a preferred projection by drawing on their knowledge of the study

TABLE 6.24	Per Capita Income: Decatur and DeKalb County, 2008–2013 and 2022

Area	Per Capita Income						
	2008[a]	2009[a]	2010[a]	2011[a]	2012[a]	2013[a]	2022
(1)	(2)	(3)	(4)	(5)	(6)	(7)	(8)
Decatur	37,130	38,972	42,926	41,632	43,477	43,032	53,542
DeKalb County	26,064	28,412	28,843	28,760	28,810	28,971	33,060
Ratio	1.42	1.37	1.49	1.45	1.51	1.49	1.62

Source:
[a] US Bureau of the Census (n.d.a).

TABLE 6.25	Adjusted Constant-Share Employment Projections: Decatur, 2022

NAICS Code	NAICS Sector	DeKalb County Employment		Decatur Employment		Percent Growth	
		2013[a]	2022	2013[b]	2022[c]	DeKalb County	Decatur
(1)	(2)	(3)	(4)	(5)	(6)	(7)	(8)
54	Professional Services	18,941	21,670	1,318	1,644	14.4	24.7
61	Educational Services	22,254	40,750	1,658	3,310	83.1	99.7
62	Health Care	38,691	45,440	2,579	3,302	17.4	28.1
72	Accommodation	24,821	21,170	1,642	1,527	−14.7	−7.0
—	Other Sectors	155,487	207,310	5,061	7,357	33.3	45.4
—	**Total**	**260,194**	**336,340**	**12,257**	**17,140**	**29.3**	**39.8**

Source:
[a] US Bureau of the Census (2015a).
[b] US Bureau of the Census (2015b).
[c] Georgia Department of Labor (n.d.).

area's characteristics and history and their reasoned judgment, which they typically obtain through repeated observation and practice. A middle-of-the-road projection is often preferable to an extreme one and some statistical evidence suggests the superior accuracy of intermediate forecasts.[9] For example, the constant-share projection of 15,720 is near the middle of the results in the table. Averaging multiple projections can also provide a mid-range option. In this case, averaging the four projections yields a similarly central projection of 15,835.

Another strategy does not use technical criteria or judgment to select a projection. Instead, it retains the collection of reasonable projections, presenting them all with the assumptions and choices they embody as part of a reflective process of analysis and understanding. This process of scenario analysis is described in the last chapter of this book.

TABLE 6.26 **Trend, Constant-Share, Shift-Share, and Share-Trend Employment Projections: Decatur, 2022**

NAICS Code	NAICS Sector	2013 Employment[a]	Trend Projection[a]		Constant-Share Projection[b]		Shift-Share Projection[c]		Share-Trend Projection[d]	
			2022 Employment	Percent Change	2022 Employment	Percent Change	2022 Employment	Percent Change	2022 Employment	Percent Change
(1)	(2)	(3)	(4)	(5)	(6)	(7)	(8)	(9)	(10)	(11)
54	Professional Services	1,318	1,556	18.1	1,508	14.4	1,297	–1.6	1,907	44.7
61	Educational Services	1,658	1,915	15.5	3,036	83.1	3,136	89.1	3,219	94.2
62	Health Care	2,579	1,858	–27.9	3,029	17.4	2,494	–3.3	2,408	–6.6
72	Accommodation	1,642	2,452	49.4	1,400	–14.7	1,592	–3.0	1,990	21.2
—	Other Sectors	5,061	5,372	6.1	6,748	33.3	8,131	60.7	8,292	63.9
—	**Total**	**12,257**	**13,153**	**7.3**	**15,720**	**28.3**	**16,650**	**35.8**	**17,817**	**45.4**

Source:
[a] Table 6.16.
[b] Table 6.17.
[c] Table 6.19.
[d] Table 6.23.

Using Economic Projection Methods

Using Ready-Made Projections

Planners in the United States who work with or on behalf of cities or other sub-state areas often do not use the trend and share employment projection methods described in this chapter. Instead, they use estimates and projections generated with more complex projection methods such as single-equation multivariate regression or multiple-equation econometric models such as input–output that are beyond the scope of this text.[10] Planners rarely apply these methods themselves because the specialized training, data, software costs, and demands of ongoing model maintenance are prohibitive. Instead, they adjust employment and other economic projections that are developed for larger jurisdictions or obtained from private vendors.[11]

Data vendors in the United States provide software or online tools that make producing projections for user-defined areas easy, particularly at the county scale and above.[12] These systems provide estimates for data gaps associated with data suppression rules, making the local analyst's job much easier. Graphical displays and mapping applications are growing in sophistication and ease of use, supporting in-depth exploration of the analysis results. Several consulting firms also provide estimates and forecasts for cities, counties, regions, and states, often by applying their proprietary models and data. Economies of scale and specialization mean that using vendors and consultants generally makes more sense than building sophisticated projection capacity in-house, except for the largest cities, regional organizations, and states.

It is, of course, essential that analysts use vendor and consultant-generated projections wisely. Off-the-shelf software and fee-based online tools make it easy to produce volumes of data and hundreds of projections for a given region. However, the value of any projection depends on the quality of the data, assumptions, and scenarios that support it. Developing useful assumptions and scenarios requires an in-depth knowledge of local conditions and close attention to the plausibility of the projections as they relate to an area's current conditions and anticipated futures.

Two precautions should be taken in applying externally produced economic projections. First, to the extent possible, analysts should evaluate the choices, assumptions, and data inputs used to generate the projections. Information on the methods and data used to prepare projections are generally available from public data providers such as state agencies. Unfortunately, private data vendors often are reluctant to specify their methods and data in detail, although they must describe their approaches in general terms to satisfy current and prospective customers.

Second, as is true for internally produced projections, planners and analysts should modify the employment projections to incorporate local data, trends, perspectives, and purposes. This is particularly important if the assumptions grounding external projections do not match local conditions. In these cases, planners can improve the accuracy and utility of externally produced projections by applying adjustments such as the ones discussed in the previous sections of this chapter. Even when the most appropriate modifications are uncertain, testing various alternatives to illustrate the range of the more probable outcomes provides valuable information and experience.[13] Treating the development of economic projection and forecasts as an exercise of exploring scenarios and options, rather than as an instrument for achieving a single best answer, is generally the preferred strategy.

Aggregation Level

Planners analyzing local areas in the United States normally avoid projecting employment for individual industries or detailed sectors because data quality and precision limitations force a trade-off between the geographic and sectoral detail of their analyses. The accuracy and consistency of projections are severely limited for smaller geographic areas and for more disaggregated economic units. In addition, the factors that determine the employment in particular industries are interconnected at spatial scales that extend far beyond particular communities. As a result, most local employment projections—and particularly the projections for relatively small municipalities such as Decatur—focus on the entire economy or on broad sectors such as manufacturing or services. Industry-specific employment projections are more commonly prepared for metropolitan and state-planning efforts, providing ready-made projections that can be apportioned or adjusted for considering their component localities.

Economic Analysis and Other Planning Information

Local economies are intertwined with many other physical and governmental characteristics of a location: the transportation systems that provide mobility and accessibility for people and commodities, streams of government revenue and public expenditures for infrastructure and services, and the spatial distribution and use of land and natural resources. As a result, often it is beneficial to investigate the features of an economy in concert with a variety of other information, as is done in the final chapter of this text. However, an investigation of the local economy should also stand on its own in describing the economic features and conditions of a place. Economic information often is collected and interpreted independently from other local features due to time, effort, resources, and competency constraints. Even when the analytical focus is solely economic, planners should apply multiple methods and evaluate several complementary data sources to develop perspective and uncover useful information about the complex and multifaceted nature of a local or regional economy.

Economic Futures and Scenario Planning

Several rationales justify preparing and considering an assortment of projections of local economic variables rather than forecasting a single most likely future. On the technical side, the imperfect nature of the information and methods available to conduct and evaluate projections make it difficult to select one as the most probable. Offering a selection of projections along with the choices that underlie them, perhaps accompanied by a ranking or other account of their relative likelihood, also is more informative by providing a more accurate reflection of the information generated by the projection exercise. Consumers of forecasts and projections are more fully informed when they are presented with a range of possible outcomes calculated using different data sources, assumptions, and methods, even if they then choose to concentrate on a particular outcome. When the planner or analyst eventually must present a unitary result, the multiple projections still serve a valuable analytical purpose: the process of deliberating among several well-developed options forces the analyst to acknowledge and evaluate the projection's underlying decisions and judgments.

Scenario planning, described in the final chapter of this book, is an alternative and often complementary approach to formal forecasting for considering future possible states and options.[14] In its simplest application, the process of preparing scenarios can be considered an extension of the strategy of crafting multiple projections to yield a range of outcomes. By incorporating a range of assumptions and decisions, scenarios increase the breadth and reach of the analysis process. Considered as a set of illustrative examples, scenarios can be a powerful communicative tool for conveying complexity without overwhelming audiences.

Of course, local economic scenarios need not be restricted to employment. Dimensions of the local economy such as productivity, income and wealth distribution, and entrepreneurship can be the focus instead of, or in relation to, various aspects of employment. Fully developed local economic scenarios also can integrate economic futures with other aspects of community planning and development, including population and demographics, land use, transportation, housing, and education.

Taken further, scenario planning does not attempt to suggest precise states to be reached at a point in time. Instead, planning scenarios depict processes of change that identify plausible paths to a wide range of potential futures and reveal the implications of different assumptions and choices. Well-conceived scenarios can expose the hidden assumptions and preferences that individuals—including analysts—bring to their consideration of economic futures. Scenario planning also can support engagement and meaningful participation by facilitating discussion about the forces that an organization or community can—and cannot—influence and the desirability of alternative economic futures.

Notes

[1] The geometric mean is calculated by taking the nth root of the product of a set of data. That is, the geometric mean of n numbers a_1 to a_n is $\sqrt[n]{(a_1 \times a_2 \times \cdots \times a_n)}$. The geometric mean is preferable for computing the central tendency of data that have a skewed distribution—a distribution with one tail substantially longer or "weightier" than the other. Skewed distributions appear regularly among diverse phenomena, both natural and human-made, including many that occur as equilibrium states of stochastic processes. The distribution of firm sizes is an example of a skewed distribution (Ijiri and Simon 1974).

[2] Isserman and Westervelt (2006) describe a sophisticated method for estimating the roughly 1.5 million suppressed employment values in the nation's CBP employment data. Unfortunately, these estimates and the software used to prepare them are no longer available.

[3] Unfortunately, there is no reliable way to estimate the employment in open-ended employment categories such as 1,000+ for the CBP data. This is because the largest employment value is unknown and must be assumed. When estimating this value, analysts should remember that industries are usually clustered at the lower end of an employment category and assume a low value for the maximum employment value.

[4] See, e.g., descriptions of the methods used by the US Bureau of Labor Statistics (n.d.a, n.d.b).

[5] County Business Patterns reports information for business establishments, not for individuals. Some people may hold multiple full-time or part-time jobs, and each job would be included in the CBP data. As a result, fewer than 12,257 people may have been employed in Decatur in 2013.

[6] Other versions of the shift-share method are described in Stevens and Moore (1980) and Loveridge and Selting (1998).

[7] The Industry Mix term is hard to interpret and explain for individual components and the analysis of this component is generally conducted at the aggregate level. Positive Industry Mix terms for an industry may reflect either: (1) a local concentration in an industry which grew at a faster rate (or shrank more slowly) than the overall economy or (2) a local under-concentration in an industry which declined more rapidly (or grew more slowly) than the nation's economy. Negative figures reflect either: (1) a local concentration in an industry which performed more poorly than the nation as a whole or (2) or a local under-concentration in an industry whose national performance exceeded the nation's.

[8] Projections are prepared for 2022 so they correspond with the share projections that are described in the following sections which use the 2022 employment projections prepared by the Georgia Department of Labor.

[9] See Armstrong (1985), especially chapter 6, and Armstrong (n.d.).

[10] Goldstein (1990, 2005) describes these techniques in the context of economic projections. For the basics of forecasting using regression techniques, see Pindyck and Rubinfeld (1997); for input-output modeling, see Miller and Blair (2009).

[11] Most US states publish total employment forecasts at the state and county scales. Many states produce statewide forecasts that disaggregate by selected levels of industry or sectoral detail. Some states estimate forecasts at more detailed sectoral levels for substate regions such as economic planning zones or counties.

[12] Emsi Inc. (http://www.economicmodeling.com/) and Regional Economic Models Inc. (http://www.remi.com/), are two such vendors. Other organizations, including Moody's Analytics (https://www.economy.com/) and Oxford Economics (http://www.oxfordeconomics.com/), produce on-demand local forecasts and other analyses.

[13] This strategy is at the heart of a family of mathematical and statistical approaches that includes sensitivity analysis, uncertainty analysis, probability bounding, and robustness checking or robustness testing.

[14] Scenario planning is described in Hopkins and Zapata (2007) and Klosterman (2013).

Spatial Analysis Methods 7

The analysis techniques described in previous chapters are important because they consider the major causes of local growth and change. However, they are incomplete because they do not explicitly recognize that human and natural activities are located on the Earth's surface. The techniques described in this chapter add this dimension by describing methods for understanding the location of demographic and economic processes and the characteristics of the locations in which they occur.

The chapter contains six major sections. The first section describes geographic information systems (GIS) and basic **spatial analysis** concepts and terms. The second section describes five approaches for locating **spatial features** on the Earth's surface and maps. The third section describes the vector and raster data models that are widely used with spatial analyses methods. The fourth section describes methods for analyzing the attribute information that describe spatial features. The fifth section describes methods for analyzing and representing spatial relationships. The final section describes widely used spatial analysis operations.

Concepts and Terminology

Geographic Information Systems

Geographic information systems (GIS) are the primary tools that planners and policy analysts use to consider spatially related phenomena. GIS are computer systems for capturing, storing, querying, analyzing, and displaying information about natural and human-built entities and activities that can be located on the Earth's surface. They are particularly powerful analysis tools because they consider cartographic, attribute, and topological data about spatially referenced entities and activities.[1]

Cartographic data describe *where things are*, that is, their location relative to a reference system, and provide the information needed to prepare a map. For example, the cartographic data required to display lines on a map include their location (beginning and ending points), shape (curved or straight), line type (solid, dotted, or dashed), and color. Attribute data describe *what things are*, for example, their magnitude and classification. **Attribute data** are normally stored in a database and displayed in a **map legend**. For example, the attribute data for a zoning map are the zoning codes for different parts of a city. **Topological data** describe spatial relationships, that is, *where things are located relative to other things.* Topological relationships include: (1) distance (e.g., that Cleveland is 345 miles from Chicago), (2) adjacency (e.g., the fact that Ohio and Illinois are both next to Indiana but are not next to each other), (3) containment (e.g., that Chicago is in Illinois, not Ohio), and (4) intersection (e.g., whether one road intersects with another road).

Maps, Layers, and Features

Geographic information systems cannot capture all the complexity of the real world so must use simplified models to represent the aspects of reality that are assumed to be most important for a particular application. Reflecting their intellectual roots in cartography, GIS reduce the complexity of the real world into maps, layers, and features.

Maps. A **map** is a two- or three-dimensional model of all or a part of the Earth and features on the Earth's surface that can be reproduced on paper or displayed on a computer screen. People have been preparing maps for centuries and an entire field of study, cartography, is devoted to the art and science of mapmaking.[2]

Layers. A **layer** is a collection of features that are related to a common **theme** and represented by the same class of geometric entities, for example, points, lines, or areas. For example, a city can be represented by several map layers: by a street map showing the location of streets, by a property map showing the location of land **parcels**, or by a land cover map showing what is on the Earth's surface.

Features. A **feature** is an entity or activity that can be located on a map at a given map scale. Features are displayed in one or more map layers. For example, parcels are features on a property layer and the location of different crimes are features on a crime layer. It is important to recognize that entities can only be represented as features on maps that have an appropriate scale. For example, the population of a county cannot be displayed on a neighborhood property map and parcel boundaries will be too small to be displayed on a state map.

Feature Geometries

Geographic information systems generally represent features with five feature geometries: (1) points, (2) lines, (3) areas, (4) networks, and (5) surfaces.

Points. **Points** are used to represent features that are too small to be visible at a given map scale. Thus, for example, points can be used to represent buildings that are too small to be depicted on a county map and cities that are too small to be displayed on the map of a country. They can also be used to represent features that have no width or length such as parcel centroids.

Lines. **Lines** are used to represent features that have length but no width or are too narrow to be visible at a particular map scale. Thus, for instance, property lines have no width and roads and streams will generally not be wide enough to be visible on a county map.

Areas. **Areas** are used to represent features that are large enough to have length and width at a given map scale. Thus, for example, a building can be represented as an area on a parcel map, a land parcel can be represented as an area on a neighborhood map, and a forest can be represented as an area on a county land cover map.

Networks. **Networks** are used to represent linear features that transmit goods, people, energy, and information. Roads, rivers, pipelines, and electricity transmission lines are all familiar examples of networks.

Surfaces. **Surfaces** are used to represent continuous features that cover an area and include measurements along a third dimension. Elevations, air pollution, and noise levels are all surfaces.

Map Scale

One of the most important characteristics of a map is the **map scale**, that is, the size relationship between the features displayed on the map and on their counterparts on the Earth's surface. For example, as the previous discussion illustrates, whether a real-world entity is represented as a point, line, or area is dependent on the map scale used to represent the features. The concept of scale is particularly important because it has three different and somewhat contradictory meanings.

"Scale" can first refer to the level of spatial detail that is displayed on a map. On this meaning a fine-scaled map will include smaller objects such as land parcels and buildings and a course-scaled map will include large objects such as counties or states.

A second meaning relates scale to the geographic extent of a map. On this meaning, a large-scale map covers a large area such as a state or country and a small-scale map covers a small area such as a neighborhood. On this interpretation, a small-scale map corresponds to a fine-scale map because it covers a small area and shows small features. Conversely, a large-scale map corresponds to a course-scale map because it covers a large area and contains large features.

Cartographers use the word "scale" in a third way to refer to a map's **representative fraction**, that is, the ratio between distances on a map and the distances on the Earth's surface the map represents. For example, a representative fraction scale of 1:1,000 indicates that 1 inch on a map represents 1,000 inches or 83.3 feet on the ground. A representative fraction of 1:1,000,000 indicates that one inch on a map represent one million inches or 83,333 feet on the ground. On this meaning, a large-scale map has a large representative fraction (e.g., 1:1,000) and covers a small area at a high degree of detail; a small-scale map has a small representative fraction (e.g., 1:1,000,000) and covers a large area with little detail. On this meaning a large-scale map corresponds to a fine-scale map and a small-scale corresponds to a course-scale map.

The second and third definitions of map scales provide opposite interpretations of the terms **small scale** and **large scale**, causing a great deal of confusion between cartographers and noncartographers. To avoid this confusion, it is advisable to use **fine scale** for maps that cover a small area and show smaller objects with a great deal of detail and **course scale** for maps that cover a large area and show larger objects with little detail.

Representative fractions have the same units in the numerator and denominator. This means they are "unitless" because the ratio can employ any unit. For example, one inch on a course-scale map with a representative ratio of 1:24,000 is equal to twenty-four thousand inches on the ground; one foot on the map represents twenty-four thousand feet on the ground.

Maps used in the architecture and engineering fields often use "graphic" or "verbal" scales, for example, "1 inch = 1,200 feet." Graphic scales do not have the same units in the numerator and denominator. As a result, converting a graphic scale to a representative fraction requires converting the denominator to the same units as the numerator. For example, a graphic scale of 1 inch = 100 feet is converted to a representative fraction of 1 inch = 100 feet × 12 inches/foot or 1:1,200. The process is reversed when converting a representative fraction to a graphic scale.

Map Accuracy and Metadata

The map scale affects the **accuracy** at which features are displayed on a map.[3] **Spatial data** are never perfectly accurate; they always include some errors in the location of map features or in the attributes associated with map features. While map users can rarely improve a map's accuracy, they should be aware of, and account for, the errors it inevitably contains.

For example, the US Geological Survey (USGS) produces widely used maps at two scales: 1:24,000 and 1:100,000. A road drawn as a one millimeter (mm) or 0.04-inch-wide line on a fine-scale 1: 24,000 map corresponds to 24,000 mm, 24 meters or 78.7 feet on the ground. A 1 mm line on a course-scale 1:100,000 map corresponds to 100 meters or 328.1 feet on the ground. As a result, combining a fine-scale USGS map with a course-scale USGS map ensures only that the road is located somewhere in a 428-foot-wide swath on the ground.

Fortunately, many maps provide metadata, data about data, that document the accuracy, contents, publisher, currency, and lineage of spatial data. Metadata are required for all spatial data produced by the US government and many states. The International Standards Organization has developed and promulgated similar standards for map data.[4]

Map Dimensions

Two-dimensional (2-D) objects that have length and width, but no height can be represented on a map by x and y coordinates on a coordinate system. Three dimensional (3-D) objects have length, width, and height and require x, y, and z values that define a volume. Continuous variables such as elevation and depth can almost always be defined by a single height value at any location. Exceptions include unusual features such as vertical or overhanging cliffs and caves that have more than one height values for a single x-y location. As a result, terrain is normally represented by a single z value for each x and y coordinate pair, or what is often referred to as "2.5-D," and full three-dimensional (3-D) representations are reserved for representing unusual features such as cliffs and caves. 2.5-D and 3-D images arc useful for representing three-dimensional features such as buildings.

Locating Spatial Features

Geographic information systems are powerful tools for representing and analyzing entities and activities that take place on the Earth's surface. As a result, it is essential that they accurately locate spatial features on the Earth surface and on maps showing a portion of the globe. The most widely used systems for locating spatial features are: (1) postal addresses, (2) linear referencing systems, (3) cadasters, (4) geographic coordinate systems, and (5) projected coordinate systems.[5]

Postal Addresses

Postal addresses that include a street address, an optional apartment number, a city name, a state name, and a **postal code** are the most familiar system for locating human activities. Postal addresses were developed to facilitate mail delivery in the eighteenth century and are used in conjunction with national mapping systems such as the Topologically Integrated Geographic Encoding and Referencing (TIGER) system in the United States. GIS-based geocoding tools allow address-based features to be located in space and displayed on a map, providing the basis for the ubiquitous online mapping and route-finding systems (Klosterman 1991). However, postal addresses are not useful for locating natural features such as mountains and rivers that do not have a street address. They are also not useful when buildings are not numbered consecutively along a street, as is the case in Japan where street numbering indicates the date of construction, not the location along a street.

Linear Referencing Systems

Linear referencing systems identify locations along a network by measuring the distance from a defined point in each direction along a defined path along the network. For example, the location of an automobile accident can be reported as a measured distance from a street intersection in each direction along a named street. Linear referencing systems are widely used for applications that use linear networks such as highways, railroads, pipelines, and electric transmission lines. GIS provide specialized tools for creating, maintaining, displaying, and analyzing linear reference systems.

Cadasters

A **cadaster** is a map of land ownership that provides a record of land ownership to facilitate land taxation. Land parcels in a cadaster are usually identified by a unique number or code, often called a parcel identification number or PIN. The US Public Land Survey System, developed in the early nineteenth century to map the western United States, is the dominant cadaster for the US west of Ohio and in western Canada. The Public Land Survey System divides an area into townships made up of six-by-six-mile squares. Townships are divided into one-mile-by-one-mile sections. Sections are 640 acres in size and are divided into four quarter sections of 160 acres each. Each subunit is labeled by a unique identifier.

Geographic Coordinate Systems

Geographic coordinate systems recognize that the Earth is roughly spherical and define locations on its surface in terms of spherical coordinates of latitude and longitude.

Longitude is defined by slices through the Earth that pass through the North and South Poles, like the segments of an orange. A slice that passes through the poles and a marker on the grounds of the Royal Observatory in Greenwich, England, defines the Prime Meridian which is assigned a value of 0 longitude. The angle between this slice and a location to the east or west defines a location's latitude. Locations west of the Prime Meridian are assigned negative longitude values up to −180 degrees. Locations east of the Prime Meridian are assigned positive longitude values up to +180 degrees. The degrees of longitude can be divided into sixty minutes that are divided into sixty seconds; instead they are generally represented by their decimal parts. Lines of equal longitude are represented on a map or globe by lines that pass through the Earth's poles, perpendicular to the Equator.

Latitude is defined by lines that pass through the Earth's center and a location north or south of the Equator. The angle made between this line and the plane of the Equator defines its latitude. Latitude varies from 90 degrees south (the South Pole), represented by negative values, and 90 degrees north (the North Pole), represented by positive values. Lines of equal latitude are represented on a map or globe by lines parallel to the Equator. Like longitude, degrees of latitude can be divided into seconds and minutes but are generally represented with decimals. Latitude is often symbolized by the Greek letter psi (φ); longitude is often symbolized by the Greek letter lambda (λ).

Geographic coordinate systems are complicated by the fact that the Earth is not a perfect sphere; the distance between its poles is roughly 1/300 less than its diameter at the Equator. An extensive effort has been devoted over the last two hundred years to define a mathematical model that best represents the Earth's shape. One model, known as WG84 (the World Geodetic System of 1984), is now generally accepted, along with its North American equivalent, the North American Datum of 1983 (NAD83). Many GIS datasets still adhere to earlier standards such as the North American Datum of 1927 (NAD27), but today's GIS allow spatial data to be easily converted from one coordinate system to another.

Projected Coordinate Systems

Geographic coordinate systems provide the most comprehensive system for defining locations on the Earth's surface and calculating the distance between point locations. However, paper maps and aerial photographs are flat, which often makes it easier to work with **projected coordinate systems** that project the Earth's curved surface onto a two-dimensional plane. For example, the Cartesian coordinate system assigns x and y coordinates to points on a flat surface defined by horizontal and vertical axes. Map projections transform locations on the Earth's surface identified by latitude and longitude (φ, λ) into equivalent positions in Cartesian coordinates (x, y).

Projections onto a flat surface necessarily distort the Earth's curved surface, making it impossible to simultaneously represent the shape, sizes, and direction of spatial features accurately on a map. Thousands of projections have been developed to preserve certain properties for particular uses. The Universal Transverse Mercator (UTM) projection is often used for military applications and for national maps. The UTM system consists of sixty zones; each zone corresponds to a half cylinder wrapped along a line of longitude.

Universal Transverse Mercator coordinates are very useful for spatial analysis because they allow distances between points in the same zone to be measured with less than 0.04 percent error. However, this accuracy is not sufficient for surveying and other precise applications. As a result, in the 1930s each state in the United States adopted its own projection system, generally known as the State Plane Coordinate System (SPCS), to support high-accuracy applications. Some large states have developed SPCSs for portions of their states; for example, California has six SPCS zones and Ohio has two. Many other countries have adopted their own projected coordinate systems. Modern GIS can easily convert geographic coordinates into the appropriate projected coordinate system for a location.

Vector and Raster Data Models

Geographic information systems use two main models to represent points, lines, and polygons: the vector data model and the raster data model.[6] GIS also use **triangulated irregular networks** (**TINs**) to represent surfaces such as terrain. TIN models comprise x, y, and z (elevation) coordinates for points located on the surface being modeled. Three adjacent points define a triangle; the face of the triangle represents the surface; and the edges of the triangle represent ridges and break lines in the surface.

Vector Data Model

The **vector data model** represents points such as trees by x and y coordinates defined with respect to a coordinate system. Lines such as streets are represented by line segments connecting two or more x, y coordinate pairs. Areas such as buildings are represented by polygons, made up of line segments connecting three or more x, y pairs. The line segments for a polygon are closed, that is, the first coordinate pair is also the last coordinate pair.

The vector data model is most appropriate for displaying **discrete features** that are unique, identifiable, and can be counted. Discrete features include entities that have distinct boundaries such as buildings, locations such as addresses, linear features such as roads and rivers, and political and administrative units such as cities, census tracts, and land parcels. The vector model is often preferred because it generally produces more attractive maps. However, it is not well suited for representing continuous variables such as pollution concentrations or natural features with indistinct boundaries such as land cover and soil types. The raster data model is generally preferred for these uses.

Raster Data Model

The **raster data model** divides an area into an array of equal-sized grid cells. Each cell is associated with one or more attributes. Points are represented in a raster data model by a single cell, lines are represented by a collection of adjacent cells with a common value, and areas are represented by a group of adjacent cells with a common value.

The raster model is particularly well suited for representing **continuous features** that vary continuously over space such as temperature and noise levels and phenomena with indistinct boundaries such as natural vegetation areas. The raster data model is also desirable for displaying map images such as satellite imagery or aerial photographs and for analyzing data that are provided in a uniform grid of elevation values such as digital elevation models. However, raster data models are not appropriate for representing point or line features or entities that have distinct boundaries such as roads, parcels, and administrative units. In addition, raster maps are generally less attractive than vector maps.

Modern GIS can easily handle both vector and raster data and quickly and easily convert from one type of data to the other. For example, spatial data may be obtained as raster data from a remotely sensed image, converted to a vector model for locating roads and buildings, and converted back to a raster for display on the Web. Raster aerial photos or space imagery can also be used to provide a photo-realistic background for a vector data analysis.

As pointed out above, raster data are particularly useful for dealing with remotely sensed imagery and continuous features that do not have distinct boundaries. However, city and regional planners deal primarily with features from the built environment such as buildings and roads and work with information for political and administrative units such as cities, census enumeration areas, school districts, and traffic analysis zones. As a result, the discussion of spatial analysis techniques in this chapter will only consider the vector data model.[7]

Analyzing Spatial Data

Geographic information systems are particularly useful because they can be used to prepare maps that allow planners, public officials, and the public to answer spatial queries about the location of entities and activities that can be answered by consulting a map. Maps can be used to determine *what is at a location*, for example, the name of a county on a state map or the parcel identification number for a parcel on a cadastral map. They can also be used to determine *where a feature is located*, for example, where a street is located on a city's street map.

Maps can also be used to identify spatial patterns that help identify policies that should be implemented or actions that should be taken. For example, a crime map may identify areas that require additional policing and a map of dilapidated housing units may identify areas for targeting home improvement loans. They can also be used to help identify the causes for spatial patterns. Thus, for example, an ecologist may look at the distribution of plant communities to see if the patterns are related to terrain, rainfall, or other factors.

Quantities, Categories, Ratios, and Ranks

Before preparing a map, it is important to identify the most appropriate way to analyze the available information and present it clearly. In particular, it is important to recognize whether the data should be displayed as quantities, categories, ratios, or ranks.

Quantities **Quantities**, the number or value of the features being considered, is the most obvious measure to use in preparing a map. Quantities include the number people for census tracts, the size of land parcels, and the average daily traffic for road segments. Quantities are clearly useful for showing the values of map features (e.g., the size of land parcels) or comparing the value of different features (e.g., the population of different census tracts).

However, maps that display quantities may be misleading if the size of the areas being mapped is substantially different. Mapping quantities can also obscure important relationships and patterns that are revealed only when the data are standardized against a meaningful common denominator.

For example, map 7.1 displays the number people living in poverty in 2010 in Georgia's 159 counties. It is almost identical to map 7.2 which shows the total population of each county in 2010. This is hardly surprising because we would expect more poor people to live in counties that have larger populations.

Now consider map 7.3 which shows the proportion of the population in each county that is poor. The pattern here is substantially different from map 7.1, showing that a large proportion of the population living in the central part of the state is in poverty. This map is much more informative than the map of the total population in poverty displayed in map 7.1 and raises interesting questions of why the population here is poorer than it is in other parts of the state.

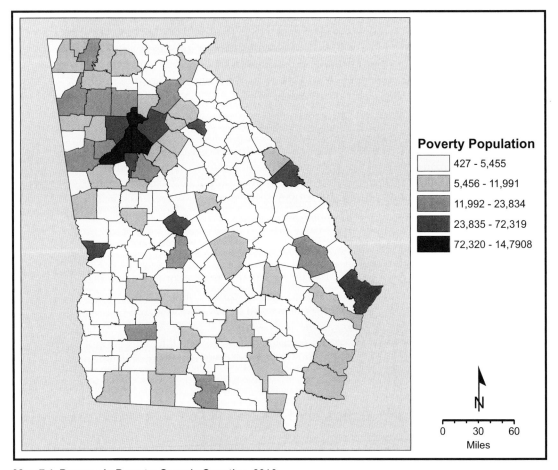

Map 7.1 Persons in Poverty: Georgia Counties, 2010

Source: US Bureau of the Census (2013c).

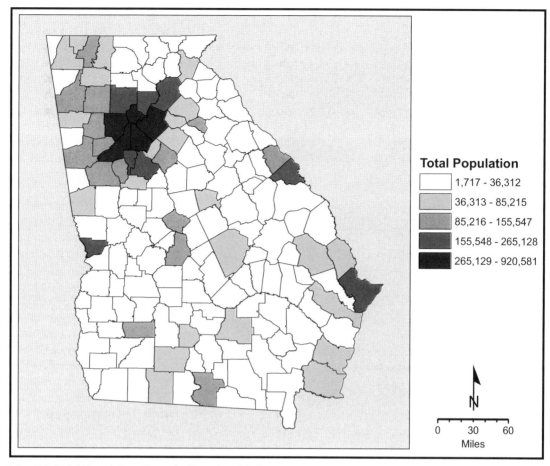

Map 7.2 Total Population: Georgia Counties, 2010

Source: US Bureau of the Census (2013c).

Categories Categories are groups of similar things. All features with the same value for a category are alike in one aspect and different from features in other categories with respect to that aspect. For example, a land use map displays the land uses (residential, commercial, agricultural, and so on) for different land parcels and a street map may show different types of streets (expressways, major arterials, and local streets). The categories used in a map are identified in the legend and with different map symbols (colors, line widths, hatching patterns, and so on).

Ratios Ratios record the relationship between two quantities and are computed by dividing one quantity by another. Ratios are particularly useful when values are summarized by area because they may even out differences between large areas with many features and small areas with few features. The three most common ratios are averages, proportions, and densities.

Averages Averages are familiar to planners and include measures such as the average number of people per household, the average household income for block groups, or the average age for a population group. Averages are computed by dividing the value for one variable (e.g., the total income in census tracts) by the value for a second variable (e.g., the total number of people in each tract).

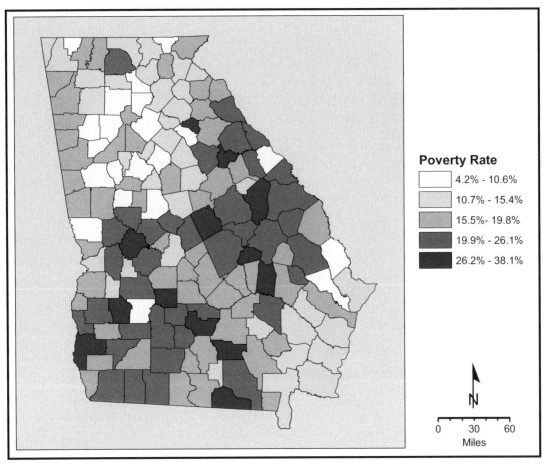

Map 7.3 Poverty Rate: Georgia Counties, 2010

Source: US Bureau of the Census (2013c).

Proportions Proportions record the portion of a whole that a quantity represents and are computed by dividing quantities that measure the same variable. For example, dividing the number of people between the ages of twenty and twenty-five by the total population in a census tract shows the proportion of the tract's population between twenty and twenty-five. As map 7.3 illustrates, proportions are particularly useful for comparing places with few features to areas with many features.

Proportions are often presented as percentages (the proportion multiplied by 100) or rates (the proportion multiplied by 1,000). This is done because people are generally more comfortable thinking about percentages (e.g., "22 percent") than about ratios ("twenty-two hundredths"). Similarly, for rare events such as the number of deaths in an age cohort, it is more convenient to talk about rates per thousand (e.g., "3.5 deaths per thousand" instead of "three and a half thousandths").

Densities Densities are useful for showing the distribution of features when the size of the areas varies greatly. Densities are computed by dividing the value of a variable for an area by the size of the area. Density is good for showing the distribution of values when the size of the areas being compared varies greatly. For example, census tracts have roughly the same sized population so mapping the total population for census tracts may not be very informative. However, some of the census tracts may be small (and densely settled) and others may be large (where the same number of people is spread out over a larger area), so a density map may be much more informative.

Ranks Ranks show relative values, that is, put features into an order from high to low. Ranks can be represented by labels, for example, low, moderate, and high, or by ordinal numbers, for example, from 1 (low) to 9 (high). Ranks are useful when direct measurement is difficult or impossible. For example, it may be difficult to quantify the scenic value of landscapes but possible to rank them from more to less attractive. Ranks are also useful when a quantity represents a category or combination of factors. For example, as shown in chapter 8, suitability analysis techniques can be used to rank locations by their relative suitability for accommodating different land use demands.

Creating Classes

After choosing to use quantities, categories, ratios, or ranks, a mapmaker must decide whether the map should represent individual values or group the values into classes. Maps that show individual values (e.g., the population of each county in a state) are useful for presenting a detailed picture of the data. However, they may be difficult to interpret if there are many different values that make it difficult to identify features with similar values. In these cases, it is often desirable to group the values into classes. Classes are also useful when a map will be used with the public because it allows map readers to compare areas quickly and easily. As a result, maps showing quantities and ratios for many features generally group the values into classes. Maps that show ranks, for example, the suitability of different locations for a land use, generally display individual values.

Classes group features with similar values together by assigning them the same area fill (color and hatching pattern) or line type (width, color, and dashes), allowing users to quickly and easily identify features with similar values. However, the way the values are classified determines which features fall into each class and how information is portrayed to the reader. Five methods can be used to group values: (1) equal intervals, (2) quantiles, (3) natural breaks, (4) manual classifications, and (5) standard deviations. As the following discussion illustrates, the method used to group values can have a tremendous impact on the way information is reported on a map.

Equal Interval Classification Method The **equal interval classification method** divides values into classes that have an equal range, that is, the difference between the high and low value is the same for each class. For example, if three classes are used to group features whose values range from 1 to 300, the equal interval method will use class ranges of 1–100, 101–200, and 201–300.

The equal interval method is good for presenting information to a nontechnical audience because it is easy to understand that the data have been divided into equally broad ranges. It is also useful for displaying continuous values such as temperature and familiar values such as percentages. However, it is much less informative if the data values are not evenly distributed between equally broad classes.

For example, map 7.4 uses the equal interval classification system to display the population of Georgia's 159 counties. The map divides the counties' population values into five intervals, each of which has a range of 183,733 people. The ranges are computed by subtracting the smallest value, 1,717, from the largest value, 920,581 and dividing by five.

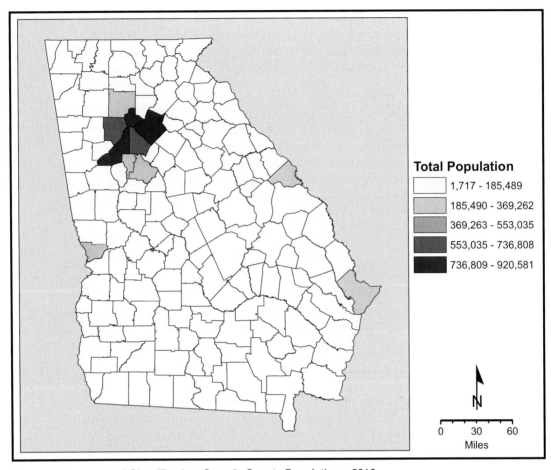

Map 7.4 Equal Interval Classification: Georgia County Populations, 2010

Figure 7.1 reports that the equal interval method includes 2 counties in the top class, 2 counties in the second class, no counties in the third class, 4 counties in the fourth class, and 149 counties in the lowest class. The map is correct but not at all helpful because it suggests that with ten exceptions, all of Georgia's counties have the same population. This is clearly not correct, and the map does an extremely poor job of representing the population distribution among Georgia's counties.

Quantiles Classification Method The **quantiles classification method** groups features into classes that contain the same number of features. For example, if five classes are used to group twenty-five values, the quantiles method will place five values in each class. The quantiles method is useful when the values are evenly distributed between classes and the areas being compared are roughly the same size. However, it can be less informative if the areas being compared vary greatly in size. It can also be misleading by exaggerating differences between closely grouped features and minimizing the differences between adjacent but widely different values.

Map 7.5 uses the quantiles method to divide the population of Georgia's counties into five classes: four classes with thirty-two counties and one class with thirty-one counties. The map is more informative than the first map because it suggests that the population of the state's counties varies widely. However, it is still very misleading.

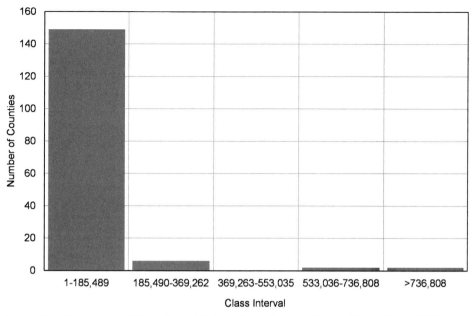

Figure 7.1 Equal Interval Class Break Distributions: Georgia County Populations, 2010

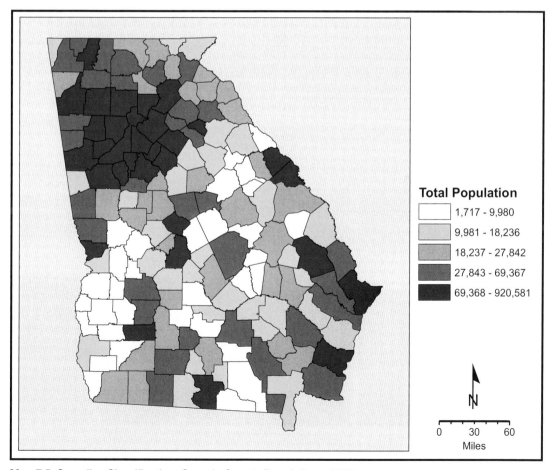

Map 7.5 Quantiles Classification: Georgia County Populations, 2010

Figure 7.2 reports that the quantiles method places the thirty-one counties with the largest population values in a single class. This suggests that one-fifth of the counties have roughly the same population as the highest population county, Fulton County (containing Atlanta), which has more than thirteen times as many people as Barrow County, at the bottom of its class.

Natural Breaks Classification Method The **natural breaks (or Jenks) classification method** identifies natural grouping in the data and creates classes that minimize the differences between the members in each class and maximize the differences between classes. This method was developed by George Jenks (1916–1996) and is good for mapping values that are not evenly distributed because it places nearby values in the same classes. However, the natural breaks categories are specific to a data set, making it difficult to compare maps that use this method.

For example, map 7.6 uses the natural breaks method to display the population of Georgia's 159 counties. The map clearly demonstrates the distribution of the state's population with several counties with large populations near Atlanta, a few counties with moderately large populations scattered around the state, and many counties with small populations distributed evenly throughout the state. Figure 7.3 correctly reports that many Georgia counties have small populations and progressively fewer counties have larger populations.

Manual Classification Method The **manual classification method** allows users to create their own classes with break points that they feel most accurately represent important divisions in the data. The breaks may reflect the clustering the data, corresponding to results of the natural breaks method with modifications to highlight particular characteristics in the data. The breaks may also reflect the type of data being considered and the information the map attempts to convey. For example, it would be appropriate to create categories that identify tracts in which at least 35 percent of the residents is below the poverty level in preparing a map for locating Empowerment Zones.

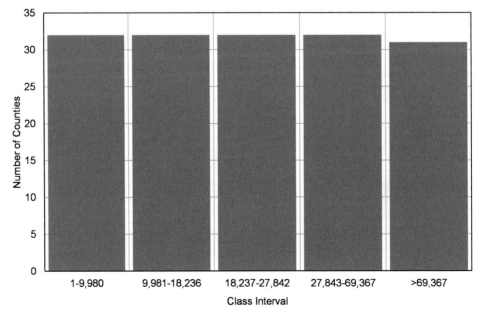

Figure 7.2 Quantiles Class Break Distributions: Georgia County Populations, 2010

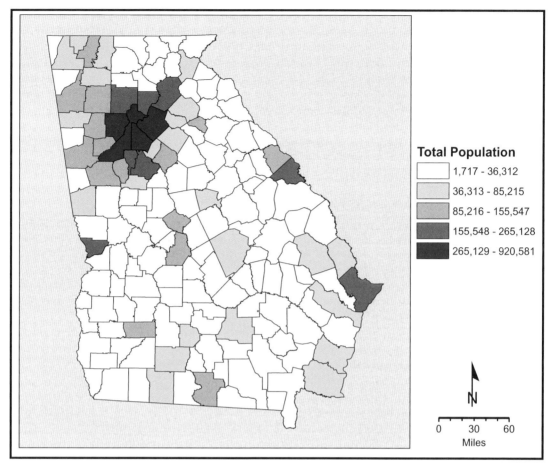

Map 7.6 Natural Breaks Classification: Georgia County Populations, 2010

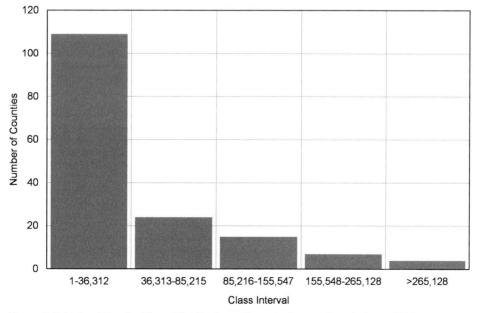

Figure 7.3 Natural Breaks Class Distributions: Georgia County Populations, 2010

Standard Deviation Classification Method The **standard deviation classification method** creates classes based on their distance from the mean value for all the features. The classes are created by computing the average of all the values and then creating class breaks above and below the mean based on the number of standard deviations the user designates. The standard deviation method is useful for identifying features that are above or below the average value. It is also useful for displaying data that have many values near the mean and fewer values further from the mean (i.e., a bell-shaped curve or normal distribution). However, the method can be misleading if very high or small values skew the mean so that most features fall in the same class.

Choosing a Classification Scheme

As the Georgia county population mapping example illustrates, the method used to group map values can dramatically affect the user's understanding of the underlying data. The best scheme to be used for a data set is determined primarily by the way in which the feature values are distributed across their entire range. The easiest way to determine this is to prepare *histograms* such as figures 7.1, 7.2, and 7.3 that display the attribute values on the horizontal axis and the number of features with each value on the vertical axis.

In general, the natural breaks method should be used if the data are unevenly distributed, that is, many features have the same or similar values and there are gaps between groups of values. The equal interval method should be used if the data are evenly distributed continuous variables and for presenting familiar concepts such as percentages to nontechnical audiences. The standard deviation methods should be used if the data are evenly distributed, and it is important to demonstrate how values are related to the mean value. The quantiles method should be used when the data are evenly distributed and it is useful to emphasize the relative differences between values.

Analyzing Attribute Data

Planners, public officials, and the public are obviously interested in using cartographic data to map things and answer spatial queries about the location of entities and activities displayed on a map. However, they are also interested in understanding the characteristics of things, that is, their attributes. Thus, for example, they may want to know the land use for a parcel of land, a city's racial breakdown, or how one area's population compares to another area's population. GIS are particularly useful for answering these questions because they contain sophisticated **database management systems (DBMS)** for creating, updating, querying, and maintaining attribute databases.

Attribute Tables

Attribute data are normally stored in tables.[8] As shown in table 7.1, attribute tables can be conceptualized as being comprised of rows (or **records**) and columns (or **fields**). Each row represents a spatial feature in the database; each column contains information on a characteristic of the features in the database; and the intersection of a row and column records the value of a characteristic for a feature.

For example, table 7.1 records the 2010 population data for ten Georgia counties. Each row contains the data for a county; for example, the third row (highlighted in yellow) contains data for Bacon County. Each column contains a

TABLE 7.1	Total Population: Georgia Counties, 2010		
Name	County	County FIPS Code	2010 Population
(1)	(2)	(3)	(4)
Appling County	001	13001	18,236
Atkinson County	003	13003	8,375
Bacon County	005	13005	11,096
Baker County	007	13007	3,451
Baldwin County	009	13009	45,720
Banks County	011	13011	18,395
Barrow County	013	13013	69,367
Bartow County	015	13015	100,157
Ben Hill County	017	13017	17,634
Berrien County	019	13019	19,286

Source: US Bureau of the Census (2013a).

characteristic for all the counties; for instance, the fourth column (highlighted in blue) contains each county's 2010 population. The intersection of a row and column contains a piece of data for a county; for example, the intersection of the third row and the fourth column contains the 2010 population for Bacon County (highlighted in blue).

Geographic information systems are particularly powerful tools because they automatically link cartographic databases and attribute databases allowing users to quickly and easily perform queries that utilize both databases. For example, they can use spatial queries to select one or more features on a map and view the information on the associated feature(s) stored in the attribute database. Similarly, they can use attribute queries to select one or more records in the attribute data table and see where these features are located on a map. Spatial and attribute queries made from an entire database are "selected sets" that can be used to answer a question or as the basis for further analysis.

The link between the features displayed on a map and their attributes in the attribute table also allows information in the attribute table to be graphically displayed on a map. For example, a GIS map showing a city's streets could use line widths to represent different road classes, colors to represent the peak hour congestion on each link, and line types (e.g., solid, dashed, and dotted) to represent the condition of the road surfaces. Similarly, a three-dimensional GIS display of block group populations could use the height of blocks to represent their densities and their colors to represent the proportion of the population in an income category.

Types of Attribute Tables

Geographic information systems use two main kinds of attribute tables. **Feature attribute tables** are created and maintained by the GIS software and provide a direct connection between the cartographic and attribute data tables. This connection is maintained by unique feature codes in both tables that are created by the GIS program and cannot be modified by the user.

Nonspatial attribute tables are not created by a GIS system and are not automatically connected to the cartographic database. These tables may take many forms including Excel tables, delimited text files, and files maintained by dedicated DBMS such as Oracle or IBM DB2.

Nonspatial attribute tables often contain fields that can be used to connect a nonspatial attribute table to a feature attribute table, allowing the **nonspatial data** to be displayed on a map. For example, a user may have a feature attribute table that contains the boundaries for the block groups in a county and a nonspatial attribute table that contains detailed population and housing data for the block groups. Both tables may have a field containing a unique identifying code for each block group such as the US federal government's **Federal Information Processing Series (FIPS)** code described in appendix A. The common field can be used to join the two tables, allowing the detailed population and housing data to be displayed on a map.

Attribute Measurement Scales

It is important to recognize that there are many kinds of attribute data that can be used for very different kinds of operations. Common types of attribute data include: (1) text (or characters) that can be used as labels, (2) numbers, (3) dates, and (4) binary large objects (BLOBs) that are used to store images and multimedia information. Numbers can be either **integers** (for numbers without decimal points) or **floating points** (for numbers with decimal points). Attribute data can also be defined by their measurement scale: nominal, ordinal, interval, or ratio.

Nominal Scale Data The simplest form of attribute data are **nominal scale** data that assign labels to objects, allowing an object to be distinguished from other objects of the same class. Examples of nominal scale numeric values include driver's license numbers, ZIP code numbers, and street addresses. In all these cases a number is used to label an object, allowing it to be identified clearly. Nominal attributes can include not only numbers but also labels, letters, and even colors (e.g., land use types can be represented by different colors). And while nominal attributes may be numbers, they cannot be used for numeric operations such as addition, subtraction, or division. For instance, it makes no sense to add drivers' license numbers or compute the average ZIP code for an area.

Ordinal Scale Data Attributes are measured on an **ordinal scale** if their values have a natural order. For example, soils can be classified as "Class 1" (best) "Class 2" (second best), and so on, to reflect their appropriateness for a particular purpose. Similarly, an ordinal scale can be used to identify the order in which runners finished a race, that is, who finished first, who finished second, and so on. Ordinal numbers cannot be meaningfully added, subtracted, multiplied, or divided. For example, a Class 2 soil is not twice as good as a Class 1 soil. It makes no sense to compute averages for ordinal scale values, for example, the average order in which runners finished. However, the median or middle value provides a useful central value measure. For example, the **median** value identifies runners who finished among the first half of the race contestants.

Interval Scale Data Attributes are measured on an **interval scale** if the interval between observations makes sense. Temperature measured on a Celsius or Fahrenheit scale is an obvious example of an interval scale because the interval between ten degrees and fifteen degrees on either scale is the same as the interval between twenty degrees and twenty-five degrees on that scale. Interval-scaled numbers can be added, subtracted, or used to compute averages. For instance, it makes sense to say that today's high temperature is five degrees warmer than yesterday's and to compare the average temperature for two dates or locations. However, it makes no sense to multiply or divide interval numbers because they are not measured with respect to a meaningful zero value. That is, it makes no sense to say today is twice as hot as yesterday because today's high temperature is sixty degrees and yesterday's high was thirty.

Ratio Scale Data Values measured on a **ratio scale** such as weight, height, and population density have a meaningful zero value and allow the full range of mathematical operations. Thus, for example, it makes sense to say that a person who weighs 100 kg is twice as heavy as a person who weighs 50 kg or that an area with a density of twenty-five persons per hectare is half as densely settled as an area with fifty persons per hectare. Temperature measured on the Kelvin (K) temperature scale (−273 °C or −459 °F) is measured on a ratio scale because it is based on a meaningful zero value, absolute zero characterized by a complete absence of heat energy.

Categorical, Numeric, and Interpreted Data As their name suggests, **categorical data** (measured on nominal or ordinal scales) allow observations to be assigned to categories but do not allow arithmetic operations such as computing means and ratios. **Numeric data** are measured on interval or ratio scales that allow a much broader range of arithmetic operations. The two types of data are clearly different, and it is normally not possible to convert categorical data into numeric data. However, suitability analyses such as the methods described in the next chapter assign (categorical) suitability ratings to different suitability factors and combine these ratings to determine an overall suitability score measured on an ordinal scale. Assigning suitability scores requires expert opinion, for example, on the relative weights to be given to different suitability factors. As a result, the suitability scores are **interpreted data** based on expert judgment.

Boolean Queries

Geographic information systems include flexible and powerful tools for querying databases, that is, specifying questions of a database that satisfy specified conditions. Databases can be queried with respect to their location (**spatial queries**) and their attributes (**nonspatial** or **attribute queries**). These queries are generally expressed in a standard database query language known as **Standard or Structured Query Language** (**SQL**) and are most clearly understood as Boolean queries visualized as Venn diagrams.

    Boolean queries are named after George Boole (1815–1864), an Irish mathematician who used binary logic (true/false, yes/no) to combine logic and mathematics into a coherent system that is ideal for use with computers. Boolean operators—AND, OR, NOT, and XOR—can be combined with arithmetic operators—addition, subtraction, multiplication, and division—and logical operators—equals (=), greater than (>), less than (<), and not equal to (<>)—to define complex and precise database queries. The results of a Boolean query are either true or false, which makes them very useful for GIS analysis.

Venn Diagrams

Venn diagrams are named after their inventor, John Venn (1834–1923), a priest, philosopher, and logician who taught at Cambridge University, England. Venn diagrams use overlapping circles to represent logical relationships between two or more classes or sets of entities. Venn diagrams can be used to visualize attribute queries, spatial queries, or queries that jointly consider features' locations and attributes.

    For example, an **attribute query** could use one circle to represent all the census tracts in a city that have a majority Hispanic population; a second circle could represent all the tracts with a low-income population; the overlapping area (the intersection) would represent the tracts in the city that have a majority Hispanic and a low-income population.

A spatial query could use one circle to represent all the tracts in a county; a second circle would represent all the tracts in the state within a quarter mile of hazardous waste site; the intersection represents the tracts in the county that are within one-quarter mile of a hazardous waste site. A combined query could use one circle to represent all the tracts in a state that have a majority Hispanic population (an attribute query); a second circle represents all the tracts within a quarter mile of hazardous waste site (a spatial query). The intersection represents all the tracts that have a majority Hispanic population and are within one-quarter mile of a hazardous waste site (combining an attribute and a spatial query).

Venn diagrams are particularly useful for representing Boolean queries that select a portion of the records in two or more data sets that satisfy specified conditions. For example, table 7.2 considers the relationship between population density and per capita income for Georgia's 159 counties. The table divides the counties on two dimensions: (1) per capita income and (2) population density. The row totals record that 128 low-density counties (D_L) have population densities below the state average and 31 high-density counties (D_H), have population densities greater than the state average.[9] The column totals indicate that ninety-seven low-income counties (I_L) have per capita incomes below the state average, and sixty-two high-income counties (I_H) have per capita incomes greater than the state average. The values in the body of the table record the number of counties with different density and income combinations. For example, ninety-four counties ($D_L - I_L$) have low densities and low incomes.

Sample Venn Diagram The relationship between the state's low-density and low-income counties is represented graphically by the Venn diagram shown in figure 7.4. The 128 low-density counties (D_L) are represented by the circle to the far left of the figure. The ninety-seven low-income counties (I_L) are represented by the second figure from the left. The figure on the right uses two overlapping circles to represents the number of counties that have a combination of density and income values. The intersection of the two circles ($D_L - I_L$) represents the ninety-four counties that have

TABLE 7.2	Per Capita Income and Density: Georgia Counties, 2010		
	Per Capita Income		
Density	**Low**	**High**	**Total**
(1)	(2)	(3)	(4)
Low	94 ($D_L - I_L$)	34 ($D_L - I_H$)	128 (D_L)
High	3 ($D_H - I_L$)	28 ($D_H - I_H$)	31 (D_H)
Total	97 (I_L)	62 (I_H)	159

Figure 7.4 Sample Venn Diagram

low densities and low incomes. The partial circle to the left $(D_L - I_H)$ represents the thirty-four counties that have low densities and high incomes. The partial circle to the right $(D_H - I_L)$ represents the three counties that have high densities and low incomes.

UNION Query The Venn diagram in figure 7.5 represents a Boolean **UNION (or OR) query** that selects all of members of overlapping sets that satisfy either: (1) one condition, (2) a second condition, or (3) both conditions. The query is illustrated in the left-hand figure by shading all three portions of the Venn diagram. Its application to the data in table 7.2 is illustrated in the right-hand figure that highlights all counties that either have either (1) low densities and high income $(D_L - I_H)$, (2) high densities and low incomes $(D_H - I_L)$, or (3) both low densities and low incomes $(D_L - I_L)$. The figure reveals that 131 Georgia counties have low densities, low incomes, or both low densities and low incomes.

INTERSECT Query The Venn diagram in figure 7.6 represents a Boolean **INTERSECT (or AND) query** that selects all of members of overlapping sets that satisfy one condition and a second condition. The INTERSECT query is illustrated in the left-hand figure by shading the intersection of the two circles in the Venn diagram. The right-hand figure indicates that ninety-four Georgia counties $(D_L - I_L)$ have both low densities and low incomes.

IDENTITY Query The Venn diagram in figure 7.7 represents a Boolean **IDENTITY (or NOT) query** that selects the members of overlapping sets that satisfy one condition but not a second condition. The IDENTITY query is illustrated in the left-hand figure by highlighting the non-overlapping portion of the Venn diagram. Its application to the Georgia county data is illustrated in the right-hand figure which indicates that only three Georgia counties $(D_H - I_L)$ have high densities and low incomes.

EXCLUSIVE OR Query The Venn diagram in figure 7.8 represents a Boolean **EXCLUSIVE OR (or XOR) query** that selects all of members of overlapping sets that satisfy one condition or a second condition but not both. The EXCLUSIVE OR query is illustrated in the left-hand figure by shading the two nonintersecting portions of the Venn diagram. Its application to the Georgia data is illustrated in the right-hand figure that highlights: (1) counties that have low densities and high incomes $(D_L - I_H)$,

Figure 7.5 UNION Venn Diagram

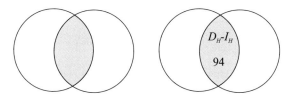

Figure 7.6 INTERSECT Venn Diagram

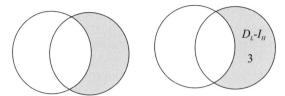

Figure 7.7 IDENTITY Venn Diagram

Figure 7.8 EXCLUSIVE OR Venn Diagram

and (2) counties that have high densities and low incomes ($D_H - I_L$), but not (3) counties that satisfy both conditions ($D_L - I_L$). The diagram reveals that thirty-seven Georgia counties have low densities or low incomes but not both.

Analyzing and Representing Spatial Relationships

As the previous discussion has demonstrated, GIS are powerful analysis tools because they incorporate cartographic data describing *where things are* and attribute data describing *what they are*. However, its ability to consider topological data describing *where things are relative to other things* is even more important and differentiates it from other computer systems.

Topology

Topology is often confused with topography. **Topography** uses maps to represent the Earth's surface, or terrain, and locate natural and man-made features on the Earth's surface.[10] Topographic maps generally represent terrain with contour lines connecting points of equal elevation above mean sea level. For example, the USGS's widely used topographic maps use contour lines to represent relief and include streams and water bodies, forest cover, built-up areas, and other features.

Topology is a field of mathematics that deals with geometric relationships that remain constant when geographic space is distorted. Consider, for example, a map of the states and major cities in the United States that is drawn on a piece of rubber. If the map is stretched, the size and shape of the states and the distances between cities may change dramatically. However, the map's topological properties such as adjacency (which features are next to each other) and containment (which features are contained within other features) will not be changed. That is, no matter how the map is stretched, Illinois and Ohio will both be next to Indiana and not next to each other (adjacency) and Chicago will be in Illinois, not Ohio (containment).

Topology is an essential feature of GIS because it allows users to analyze the spatial relationship between features. For example,

- *containment* allows an analyst to identify the public facilities in a census tract;
- *adjacency* allows an analyst to identify the owners of all land parcels adjoining a piece of land;

- *intersection* allows an analyst to identify block groups that are within a specified distance of a hazardous waste site;
- *distance* allows an analyst to determine the distance between two locations on a map; and
- *network topology* allows a person to log onto the Web and get directions for driving from one location to another.

Topology Models

Five topology models have been developed for dealing with spatial relationships in a vector GIS: (1) images, (2) computer-aided design (CAD), (3) coverages, (4) shapefiles, and (5) geodatabases.

Images Images are widely used in GIS as photographs, drawings, and remotely sensed aerial or satellite images. Images are used to provide realistic map images and to incorporate pictures such as site photographs and scanned building floor plans as attributes for geographically referenced objects. However, images are just that; they do not incorporate attribute information on the objects in the image or topological information on how the objects are related to each other in space. As a result, images are only useful in GIS for orienting users and providing graphic displays.

Computer-Aided Design The first computerized mapping systems used lines on cathode ray tubes to represent vector maps and overprinted characters on a line printer to represent raster maps. In the 1960s and 1970s, advancements in graphics hardware and drawing software allowed maps to be created with general-purpose **computer-aided design/computer-aided drafting** (**CAD**) packages such as AutoCAD. The CAD data model was very useful for producing high-quality maps. However, it was never widely used in GIS because it deals only with cartographic information and contains little, if any, attribute information, and no topological information.

Rasters There are two main types of raster or grid models: georeferenced rasters and non-georeferenced rasters. **Non-georeferenced rasters** are images that are not tied to locations in space and do not store attribute information on the objects in the image or topological information on how the objects are related to each other. As a result, they can only be used to display images that can be superimposed on other GIS layers.

Georeferenced rasters are made up of cells of a known size that are tied to locations in space and stored as a matrix of numeric values. For example, the USGS **Digital elevation models** (**DEMs**) consist of thirty-meter-by-thirty-meter cells that cover the United States and store the cells' elevation above sea level. Locations within a raster layer can be identified by their row and column values. Topological relationships are not stored in rasters, but grid-based GIS systems provide an array of tools for considering spatial relationships.

Coverages In 1982, Esri, Inc., formerly called the Environmental Systems Research Institute, released its first commercial GIS software, ARC/INFO. Designed to run on desk-sized minicomputers, ARC/INFO was the first modern vector GIS. ARC/INFO introduced the **coverage or georelational data model** that made two major improvements to the CAD data model. First, it combined cartographic data for displaying map features and a sophisticated DBMS for handling attribute data. More importantly, the coverage data model explicitly stored topological information on the spatial relationship between map features.

Shapefiles The coverage data model's ability to simultaneously consider carto-graphic, attribute, and topological data made it the dominant vector GIS data model. However, it was large and complex, requiring more computing power than was available on the early personal computers. In 1992, Esri introduced a third data model, **shapefiles**, for personal computers that could not accommodate ARC/INFO. Shapefiles do not store topological information; instead they analyze spatial relation-ships on-the-fly, making them less desirable than coverages (or geodatabases) for spatial analysis. However, unlike the coverage model which uses proprietary formats to store attribute information, shapefiles store this information in fully documented for-mats that can be manipulated with other GIS systems and readily available software tools such as spreadsheets.

Geodatabases In 2001, Esri introduced a new data model, **geodatabases**, that store cartographic, attribute, and topological data in an integrated data system. Geodatabases are much more efficient than the other data models and allow different types of GIS data—vector, raster, TIN, location, and attribute—to be more easily integrated with non-GIS DBMS such as Microsoft Access and IBM DB2.

More importantly, the geodatabase model is an objected-oriented data model that stores features as objects with clearly defined properties and behaviors. GIS objects allow users to define "smart features" that provide much more realistic representa-tions of real-world entities than were provided by earlier GIS data models consisting of generic points, lines, and areas. Objects can have properties that correspond to the attributes or characteristics of the real-world entities they represent. They can also have methods that represent the actions associated with real-world entities. For example, land parcels could be defined as objects with properties such as their location, own-er(s), assessed value, and land use. Methods that could be associated with the parcel objects could include changes in their owner(s), their land use, or their assessed value.

Spatial Analysis Operations

Modern GIS offer dozens of sophisticated operations for analyzing spatial, attri-bute, and topological information for vector, raster, and TIN data structures. Many of these operations are useful for particular applications but less useful for other purposes. For example, network analysis is essential for transportation planning and utility management but not very useful for land use planning where overlay oper-ations are widely used. As a result, this section does not describe all of GIS spatial analysis operations. Instead, it only considers the vector analysis operations that are most useful for the analysis and projection methods described in other parts of this book.

Buffering

Buffering creates two zones around a point, line, or polygon: the buffer zone that is within a specified distance of a feature, and a second zone that is outside the specified distance.[11] **Buffers** are useful for many planning and administrative purposes. They can be used to enforce city ordinances specifying that liquor stores cannot be located within one thousand feet of a school or church. Buffers can be used to control land uses along rivers and streams to reduce nutrient, sediment, and pesticide runoffs. They can also be used to analyze the suitability of different locations for particular land uses. For example, buffers can be used to designate areas near parks and other amenities as highly suitable for residential development and areas close to disameni-ties as less desirable.

Buffering is a powerful operation that provides several options. The buffers around different features do not have to be equally large. For example, two-hundred-foot noise buffers could be defined for highways and hundred-foot buffers created for secondary streets. Buffers can also be created around more than one feature. The boundaries for overlapping buffers can be dissolved to create a single buffer zone that identifies all areas that are within a specified distance of one or more features, for example, all locations within five miles of a fire station or an expressway. Overlapping buffers can also be retained to identify areas that are within a specified distance of more than one feature and which feature is closest to a given location.

Buffering usually creates a buffer polygon that includes all locations within the buffer distance. The buffer polygon can be used to identify other features that lie inside or outside of the buffer. Many GIS systems allow buffers to be created "on the fly," to, for example, select features in one layer that are within a specified distance of the features in another layer.

Overlays

An overlay operation combines, erases, modifies, or updates the geometries and attributes of two or more input layers to create an output layer that reflects the spatial relationship between the input layers. Many different overlay operations are available for relating points, lines, and polygons. The most important of these operations will be described for analyzing the spatial features in figure 7.9. The figure contains four spatial features: (1) a polygon layer consisting of four parcels, labeled P_1, P_2, P_3, and P_4, (2) a second polygon layer showing a floodplain colored in green, (3) a point layer containing two buildings, labeled B_1 and B_2, and (4) a line layer in red representing a road passing through the floodplain and two of the parcels.

Polygon-Polygon Overlays

Polygon-polygon overlays combine two or more polygon layers to create a new polygon layer containing the attributes for one or more of the input layers. The polygon–polygon overlays most widely used by planners are (1) Union, (2) Intersect, (3) Clip, (4) Erase, and (5) Spatial Joins.

Polygon Union Overlay A **Polygon Union overlay** is the spatial equivalent of the Boolean UNION or OR operator that combines two or more layers to create a new layer that preserves the features and attributes for all the input layers. For example, a Union overlay for the parcel and floodplain layers in figure 7.9 generates a new layer consisting of the seven polygons in figure 7.10 that are in the parcel layer or in the floodplain layer. The attribute table contains seven records. Each record contains the information for both layers, for example, identifying the parcel in which each polygon is located and the fact that polygons 3, 5, and 7 are in the floodplain and the remaining polygons are not.

Polygon Intersect Overlay A **Polygon Intersect overlay** is the spatial equivalent of the Boolean INTERSECT (or AND) operator because it computes the geographic intersection of two or more polygon layers to create a new polygon layer that preserves only locations in the overlapping input layers. For example, an Intersect overlay for the two polygon layers in figure 7.9 yields the three green polygons in figure 7.10 that are in the parcel layer and the floodplain layer. The attribute table contains all the information from the parcel and floodplain layers for the three polygons.

Polygon Clip Overlay A **Polygon Clip** overlay extracts the features from an input layer that overlay the features in the clip layer. That is, the clip layer acts as a cookie cutter that cuts out a subset of the features in the input layer. This is particularly useful for defining a study area that contains a geographic subset of the features in a larger area. Using the floodplain polygon layer to clip the parcel layer in figure 7.9 yields the three green polygons in figure 7.10 that are in the floodplain yellow. The attribute table contains the attributes for only the parcel layer, unlike the attribute table for the Intersect overlay that has the information for both layers.

Polygon Erase Overlay A **Polygon Erase overlay** creates an output layer that contains only the features in the input layer that lie outside the boundaries of the erase layer. For example, using the floodplain polygon layer as an erase layer for the parcel layer in figure 7.9 yields the four yellow polygons in figure 7.10 that are not in the floodplain layer. The attribute table contains the attributes for only the parcel layer.

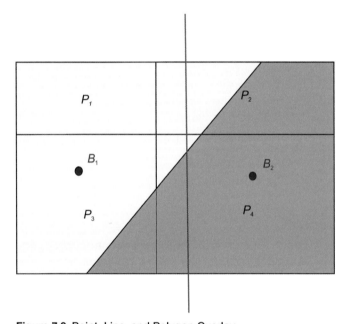

Figure 7.9 Point, Line, and Polygon Overlay

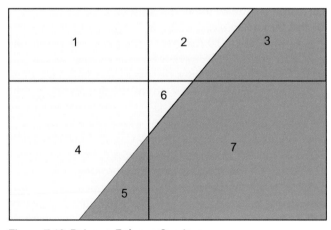

Figure 7.10 Polygon-Polygon Overlays

Polygon Spatial Join A **Polygon Spatial Join** joins the part of an input layer that shares locations with the join layer to create an output layer containing the shared locations and the associated attributes from the input and join layers. Spatial joins can be defined for many spatial relationships between all combinations of points, lines, and polygons. Two spatial relationships will be illustrated for the parcel and floodplain layers in figure 7.9. Joining the features in the parcel layer that *intersect with* the floodplain polygon generates a new polygon layer containing parcels P_2, P_3, and P_4 in figure 7.9, which are partially within the floodplain layer. Joining the features in the parcel layer that *have their centroids in* the floodplain polygon generates a polygon layer containing parcels P_2 and P_4, which have centroids in the floodplain layer. In both cases the output layer contains attributes from both polygon layers.

It is important to understand the difference between the Polygon Union, Polygon Intersect, and Polygon Spatial Join overlays:

- The Polygon Union overlay creates a new output layer that contains all the spatial features and attributes for the input layers. In the example above, the Union overlay generates a new layer contains the seven polygons shown in figure 7.10.
- The Polygon Intersect overlay generates a new layer containing only the overlapping portion of the input layers. In the example above, the Polygon Intersect overlay generates an output layer containing only the portion of the parcels that is inside the floodplain.
- The Polygon Spatial Join operation selects the entire features from the input layer that have the specified spatial relationship with the join layer. Thus, in this example, the Polygon Spatial Join generates a new polygon layer containing the entire parcels that intersect with, or have their centroids in, the floodplain.

All three operations create a new output layer containing the attributes from both layers, unlike the Polygon Clip and Polygon Erase overlays whose output layers contain only the attributes for the input layer.

Point–Polygon Overlays

Point-in-Polygon Overlay A **Point-in-Polygon overlay** combines a point layer and a polygon layer to create an output layer that contains all the points that are inside the polygon layer. The output layer contains the attributes for the polygons in which the points are located. For example, a point-in-polygon overlay for the point and floodplain layers in figure 7.9 yields point B_2, which is inside the floodplain layer. Its attribute table includes information from the point and the floodplain layers.

Point Clip Overlay A **Point Clip overlay** for a point layer (the input layer) and a polygon layer (the clip layer) selects all the points that are inside the clip layer. For example, a Clip overlay for the point and floodplain layers in figure 7.9 yields point B_2, which is inside the floodplain layer; its attribute table includes only information from the point layer.

Point Erase Overlay A **Point Erase overlay** for a point layer (the input layer) and a polygon layer (the erase layer) selects the points that are outside the erase layer. For example, a Clip overlay for the point and the floodplain layers in figure 7.9 yields point B_1, which is outside the floodplain layer; its attribute table includes only information from the point layer.

Point Spatial Join A **Point Spatial Join** joins the records from an input layer that satisfy a specified spatial relationship with the join layer to create an output layer containing the attributes for both layers. If the parcel layer in figure 7.9 is the input layer and the point layer is the join layer, a Point Spatial Join generates a polygon layer containing parcels P_3 and P_4, which contain points B_1 and B_2. If the point layer is the input layer and the floodplain layer is the join layer, a Point Spatial Join generates a point layer containing point B_2, which is inside the floodplain layer. In both cases, the output layer contains attributes from the polygon and point layers.

Line–Polygon Overlays

Line-in-Polygon Overlay A **Line-in-Polygon overlay** combines a line layer and a polygon layer to create an output layer containing the portion of the line layer that overlaps the polygon layer. For example, an Intersect operation for the line and parcel layers in figure 7.9 yields a new line layer with two segments that overlay parcels P_2 and P_4. The layer's attribute table contains information from the line and parcel layers.

Line Clip Overlay A **Line Clip overlay** combines a line layer (the input layer) and a polygon layer (the clip layer) to create an output layer consisting of the portion of the line layer that is inside the polygon layer. For example, a Clip overlay for the line and floodplain layers in figure 7.9 yields a new line layer for the portion of the line that is inside the floodplain. The layer's attribute table contains only information from the line layer.

Line Erase Overlay A **Line Erase overlay** combines a line layer (the input layer) and a polygon layer (the erase layer) to create an output layer made up of the portion of the line layer that are outside the polygon layer. For example, a Line Erase for the line and floodplain layers in figure 7.9 generates a new line layer consisting of the portion of the line that is outside the floodplain. The layer's attribute table contains only information from the line layer.

Line Spatial Join If the parcel layer in figure 7.9 is the input layer and the line layer is the join layer, a Line Spatial Join generates a polygon layer containing parcels P_3 and P_4. If the line layer is the input layer and the floodplain layer is the join layer, a Line Spatial Join generates an output layer containing the entire line. In both cases the output layer contains the attributes from the line and polygon layers.

Areal Interpolation

Polygon-on-polygon analysis is often used for **areal interpolation** that assigns data reported for one or more sets of geographic areas (the **source zones**) to an overlaying and incompatible set of geographic areas (the **target zones**). Areal interpolation can be used to convert the data from one year's census boundaries to a different set of boundaries for another census. It can also be used to convert data aggregated by census enumeration areas to the boundaries of incompatible government and administrative boundaries or to the boundaries of overlaying natural features such as river basins and floodplains.

Several areal interpolation techniques are described in the literature.[12] Two approaches will be considered here: area-weighted interpolation and dasymetric interpolation.

Area-Weighted Interpolation **Area-weighted interpolation** uses the area of the overlapping source and target zones to convert data from the source zones to the target zones. Given m source zones, n target zones, and a polygon overlay of the two layers, the estimated values for the target zones are computed with equation 7.1.

$$E_j = \sum_{i=1}^{m} e_{ij}, \qquad\qquad 7.1$$

where E_j = estimated value for target zone j, and e_{ij} = estimated value of the overlay polygon for m source zones i and target zone j
and,

$$e_{ij} = O_i \times \left(\frac{a_{ij}}{A_i} \right), \qquad\qquad 7.2$$

where O_i = observed value for source zone i, a_{ij} = area of the overlay polygon of source zone i and target zone j, and A_i = area of source zone i.

For example, areal interpolation can use the information for census tracts in figure 7.11 (the source zones) to estimate the population in the overlapping school districts in Figure 7.12 (the target zones). Table 7.3 reports the attribute information for the census tracts in figure 7.11: Tract 1 has a total area of forty hectares, twenty-four hectares of residential land, and four thousand residents (O_1); Tract 2 has a total area of twenty hectares, sixteen hectares of residential land, and one thousand residents (O_2). Table 7.4 reports the attribute information for the school districts layer in figure 7.12: District 1 has an area of twenty hectares; and District 2 has an area of forty hectares.

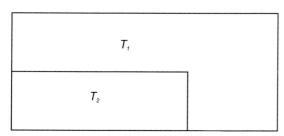

Figure 7.11 Census Tract Layer

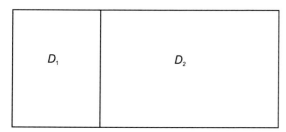

Figure 7.12 School District Layer

TABLE 7.3		Census Tract Layer Attribute Data		

		Area (ha)		
Polygon	Tract	Total	Residential	Population
(1)	(2)	(3)	(4)	(5)
1	T_1	40	24	4,000
2	T_2	20	16	1,000
Total	–	60	40	5,000

TABLE 7.4	School District Attribute Data	

Polygon	District	Area (ha)
(1)	(2)	(3)
1	D_1	20
2	D_2	40
Total	–	60

Figure 7.13 shows for the Polygon Union overlay of the layers in figures 7.11 and 7.12. Table 7.5 records the census tract, school district, total area and residential area in each overlay polygon. For example, polygon 1 (designated $T_1 - D_1$ in the figure), is the overlay of Tract 1 and District 1 and has a total area of ten hectares (16.7 percent of the total area) and a residential area of eight hectares miles (20 percent of the total residential area).

The area-weighted interpolation method uses the source and overlay polygons' total area to estimate the population in the target zones. For example, given equation 7.2, the estimated residential population in the four overlay polygons can be computed as follows:

$$e_{11} = O_1 \times (a_{11} / A_1) = 4,000 \times (10 / 40) = 1,000,$$
$$e_{12} = O_1 \times (a_{12} / A_1) = 4,000 \times (30 / 40) = 3,000,$$
$$e_{21} = O_2 \times (a_{21} / A_2) = 1,000 \times (10 / 20) = 500,$$
$$e_{22} = O_2 \times (a_{22} / A_2) = 1,000 \times (10 / 20) = 500.$$

Given these values, the residential population in the two school districts can be computed by applying equation 7.2. That is,

$$E_1 = \sum_{i=1}^{2} e_{i1} = e_{11} + e_{21} = 1,000 + 500 = 1,500,$$

$$E_2 = \sum_{i=1}^{2} e_{i2} = e_{12} + e_{22} = 3,000 + 500 = 3,500.$$

Dasymetric Interpolation **Dasymetric interpolation** uses axillary information to improve the target zone estimates produced by the area-weighted interpolation method. The dasymetric interpolation method draws on dasymetric mapping, an alternative to the more familiar choropleth maps.

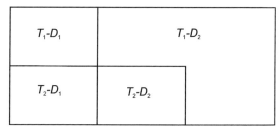

Figure 7.13 Census Tract-School District Union Layer

TABLE 7.5	Census Tract-School District Union Attribute Data					
			Total Area		Residential Area	
Polygon	Tract	District	Hectare	Percent	Hectare	Percent
(1)	(2)	(3)	(4)	(5)	(6)	(7)
1	T_1	D_1	10	16.67	8	20.00
2	T_1	D_2	30	50.00	16	40.00
3	T_2	D_1	10	16.67	8	20.00
4	T_2	D_2	10	16.67	8	20.00
Total	—	—	60	100.00	40	100.00

Choropleth maps divide an area into general purpose zones that can be used to map several variables. For example, a choropleth parcel map shows the parcel boundaries that can be used to display the parcels' land uses, zoning, assessed values, and many other variables. In contrast, a **dasymetric map** defines zonal boundaries that reflect sharp changes in the variable being mapped. For example, a dasymetric hydrology map would show watershed boundaries and a dasymetric population map would show the boundaries between different population densities.[13]

Planners often have land use information that can be used for dasymetric interpolation. For example, it is reasonable to assume that an area's residential population will only be located on residential land. This information can be used to provide more accurate target zone population estimates.

Table 7.3 records the residential area in the two source zones; table 7.5 records the residential area in the overlay zones. This information can be used with equations 7.1 and 7.2 to estimate the residential population in each school district where a_{ij} and A_i refer to the zones' residential areas, not their total area. That is,

$$e_{11} = O_1 \times (a_{11} / A_1) = 4{,}000 \times (8 / 24) = 1{,}333,$$
$$e_{12} = O_1 \times (a_{12} / A_1) = 4{,}000 \times (16 / 24) = 2{,}667,$$
$$e_{21} = O_2 \times (a_{21} / A_2) = 1{,}000 \times (8 / 16) = 500,$$
$$e_{22} = O_2 \times (a_{22} / A_2) = 1{,}000 \times (8 / 16) = 500.$$

Given these values and equation 7.2, the residential population in each school district can be computed as follows:

$$E_1 = \sum_{i=1}^{m} e_{i1} = e_{11} + e_{21} = 1{,}333 + 500 = 1{,}833,$$

$$E_2 = \sum_{i=1}^{m} e_{i2} = e_{12} + e_{22} = 2{,}667 + 500 = 3{,}167.$$

Comparing the two estimates indicates that District One's estimated population (E_1) is larger when the residential areas are used. This is true because the residential area (and presumably the residential population) in the two overlay polygons for District 1 (polygon one, $T_1 - D_1$ and polygon three, $T_2 - D_1$) have a larger share of the residential land (40 percent) than their share of the total area (33.4 percent). More accurate estimates could be obtained by using information on the quantity of land devoted to different residential types (e.g., low-density single-family, high-density single-family, and multifamily).

Notes

[1] The field of GIS is much too broad to be covered adequately in this textbook. As a result, the discussion here deals only with the aspects of GIS that are most useful for the methods described in the remainder of the book. More extensive discussions of GIS are provided by Longley et al. (2005), LeGates (2005), Clarke (2003), and Wang (2014).

[2] Guidelines for preparing informative and useful GIS maps are provided by Kent and Klosterman (2000) and Mitchell (1999). Extended discussions are provided by Robinson et al. (1995) and Monmonier (1996).

[3] Longley et al. (2005, 127–153) provides an extensive discussion of accuracy and other uncertainties associated with spatial data.

[4] For further information on these metadata standards see http://govinfo.library.unt.edu/npr/library/direct/orders/20fa.html and http://www.iso.org/iso/home/store/catalogue_ics/catalogue_detail_ics.htm?csnumber=53798.

[5] An extended discussion of geographic and projected coordinate systems is provided in Clarke (2003, 37–64).

[6] For a more complete description of the vector and raster data models see LeGates (2005, 54–90, 138–173).

[7] For extended discussions of raster GIS modeling and analysis see Tomlin (1990) and DeMers (2001).

[8] Reflecting its Latin roots, "data" is the plural form of "datum." As a result, "data" should be used to refer to more than one observation and "datum" should be used to refer to a single observation.

[9] More than half of Georgia's counties are in the low-density and low-income categories (i.e., are below the state average) because both variables are highly skewed.

[10] Topography has traditionally been concerned with recording locally detailed information, including not only relief but also vegetative and human-made features, and even local history and culture. This meaning is still used in Europe but is less common in America, where topographic maps with elevation contours have made "topography" synonymous with relief.

[11] See Mitchell (1999, 118–147) for a more complete discussion of buffering.

[12] See, e.g., Atkinson and Lloyd (2009).

[13] Eicher and Brewer (2001) and Reibel and Agrawal (2007) provide excellent discussions of dasymetric mapping and areal interpolation.

Land Suitability Analysis Methods 8

Land suitability analysis is the process of determining the suitability of one or more pieces of land for one or more uses.[1] Land suitability analysis methods use maps for one or more suitability factors to produce a set of maps showing the relative suitability of each piece of land for each land use. "Suitability" can be defined generally to identify locations that are susceptible to events such as landslides or wildfires or the best locations for locating natural features such as wildlife or plant species. However, the discussion in this chapter will focus on the analyses that planners use to prepare land-use plans and environmental impact reviews, to identify the most suitable location for siting public and private land uses and facilities, and to protect valuable natural areas.

Land suitability analysis is an example of the second of five general questions that geographic information systems (GIS) can answer: "Where is . . . ?"[2] That is, these methods use spatial and nonspatial criteria to define "what" is being looked for and then identify "where" it is located. The factors to be considered in evaluating locations include their physical characteristics, their location relative to other features, and their economic, demographic, and cultural characteristics.

Suitability criteria based on a location's characteristics (or attributes) include: (1) **categorical** or **nominal** values such as soil types, land uses, or zoning categories; (2) a range of **numeric values** (e.g., a range of parcel sizes); or (3) a threshold value (e.g., a slope value that cannot be exceeded). Criteria based on spatial relationships include: (1) how close a location is to another feature (e.g., a parcel's distance from a road), (2) whether a location is inside or outside another feature (e.g., whether a location is in a floodplain or in a city), and (3) whether a location is connected to another feature (e.g., whether a location is accessible from a city's light rail network).

Many land suitability analysis methods are described in the academic literature.[3] Land suitability analysis is often done with raster data, but this chapter will describe four widely used vector methods: (1) map overlay, (2) binary selection, (3) ordinal combination, and (4) weighted combination.

Map Overlay Method

Land suitability analysis was popularized with the publication of Ian McHarg's *Design with Nature* (McHarg 1969) but was widely used before then.[4] Charles Eliot and his associates at the landscape architecture office of Olmsted and Eliot are credited with inventing the map overlay method with "sun prints" produced on their office windows in the 1890s. Hand-drawn suitability maps were widely used during the early twentieth century but the first academic discussion of the map overlay method wasn't published until 1950 by Jacqueline Tyrwhitt (1950).

Overlay maps were used increasingly in the 1950s and 1960s but McHarg dramatically advanced previous applications by providing a theoretical basis for overlaying map information to gain an ecological understanding of an area. McHarg's work was particularly important because it provided a way to incorporate a broad array of environmental information into the planning process and protect important natural systems from development. This McHarg-inspired planning process, the "nature-first approach," begins by identifying areas essential for the protection and function of natural systems and excludes them from consideration for urban development.

McHarg's approach photographed a series of hand-drawn maps showing a study area's natural and human-made features that were printed as separate transparent images. A map was prepared for each of the **suitability factors** (e.g., soils, slopes, and land uses) to be considered in the analysis. The maps used different shades of gray to represent the distribution of different **factor types** (e.g., different slopes or soil types). More desirable features (e.g., areas with low slopes) were lighter than less desirable features. The transparent prints were superimposed on a light table to represent the composite ratings of all the factors that determined the suitability of different location for a particular land use. Lighter areas represented more suitable locations and darker areas identified less suitable areas.

Figure 8.1 uses the map overlay method to identify suitable locations for residential development. Figure 8.1(a) uses three shades of grey to identify locations relative to a road. The most suitable locations less than a quarter mile from the road are white; locations between one-quarter and a half mile away are an intermediate grey; and the least suitable locations more than half-mile from the road are a dark grey. Figure 8.1(b) uses three shades of grey to identify areas with different slopes. The most suitable locations with slopes less than or equal to 10 percent are white; locations with slopes greater than 10 percent and less than or equal to 20 percent are intermediate grey; and the least suitable locations with slopes more than 20 percent are dark grey.

Figure 8.1(c) is the overlay of the road distance and slope. The most suitable locations that are less than a quarter-mile from the road and have slopes less than 10 percent are white. The least suitable locations that have the least suitable characteristic (and the darkest shade) for one layer and an intermediate suitability (and shade) for the other layer have the darkest shade. Locations with suitabilities between these two extremes have intermediate shades.

The map overlay approach has had a tremendous influence on planning and landscape architecture and helped make land suitability analysis one of the most popular applications of GIS technology. McHarg's process of overlaying physical maps was limited by the number of maps that could be combined and the difficulty of representing variations between the factors in the map layers. Computers and GIS technology allow analysts to combine an unlimited number of map layers and adjust the values assigned to different map layers and to the features on each layer. GIS also allows analysts to represent relationships between factors on different layers, for example, to measure the distance between features on more than one map. It also allows users to easily convert one map layer into more than one suitability layer by, for example, converting digital elevation data into slope and **aspect** layers.

Binary Selection Method

The **binary selection method** uses spatial and nonspatial (attribute) data to identify locations that satisfy one or more suitability criteria. It is a "binary" method because it uses binary (yes/no) rules to divide the suitability factor types and locations into two groups: suitable and unsuitable. Suitable locations satisfy all the suitability criteria; unsuitable locations do not satisfy one or more suitability criterion.

The binary selection method is equivalent to the INTERSECT query and the INTERSECT overlay described in chapter 3. It is particularly useful when the criteria for selecting suitable locations are clearly defined and all the criteria must be satisfied for a location to be suitable. This is often the case when the suitability criteria are defined by regulations such as finding suitable locations for new public facilities. The binary selection method is also appropriate for determining whether features such as land parcels or roads satisfy a number of selection criteria.

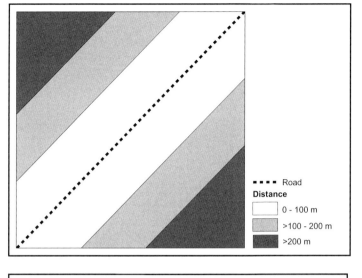

(a) Distance to Road

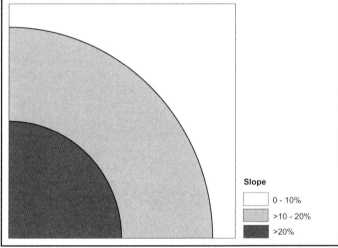

(b) Slope

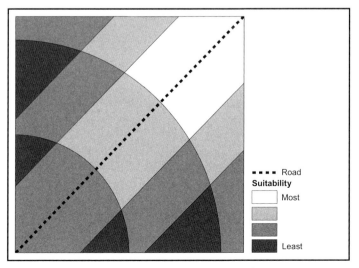

(c) Map Overlay

Figure 8.1 Residential Suitability Analysis: Map Overlay Method.

Figure 8.2 applies the binary selection method to the residential suitability analysis in figure 8.1. The figure reflects three criteria for selecting suitable locations: (1) all distances from the road are suitable, (2) locations with slopes less than or equal to 20 percent are suitable, and (3) locations with slopes greater than 20 percent are not suitable. The binary selection method ranks all locations that satisfy a suitability criterion as one and all locations that do not satisfy a suitability criterion as zero. Figure 8.2(a) reflects the first criterion by ranking the three distance buffers as one. Figure 8.2(b) reflects the second and third criteria by ranking locations with slopes less than 20 percent as one and ranking locations with slopes greater than 20 percent as zero.

The binary selection **suitability score** for location i (s_{BSi}) for an analysis of n suitability factors is computed by applying equation 8.1:

$$s_{BSi} = r_{i1} \times r_{ri2} \times \cdots \times r_{in},\qquad\qquad 8.1$$

where s_{BSi} = binary selection suitability score for location i and r_{in} = suitability ranking for location i and suitability factor n.

Any number multiplied by zero is zero. As a result the binary selection suitability score for any location is zero if the suitability rank for one or more suitability factor is zero. The binary selection suitability score for any locations is one only if the suitability ranking for all of the suitability factors is one.

The binary selection suitability scores in figure 8.2(c), the overlay of the first and second layers, use equation 8.1 to determine the combined effect of considering both suitability criteria for locating new residential development. Locations that are suitable for both the distance and slope criteria have a value of 1. For example, the binary selection suitability score for locations in the upper right corner of figure 8.2(c) are equal to the suitability ranking for distance (one) multiplied by the suitability ranking for slopes less than 10 percent (one), or one. Locations that are not suitable on one or both criteria have a value of 0. For example, the binary selection suitability scores for locations in the lower left corner of figure 8.2(c) are equal to the suitability ranking for distance (one) multiplied by the suitability ranking for slopes greater than 20 percent (zero) or zero.

Figure 8.3 illustrates the implications of applying different assumptions to consider the suitability of different locations for commercial development. The maps reflect four suitability criteria: (1) locations less than one-half mile from the road are suitable, (2) locations greater than one-half mile from the road are not suitable, (3) locations with slopes less than or equal to 20 percent are suitable, and (4) locations with slopes greater than 20 percent are not suitable. Figure 8.3(a) reflects the first and second criteria by assigning a value of 1 to all locations less than one-half mile from the road and a value of 0 to locations more than one-half mile from the road. Figure 8.3(b) reflects the third and fourth criteria by assigning a value of 1 to locations with slopes less than 20 percent and a value of 0 to locations with slopes greater than 20 percent.

Figure 8.3(c) combines these two layers to reflect the combined effect of considering both suitability criteria. All locations within one-half mile of the road with slopes less than 20 percent have a value of 1, indicating that they are suitable. Locations that do not satisfy one or both criteria, that is, are more than one-half mile from the road or have slopes greater than 20 percent, have a value of 0, indicating that they are not suitable. The implications of applying different suitability assumptions can be determined in other ways by modifying the values in the distance or slope layers and combining them using equation 8.1.

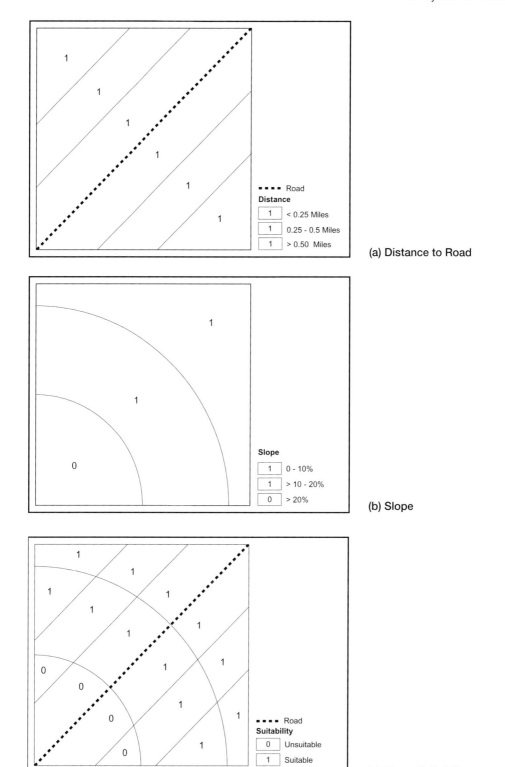

(a) Distance to Road

(b) Slope

(c) Binary Suitability

Figure 8.2 Residential Suitability Analysis: Binary Selection Method.

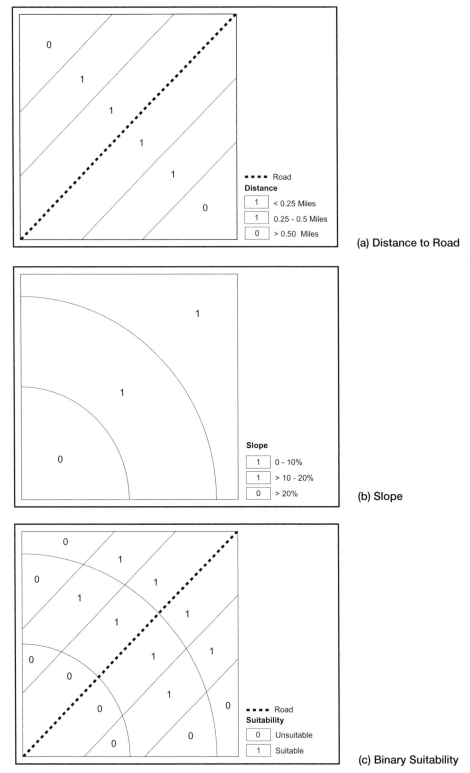

(a) Distance to Road

(b) Slope

(c) Binary Suitability

Figure 8.3 Commercial Suitability Analysis: Binary
Selection Method.

Binary Selection Analysis of Decatur Flood Risks

Predictions that climate change may produce more frequent and larger rainstorms in and around Decatur suggest that it may be useful to investigate the potential flood damage to property and buildings located near the streams shown in map 8.1. This question can be answered by using the binary selection method to identify buildings and parcels that are "suitable" for flooding. This will allow the City of Decatur to identify parcels and buildings that may be flooded and consider relocating any flood-prone critical structures such as schools, hospitals, and other public buildings.

Decatur Flood Zones As map 8.2 illustrates for a portion of Decatur, different flood zones can be used to define the flood risks for Decatur's buildings and land parcels. The Regulatory Floodway is designated by the US Federal Emergency Management Agency (FEMA) to identify areas where communities must regulate development to prevent upstream flooding. The **one-hundred-year floodplain** is designated by FEMA to identify areas with a one percent chance of flooding in any year. Similarly, FEMA's five-hundred-year flood zones have a 0.2 percent chance of being flooded in a particular year.

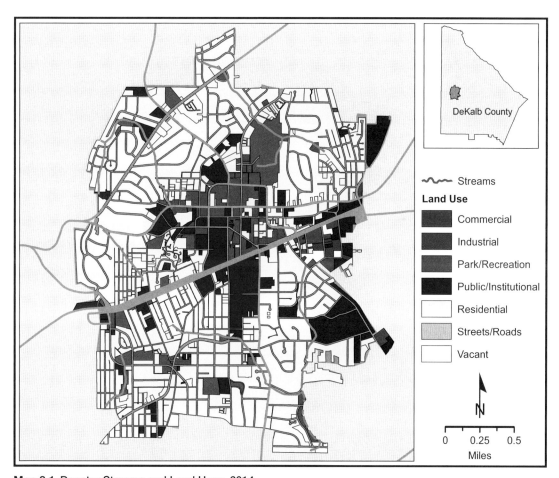

Map 8.1 Decatur Streams and Land Uses, 2014

As table 8.1 indicates, Decatur contains roughly 2,300 acres of land parcels. The portion of the city in different flood zones ranges from 156 acres for the five-hundred-year flood zone to 53 acres for the Regulatory Floodway. The binary selection method can be used to determine the total area of parcels that lie inside these three flood zones. It can also be used to identify buildings that are inside the flood zones and the characteristics of locations and buildings that might be flooded.

Decatur Flood Risks Map 8.2 indicates that large portions of many parcels, several buildings and roads, and an elementary school are located in one or more flood zones. It also reveals that there are several ways in which a parcel or building can be

TABLE 8.1	Area in Decatur Flood Zones, 2014
Land Type	**Area (Acres)**
(1)	**(2)**
Total area	2,346.4
In five-hundred-year flood zone	156.1
In one-hundred-year flood zone	133.5
In Regulatory Floodway	53.1

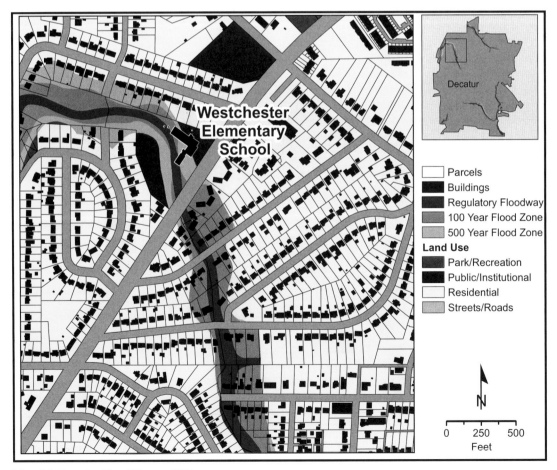

Map 8.2 Decatur Flood Zones, 2014

inside (or outside) a flood zone. Some parcels are entirely inside (or outside) a particular flood zone. Some parcels have a portion of their area in a flood zone and the remainder outside a flood zone. Other parcels have centroids inside (or outside) a flood zone. And others have more than half of their area inside or outside the flood zone. As a result, it is important to specify just how spatial criteria such as "inside the flood zone" are being defined in a particular application.

Standard GIS spatial analysis methods that use one feature to select another feature can be used to implement the different criteria for identifying parcels that are inside a flood zone. The Intersect command can be used to identify parcels that are partially (or completely) inside a flood zone. The Select by Location operation can be used to select parcels that have their centroids inside a flood zone or satisfy other location criteria such as being within a specified distance of a flood zone. A series of GIS operations can also be used to identify parcels that have more than half (or any other portion) of their area inside a flood zone.[5]

The selection methods yield substantially different results. Table 8.2 records the number and total area of parcels that satisfy different selection criteria for the three types of flood zones. For example, it reveals that 519 parcels with a total area of 353 acres (15 percent of the city's total parcel area) are at least partially inside the five-hundred-year flood zone. This area is substantially larger than the five-hundred-year flood zone in table 8.1 because a large part of some selected parcels is outside the flood zone; the "partly-within" criterion includes even the tiniest portion of a feature. A much smaller number of parcels with correspondingly smaller areas satisfy the other selection criteria.

Table 8.3 records 341 buildings in the five-hundred-year flood zone, as measured by intersecting the building footprint with the five-hundred-year flood zone boundary; this includes the 269 buildings in the one-hundred-year flood zone. A smaller number of buildings are in the other flood zones. Table 8.3 also reveals that seven public buildings are in the five-hundred-year flood zone and only one building, the Westchester Elementary School shown in map 8.2, lies inside the Regulatory Floodway. This suggests that the city should consider moving or floodproofing this building.

The preceding discussion reveals that the process of identifying land parcels and buildings that are subject to flooding is not as straightforward as it may appear. FEMA identifies three different flood zones with substantially different boundaries.

TABLE 8.2   Parcels in Decatur Flood Zones, 2014

Flood Zone	Selection Criterion	Number of Parcels	Area (Acres)
(1)	(2)	(3)	(4)
Regulatory Floodway	Partially inside	301	280.1
	Centroid inside	73	38.4
	>50% inside	0	0.0
	Totally inside	0	0.0
One-hundred-year flood zone	Partially inside	467	336.9
	Centroid inside	217	103.0
	>50% inside	213	82.8
	Totally inside	63	20.0
Five-hundred-year flood zone	Partially Inside	519	353.4
	Centroid inside	263	130.1
	>50% inside	257	100.6
	Totally inside	90	28.1

TABLE 8.3	Buildings in Decatur Flood Zones, 2014	
Building Type	**Location**	**Number**
(1)	**(2)**	**(3)**
All buildings	Inside Regulatory Floodway	85
	Inside one-hundred-year flood zone	269
	Inside five-hundred-year flood zone	341
Public buildings	Inside Regulatory Floodway	1
	Inside one-hundred-year flood zone	6
	Inside five-hundred-year flood zone	7

Communities may also want to prepare their own boundaries to identify areas where land use controls should be implemented to limit runoff from impervious surfaces that may damage nearby streams. In addition, at least four different criteria can be used to identify buildings and land parcels that are inside, or outside, the different flood zones.

These considerations suggest that analysts should work with stakeholders and concerned citizens to identify clearly the criteria to be used in identifying locations and buildings that may be flooded. They also suggest that the implications of using different flood zone definitions and selection criteria should be understood fully by everyone involved. GIS technology is extremely useful in this regard for preparing, presenting, and comparing maps, figures, and tables that can support public discussion and, ideally, agreement. The ordinal combination method described in the next section is particularly useful in this regard because it allows the suitability criteria to be ranked.

Ordinal Combination Method

The binary selection method identifies locations that jointly satisfy one or more spatial or attribute selection criteria. It assumes that all of the selection criteria must be satisfied and applies a simple binary (yes/no) criterion to define the class values for each suitability factor and the overall suitability of different locations. The **ordinal combination method** recognizes that locations can be more or less suitable for a particular land use and allows the user to rank the factor types on their suitability for that land use. It also assumes that the suitability factors are equally important and sums the factor rankings for all of the suitability factors to compute an overall ordinal combination suitability score for each location.

The ordinal combination method and the **weighted combination method** to be considered in the next section are particularly useful because they allow users to define a range of suitability values. They can be used with both **categorical data** (e.g., soil or land use types) and **continuous data** (e.g., slopes and road distances). They rank categorical data as more or less suitable for a particular land use, for example, rank soil types by their suitability for accommodating intensive development. They collapse continuous variables into suitability ranges (e.g., group the distance to roads into quarter-mile increments) or into threshold values (e.g., specify slopes from which development is prohibited).

The ordinal combination method begins by choosing the factors to be considered in selecting the most suitable locations for different land uses. The factors are defined by the user's knowledge of the study area, the available data, and the issue being considered. The suitability factors are divided into class breaks that reflect meaningful differences in the relative suitability of the factor values. Local construction regulations and industry standards can be used to define appropriate factor types.

The factor types are assigned suitability rankings, from most suitable to least suitable. By convention, the rankings are positive values with "one" representing the lowest suitability and the highest number representing the highest suitability. An odd number of classes is generally used so there is a middle value. Fewer than ten suitability classes are generally sufficient because more suitability classes are hard to discern on maps and their differences many not be meaningful in practice. Three suitability levels (High, Medium, and Low) may be enough for many applications because it is easy to grasp the difference between the classes. However, three classes may make it difficult to analyze tradeoffs and compromises between the classes because the categories are so broad. Five classes (High, Medium High, Medium, Medium Low, and Low) are often used because class differences are still readily apparent and there is room for more differentiation between classes.

A suitability value of "0" can also be used to exclude locations from consideration that are assumed to be completely unsuitable for one factor, regardless of their suitability for other factors. Exclusion areas can include water bodies, public facilities, and fully developed areas. Locations can also be excluded if they exceed critical threshold or cut off values such as areas with slopes that local development regulations specify cannot be developed. It is completely arbitrary to identify the exclusion areas with zeros. The zeros are not ordinal numbers like the class ranks; they are just labels that could be another number such as minus one or characters such as "E" or "X." Zeros are used here to correspond to the Binary Selection method that uses zero to identify unsuitable locations.

Figure 8.4 applies the ordinal combination method to the residential suitability analysis illustrated in figures 8.1 and 8.2. The figure reflects two suitability criteria: (1) locations closer to the road are more suitable than locations further away from the road and (2) locations with lower slopes are more suitable than locations with higher slopes. Figure 8.4(a) reflects the first assumption by ranking locations nearer the road higher than locations further from the road. Figure 8.4(b) reflects the second assumption by ranking locations with low slopes higher than locations with higher slopes.

The ordinal combination suitability score for location i (s_{OCi}) for an analysis that considers n suitability factors is computed by applying equations 8.2 and 8.3:

$$s_{OCi} = \sum_{j=1}^{n} r_{ij}, \qquad\qquad 8.2$$

$$s_{OCi} = 0 \quad \text{if any } r_{ij} = 0, \qquad\qquad 8.3$$

where s_{OCi} = ordinal combination suitability score for location i and r_{ij} = factor rank for location i and suitability factor j.

The ordinal combination suitability scores in figure 8.4(c) use equation 8.2 to compute the combined effect of considering both suitability factors. The most suitable locations that are closest to the road and have the lowest slopes have the highest suitability scores. For example, the ordinal combination scores (six) for the two locations in the top right-hand corner of figure 8.6(c) are the sum of the score for the distance factor (three) and the score for the slope factor (three). The least suitable locations that are furthest from the road and have the highest slopes have the lowest suitability scores. Locations at intermediate road distances with intermediate slopes have intermediate suitability scores.

Figure 8.5 applies the ordinal combination method to the commercial suitability analysis illustrated in figure 8.3. Figure 8.5 reflects three suitability criteria: (1) locations closer to the road are more suitable than locations further away from the road, (2) locations with lower slopes are more suitable than locations with higher

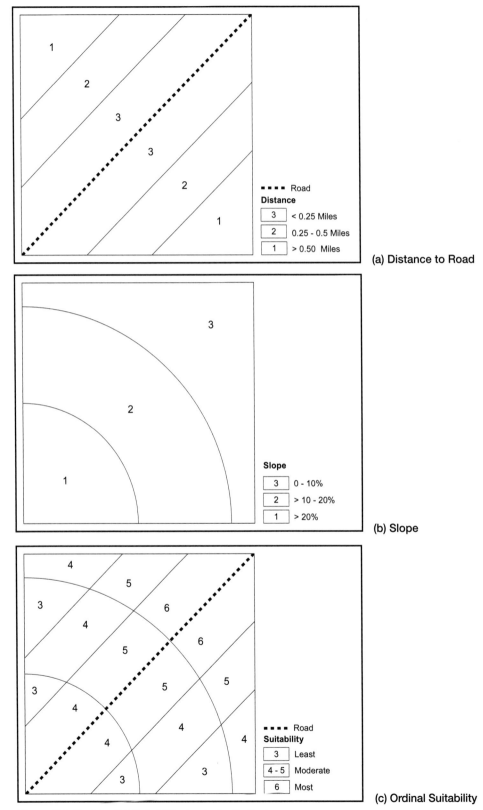

(a) Distance to Road

(b) Slope

(c) Ordinal Suitability

Figure 8.4 Residential Suitability Analysis: Ordinal
Combination Method.

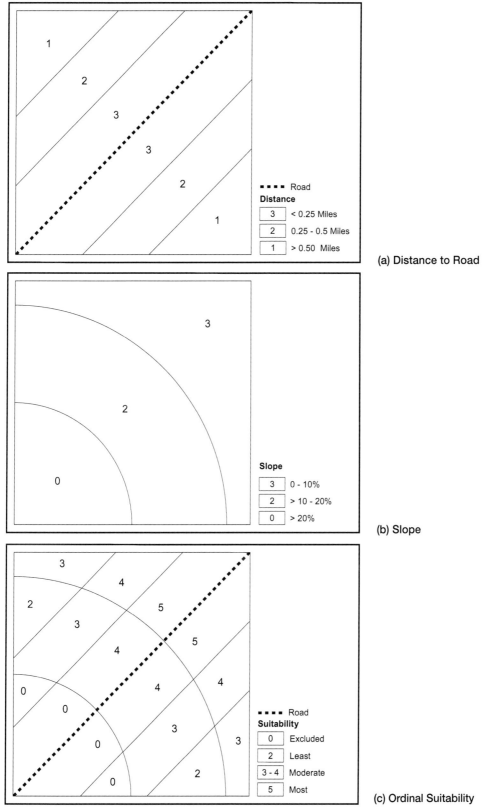

(a) Distance to Road

(b) Slope

(c) Ordinal Suitability

Figure 8.5 Commercial Suitability Analysis: Ordinal Combination Method.

slopes, and (3) commercial development is excluded from locations with slopes greater than 20 percent. Figure 8.5(a) reflects the first assumption by ranking locations nearer the road higher than locations further from the road. Figure 8.5(b) reflects second assumption by ranking locations with slopes less than ten percent higher than locations with slopes between 10 and 20 percent. Figure 8.5(b) reflects the third assumption by assigning locations with slopes greater than 20 percent a value of 0, indicating that commercial development is excluded from these locations.

Figure 8.5(c) shows the effect of combining the other two layers. The first two assumptions are implemented by applying equation 8.2 to locations where development is not excluded. Locations that are nearer the road and have lower slopes are ranked higher than location that are further from the road or have higher slopes. The third assumption is reflected by applying equation 8.3 to locations with slopes greater than 20 percent. These locations have an ordinal combination suitability score of zero, indicating that commercial development is excluded from these areas, regardless of their distance from the road.

Weighted Combination Method

The ordinal combination method ranks the classes within each suitability factor by their assumed suitability for a given land use and assumes that all the factors are equally important. The weighted combination method goes further and uses **factor weights** to reflect the user's judgment regarding the relative importance of different suitability factors. The weighted combination method is particularly useful for considering a range of development alternatives that vary the importance of the suitability factors used in the analysis.[6]

Figures 8.6 and 8.7 use the weighted combination method to consider residential and commercial development alternatives for the suitability analysis in figure 8.4. Both alternatives incorporate the suitability criteria from the ordinal combination analysis in figure 8.4. That is, they assume that (1) locations closer to the road are more suitable than locations further from the road and (2) locations with lower slopes are more suitable than locations with higher slopes.

Figure 8.6 assumes that locating new residential development near the road is more important than locating it on areas with low slopes. This assumption is reflected in figure 8.6(a), which multiplies all the distance rankings by a weight of three. The slope rankings in figure 8.6(b) are unchanged; their weights are one. Assigning a weight of three to the road-class ranks means that the differences between these ranks are three times greater than the differences between the slope-class ranks. As the result shows, this means that locations closer to a major road are always more suitable, regardless of their slope.

The weighted combination suitability score for location i (s_{WCi}) that considers n suitability factors is computed by applying equations 8.4 and 8.5:

$$s_{WCi} = \sum_{j=1}^{n}(w_{ij} \times r_{ij}),\qquad\qquad 8.4$$

$$s_{WCi} = 0 \quad \text{if any } r_{ij} = 0,\qquad\qquad 8.5$$

where s_{WCi} = weighted combination suitability score for location i, w_{ij} = factor weight for location i and suitability factor j, and r_{ij} = factor rank for location i and suitability factor j.

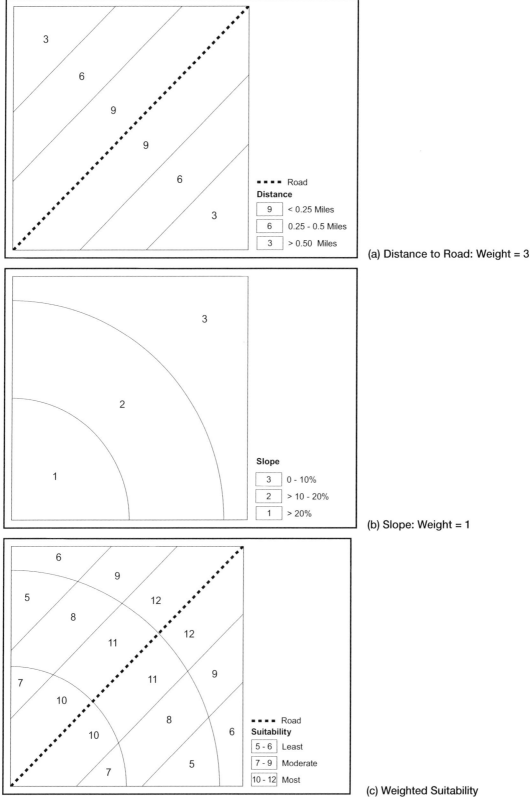

(a) Distance to Road: Weight = 3

(b) Slope: Weight = 1

(c) Weighted Suitability

Figure 8.6 Residential Suitability Analysis: Access Promotion Alternative.

Figure 8.6(c) uses equation 8.4 to compute the weighted suitability score for each location. For example, the weighted combination suitability score for the two locations at the top right-hand corner of figure 8.8(c) are computed as follows:

$$s_{WCi} = \left(w_{i1} \times r_{i1}\right) + \left(w_{i2} \times r_{i2}\right) = (3 \times 3) + (1 \times 3) = 12.$$

Figure 8.7 assumes that locating new commercial development on locations with low slopes, which reduces construction costs, is more important than road access. In this case, the distance rankings in figure 8.7(a) are weighted by one and the slope rankings in figure 8.7(b) are weighted by three. The weighted combination suitability scores in the figure 8.7(c) are the sum of the weighted distance rankings and the weighted slope rankings. The suitability scores shown in figures 8.6(c) and 8.7(c) clearly reflect the different perspectives on the relative importance of road distance and slopes in the two alternatives.

Limitations of the Ordinal and Weighted Combination Methods

The ordinal and weighted combination methods are intuitively attractive and easily applied methods for evaluating the suitability of different locations for one or more land uses.[7] However, their application makes several strong assumptions that are rarely satisfied in practice. The ordinal combination method ranks different factor types (e.g., different slopes) by their assumed suitability for locating a land use. The factor ranks are **ordinal numbers** that order the factor types from high to low. There is nothing wrong with doing this. However, adding (summing) the class ranks for the ordinal combination method and weighting them for the weighted combination method violates several important mathematical principles.

Summing suitability ranks first assumes incorrectly that the ranks are measured on an **interval scale** for which the distance (interval) between the ranks for a suitability factor are equal. That is, for example, it assumes that the difference between the suitability of low slopes and medium slopes is equal to the difference between the suitability of medium and high slopes.

Summing suitability across factor ranks also assumes inappropriately that the suitability differences between the ranks for one factor are the same as the suitability differences for a second factor. For example, it assumes that the difference between the suitability of the low and medium slopes is the same as the difference between the suitability of locations in the lowest distance classes.

Adding the ranks for two or more suitability factors also assumes that the factors are independent. This means that the ordinal and weighted combination methods cannot deal with situations in which suitability is determined by the interaction between two suitability factors. For example, high slopes on well-drained soils over clay may be disastrous for buildings along the California coast while high slopes may not be a problem for well-drained soils on different subsoils. Similarly, residential development generally prefers to be away from major roads to reduce noise and air pollution. However, roadside barriers, vegetation, and elevation changes can also reduce noise and air pollution. As a result, locations nearer highways that have appropriate barriers, elevations, or vegetation may be preferable to locations that are further from the highway that do not have these characteristics.

In some cases, the suitability ranks may not increase or decrease in a linear order. For example, locations within one-half mile of a major road may not be suitable due to the associated noise and air pollution. Locations between one-half and one mile will generally be more suitable because they have less pollution. However, locations more than one mile from the road may be less suitable than areas closer to the highway because of their reduced travel accessibility.

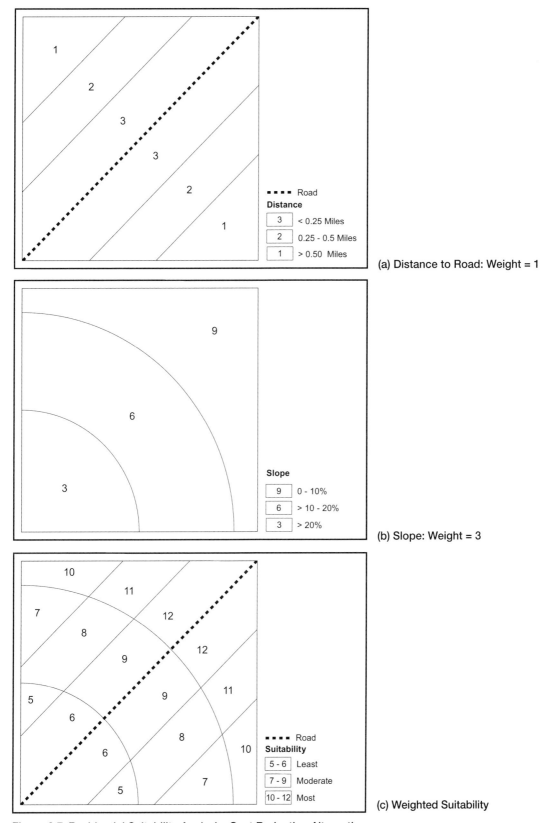

(a) Distance to Road: Weight = 1

(b) Slope: Weight = 3

(c) Weighted Suitability

Figure 8.7 Residential Suitability Analysis: Cost Reduction Alternative.

An extensive number of suitability methods have been developed that attempt to overcome the well-known limitations of the ordinal and weighted combination methods.[8] However, these methods are all much more complex than the methods discussed here and experience demonstrates that their theoretical superiority is generally overshadowed by the difficulty of applying them in practice, particularly in a public setting. The ultimate justification for using the ordinal and weighted combination method lies in "Box's Law": "All models are wrong; some models are useful" (Box and Draper 1987, 424). This principle recommends the ordinal and weighted combination methods because they have repeatedly proven to be useful for informing public debate, despite their limitations. In addition, some of these limitations can be overcome with careful analysis. For example, the soil-slope issue can be addressed by creating and ranking explicit slope-soil categories.[9]

Ordinal and Weighted Combination Analyses of DeKalb County Vacant Parcels

The ordinal and weighted combination methods will be used to consider the suitability of DeKalb County's vacant parcels for accommodating the county's projected residential and commercial growth. The ordinal combination method will be used to consider the suitability of the county's vacant parcels for additional residential and commercial development. The weighted combination method will be used to consider four alternatives for the county's residential and commercial development.

DeKalb County's Vacant Parcels

Table 8.4 indicates that DeKalb County has 56,415 vacant parcels, providing 20,727 acres of vacant land. Reflecting McHarg's design with nature emphasis on protecting important natural resources, it is assumed that some of the currently vacant parcels will be devoted to natural resource protection and enhancement. These conservation areas include: (1) vacant parcels zoned as Conservation, Park/Recreation, or Public/Institutional; (2) portions of the vacant parcels in a Water Features layer that includes water bodies, wetlands, floodplains and a hundred-foot buffer around them; and (3) other vacant parcels that are in the Protected Areas Database of the United

TABLE 8.4 **Parcels and Land Uses: DeKalb County, 2014**

Land Use	Parcels	
	Number	Acres
(1)	(2)	(3)
Commercial	6,443	11,700
Industrial	1,686	9,307
Park/Recreation	498	5,012
Public/Institutional	1,946	15,866
Residential	162,084	77,689
Vacant—Conserved	1,588	2,123
Vacant—Developable	54,827	18,604
Total	**229,072**	**140,301**

Source: DeKalb County GIS Department.

States (PAD-US), a national database on public land ownership and conservation lands in the United States. Omitting the vacant parcels that have their centroids in one of these conservation areas leaves 54,827 vacant parcels and 18,604 acres of developable land, about 13 percent of the county's total area.

Suitability Factors

Three factors will be used to consider the suitability of DeKalb County's vacant parcels: slope, distance to major roads, and distance to water features. The factors are divided into three classes, as described below.

Slope. The county's **digital elevation model (DEM)** data was used to derive three slope classes: (1) 0–5 percent, (2) 5–10 percent, and (3) more than 10 percent.

Road Buffer. GIS-generated buffers around the county's major roads were used to define three road buffers: (1) 0–0.25 miles, (2) 0.25–0.50 miles, and (3) more than 0.50 miles. The buffers do not include the county's limited access highways.

Water Features Buffer. Buffers around the water features were used to create three conservation buffer zones: (1) 0–500 feet, (2) 500–1,000 feet, and (3) more than 1,000 feet.

Table 8.5 reports the number of vacant parcels and the total area in each factor type. It indicates that roughly 60 percent of the area in the county's vacant parcels has slopes of ten percent or less, are within one-half mile of a major road, or within five hundred feet of a water feature. This information suggests that the county has ample vacant land to accommodate additional residential and commercial development.

Map 8.3 shows DeKalb County's land uses, expressways and major roads and water features. It also shows the location of the study area that will be used to illustrate the implications of the different suitability assumptions. The slope classes, road buffers, and water-feature buffers for the vacant parcels in the study area are shown in map 8.4.

| TABLE 8.5 | Vacant Parcels in Suitability Factor Classes: DeKalb County, 2014 |

Suitability Factor	Factor Class	Vacant Parcels		
		Number	Area (Acres)	Area (Percent)
(1)	(2)	(3)	(4)	(5)
Slope	0–5%	17,398	1,415	7.6
	5–10%	29,566	9,766	52.5
	Greater than 10%	7,863	7,423	39.9
	Total	**54,827**	**18,604**	**100.0**
Major Road Buffer	0–0.25 miles	26,085	7,026	37.8
	0.25–0.50 miles	12,180	4,089	22.0
	More than 0.50 miles	16,562	7,489	40.3
	Total	**54,827**	**18,604**	**100.0**
Water Feature Buffer	0–500 feet	32,318	11,136	59.9
	500–1,000 feet	16,613	5,394	29.0
	More than 1,000 feet	5,896	2,074	11.1
	Total	**54,827**	**18,604**	**100.0**

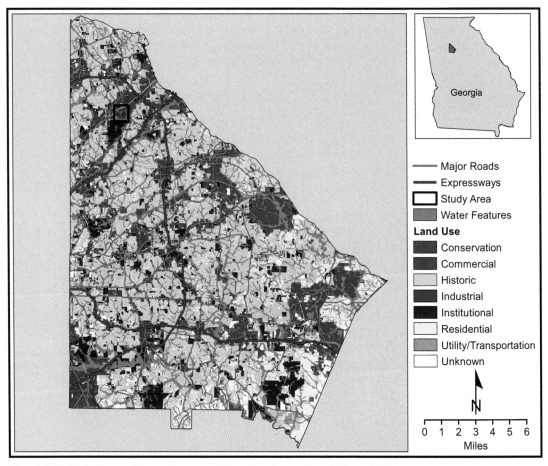

Map 8.3 DeKalb County Water Features, Major Roads, and Land Uses, 2014

Source: DeKalb County GIS Department.

Ordinal Combination Analysis

Table 8.6 shows the class rankings for two ordinal combination suitability analyses of DeKalb County's vacant parcels. The first considers the parcels' suitability for new residential development; the second considers their suitability for commercial development. The most suitable class for each suitability factor has a value of 3; the least suitable class has a value of 1. Factor types where development is excluded are assigned a value of 0.

The following assumptions underlie the two sets of class ranks.

Residential Suitability Assumptions The residential class ranks shown in column three are higher for parcels that are closer to major roads because they are more accessible and are more likely to have required infrastructure in place or nearby. The rankings are also higher for parcels with high slopes or near water features because they provide more visually attractive residential locations.

Commercial Suitability Assumptions The commercial class ranks in column four are higher for locations near major roads and exclude development from parcels that are more than one-half mile from a major road. Parcels with low slopes are assumed to be more suitable for commercial development because they will require less extensive site preparation. The commercial rankings are lower for locations near water features to protect these features from potentially harmful runoff from impervious surfaces.

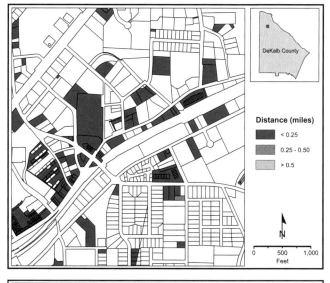

(a) Distance to Road

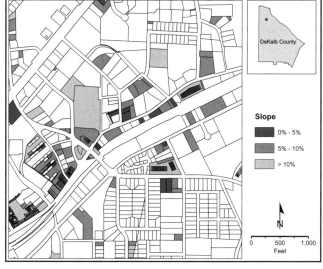

(b) Slope

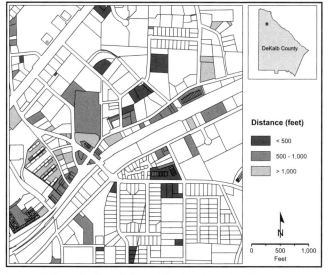

(c) Distance to Water

Map 8.4 DeKalb County Suitability Factors, 2014.

TABLE 8.6	Suitability Factor Class Ranks: Ordinal Combination Method

Suitability Factor	Factor Class	Class Rank	
		Residential Suitability	Commercial Suitability
(1)	(2)	(3)	(4)
Major Road Distance	0–0.25 miles	3	2
	0.25–0.50 miles	2	1
	More than 0.50 miles	1	0
Slope	0–5%	1	3
	5–10%	2	2
	Greater than 10%	3	1
Water Feature Distance	0–500 feet	3	1
	500–1,000 feet	2	2
	More than 1,000 feet	1	3

The ordinal combination suitability scores for each land parcel are computed by applying equations 8.2 and 8.3. Consider, for example, a parcel i that (1) is less than one-quarter mile from a major road, (2) has a slope of less than 5 percent, and (3) is less than five hundred feet from a water feature. Given equation 8.2 and the class ranks in table 8.6, the parcel's residential suitability is computed as follows:

$$s_{OCi} = \sum_{j=1}^{3} r_{ij} = 3 + 1 + 3 = 7.$$

Similarly, the parcel's commercial suitability is computed as follows:

$$s_{OCi} = \sum_{j=1}^{3} r_{ij} = 2 + 3 + 1 = 6.$$

Given equation 8.3, the commercial suitability for parcels that are more than one-half mile from a major road is zero because these parcels have a class rank of zero.

Table 8.7 reports that the residential suitability scores range from four to nine; the commercial suitability scores range from three to eight. The table groups the suitability scores into five classes: Low, Medium Low, Medium, Medium High, and High. Three and nine suitability classes are also commonly used.

Table 8.7 uses two widely used methods to group the ordinal combination suitability scores into the five classes.[10] As pointed out in chapter 7, the **Equal Interval**

TABLE 8.7	Suitability Score Class Breaks: Ordinal Combination Method

Suitability	Residential Suitability		Commercial Suitability	
	Equal Interval	Natural Breaks	Equal Interval	Natural Breaks
(1)	(2)	(3)	(4)	(5)
Low	4–5	4	3–4	3
Medium Low	6	5	5	4
Medium	7	6	6	5
Medium High	8	7	7	6
High	9	8–9	8	7–8

method divides the values into classes with the same range, that is, the difference between the high and low values is the same for each class. This method is easy to understand, making it easy to present information to nontechnical audiences but can be misleading if the values are not evenly distributed between the different groups. The **Natural Breaks method** identifies natural groupings in the data and creates classes that minimize the differences between the members of each class and maximize the differences between classes. This method is preferred for grouping values that are not equally distributed but makes it difficult to compare values from different analyses because the class breaks are specific to a particular data set.

Suitability Results As table 8.8 illustrates, the two classification methods yield substantially different information on the amount of suitable land in DeKalb County. The Equal Interval method includes nearly 70 percent of the vacant parcel area in the top three residential suitability categories; the Natural Breaks method includes more than 91 percent in these categories. Similarly, the Equal Interval method includes 15.4 percent of the land in the top three commercial suitability classes; the Natural Breaks method includes 35.7 percent of the vacant land. The Natural Breaks method will be used because the residential and commercial suitability scores are clearly not equally distributed, as shown in figure 8.8. The Natural Breaks class breaks are shown in figure 8.8.

Figure 8.9 shows the dramatically different quantities of suitable vacant land for residential and commercial development under the ordinal combination factor ranks in table 8.6. This illustrates the ordinal suitability method's ability to represent different perspectives on an area's suitability for accommodating projected land use demands.

Interpretation Table 8.8 reports that DeKalb County has substantial vacant land that is suitable for residential development under the ordinal combination assumptions in table 8.6. It also suggests that the county has much less land that is suitable

TABLE 8.8   Suitability of Vacant Parcels: Ordinal Combination Method

Land Use	Suitability	Vacant Parcels			
		Equal Intervals		Natural Breaks	
		Area (Acres)	Percent	Area (Acres)	Percent
(1)	(2)	(3)	(4)	(5)	(6)
Residential	Excluded	0	0.0	0	0.0
	Low	1,567	8.4	87	0.5
	Medium Low	4,069	21.9	1,480	8.0
	Medium	7,183	38.6	4,069	21.9
	Medium High	5,317	28.6	7,183	38.6
	High	468	2.5	5,785	31.1
	Total	**18,604**	**100.0**	**18,604**	**100.0**
Commercial	Excluded	7,489	40.3	7,489	40.3
	Low	4,479	24.1	1,125	6.0
	Medium Low	3,773	20.3	3,354	18.0
	Medium	1,936	10.4	3,773	20.3
	Medium High	768	4.1	1,936	10.4
	High	158	0.9	927	5.0
	Total	**18,604**	**100.0**	**18,604**	**100.0**

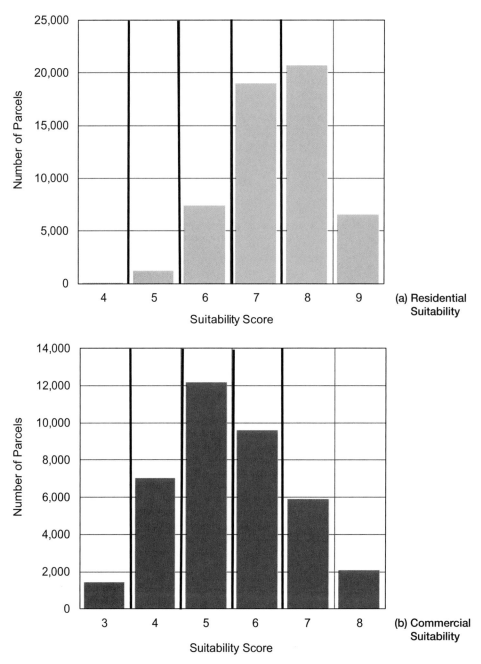

Figure 8.8 Vacant Parcels and Suitability Scores: Ordinal Combination Method.

for new commercial development. This is due in large part to the fact that commercial development is excluded from more than 40 percent of the vacant parcels because they are more than one-half mile from a major road.

Map 8.5 demonstrates the implications of the residential and commercial suitability assumptions for the study area. It shows that commercial development is not excluded from any of the vacant parcels because they are all less than one-half mile from a major road. Several of the vacant parcels on the left-hand side of map 8.5(a) are highly suitable for commercial development and less suitable for residential development. This reflects the fact that these parcels have lower slopes, which makes them highly suitable

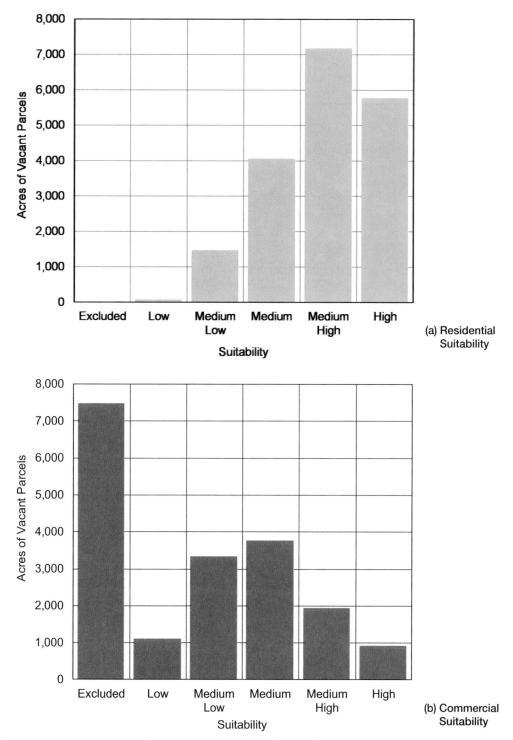

Figure 8.9 Acres of Suitable Parcels: Ordinal Combination Method.

for commercial development and less suitable for residential development. Conversely, several vacant parcels at the top of map 8.5(b) are more suitable for residential development than they are for commercial development. This reflects the fact that they are close to a water feature, making them highly suitable for residential development and less suitable for commercial development.

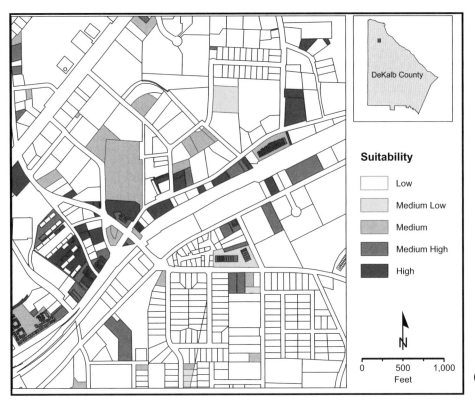

(a) Commercial Suitability

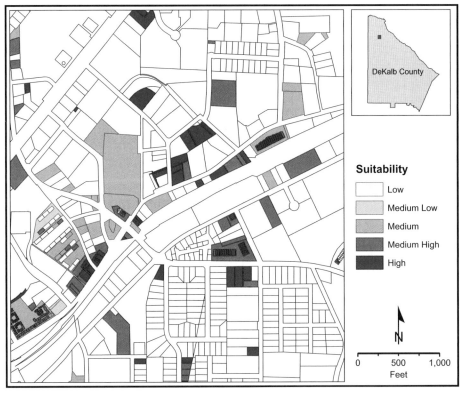

(b) Residential Suitability

Map 8.5 Suitable Parcels: Ordinal Combination Method.

Figure 8.9 and map 8.5 demonstrate the value of using GIS to apply the suitability methods and report the analysis results graphically and quantitatively. These reporting capabilities are important for supporting public discussion of different suitability assumptions.

Weighted Combination Analysis

Table 8.9 lists the assumptions of four weighted combination alternatives for developing DeKalb County's vacant parcels. The weights applied in the Conservation alternatives attempt to accommodate the projected residential and commercial growth while protecting the county's important natural features. The weights for the Development alternatives place less emphasis on protecting the county's natural resources. Both alternatives incorporate the ordinal combination class ranks from table 8.6. That is, the residential class ranks are higher for parcels that are: (1) closer to major roads, (2) have higher slopes, or (3) are nearer water features. Similarly, the commercial class ranks are: (1) higher for parcels nearer major roads, (2) zero for parcels more than one-half mile from a major road, (3) higher for parcels with low slopes, and (4) lower for parcels near water features. The factor weights in table 8.9 sum to ten, but the sum could be any other number; one and hundred are often used. In any case, the sum of the factor weights should be consistent for all of the suitability factors.

The following assumptions underlie the Conservation and Development factor weights in table 8.9.

Conservation Assumptions The Conservation alternative assign the lowest weights to the distance to a water feature factor for both residential and commercial development. This makes locations near water features less suitable for development, thereby protecting the county's lakes, rivers, and streams from potentially harmful runoff. The weights for the road distance and slope factors are twice as high as the distance to water features weights on the assumption that they are the most important considerations for locating new residential and commercial development.

Development Assumptions The Development alternative assigns the highest weight to the distance to a water feature for residential development on the assumption that locations near lakes, rivers, and streams are attractive residential locations. This alternative assigns the highest weights for commercial development to the road distance factor on the assumption that locations near roads are particularly attractive for new commercial development. The slope and distance to a water feature and the slope factors are assumed to be equally important for locating new commercial development.

TABLE 8.9 **Suitability Factor Weights: Weighted Combination Method**

Suitability Factor	Residential Development Suitability		Commercial Development Suitability	
	Conservation Alternative	Development Alternative	Conservation Alternative	Development Alternative
(1)	(2)	(3)	(4)	(5)
Road Distance	4	2	5	4
Slope	4	2	4	2
Water Feature Distance	2	6	1	3
Total	**10**	**10**	**10**	**10**

The weighted combination suitability scores for the Conservation alternative in table 8.10 are computed by applying equation 8.4. Consider, for example, a parcel i that (1) is less than 0.25 miles from a major road, (2) has a slope of less than 5 percent, and (3) is less than five hundred feet from a water feature. Given the class ranks from column three of table 8.6 and the factor weights from column two of table 8.9, the parcel's residential suitability for the Conservation alternative is computed as follows:

$$s_{WCi} = \sum_{j=1}^{3} (w_j \times r_{ij}) = (4 \times 3) + (4 \times 1) + (2 \times 3) = 12 + 4 + 6 = 22.$$

Similarly, given the class ranks in column four of table 8.6 and the factor weights in column four of table 8.9, the parcel's commercial suitability under the Conservation alternative is computed as follows:

$$s_{WCi} = \sum_{j=1}^{3} (w_j \times r_{ij}) = (5 \times 2) + (4 \times 3) + (1 \times 1) = 10 + 12 + 1 = 23.$$

Commercial development is excluded from parcels that are more than one-half mile from a major road because the class rank for these parcels is zero, as reported in table 8.6.

Alternative Comparisons Table 8.11 reports the total area of vacant parcels in six suitability classes, based on the factor ranks and factor weights for the Conservation and Development alternatives in table 8.10.[11]

The table shows that DeKalb County has ample land to accommodate its future residential development, given both alternatives' assumptions. More than 40 percent of the vacant parcel land is in the top two classes (High and Medium High) for the

| TABLE 8.10 | **Partial Suitability Scores for the Conservation Alternative: Weighted Combination** |

Suitability Factor	Suitability Class	Residential Suitability			Commercial Suitability		
		Factor Weight	Class Rank	Partial Factor Score	Factor Weight	Class Rank	Partial Factor Score
(1)	(2)	(4)	(3)	(5)	(6)	(7)	(8)
Road Distance	0–0.25 miles	4	3	12	5	2	10
	0.25–0.50 miles		2	8		1	5
	More than 0.50 miles		1	4		0	0
Slope	0–5%	4	1	4	4	3	12
	5–10%		2	8		2	8
	Greater than 10%		3	12		1	4
Water Feature Distance	0–500 feet	2	3	6	1	1	1
	500–1,000 feet		2	4		2	2
	More than 1,000 feet		1	2		3	3
Minimum Score				10			10
Maximum Score				30			25
Development Excluded				—			0

TABLE 8.11	Suitability of Vacant Parcels: Weighted Combination Method

Land Use	Suitability	Vacant Parcels			
		Conservation Alternative		Development Alternative	
		Area (Acres)	Percent	Area (Acres)	Percent
(1)	(2)	(3)	(4)	(5)	(6)
Residential	Excluded	0	0.0	0	0.0
	Low	809	4.3	1,279	6.9
	Medium Low	4,605	24.8	2,110	11.3
	Medium	5,695	30.6	3,515	18.9
	Medium High	5,260	28.3	6,894	37.1
	High	2,236	12.0	4,806	25.8
	Total	**18,604**	**100.0**	**18,604**	**100.0**
Commercial	Excluded	7,489	40.3	7,489	40.3
	Low	1,722	9.3	4,479	24.1
	Medium Low	4,340	23.3	807	4.3
	Medium	2,728	14.7	2,966	15.9
	Medium High	1,576	8.5	1,936	10.4
	High	749	4.0	927	5.0
	Total	**18,604**	**100.0**	**18,604**	**100.0**

Conservation alternative. Nearly two-thirds of the land in vacant parcels is in the top two classes for the Development alternative. The Development option identifies more than twice as much land that is suitable for residential development than the Conservation alternative. This is because the Development alternative factor weights in table 8.9 make land near water features three times more important for residential development than the Conservation alternative and the other factors, and nearly 60 percent of the county's vacant parcels are within five hundred feet of a water feature, as shown in table 8.5. Figure 8.10 reports the quantity of vacant land that is suitable for residential development for the Conservation and Development alternatives in table 8.10.

Table 8.11 and figure 8.11 show that DeKalb County has less suitable land for commercial development under these alternatives. Only 15.4 percent of the land in vacant parcels (2,863 acres) is in the top two suitability classes (High and Medium High) for the Development alternative; only 13.5 percent of the land in vacant parcels (2,325 acres) is in the top two classes for the Conservation alternative. The lack of suitable land for commercial development reflects the fact that more than 40 percent of the land in vacant parcels is excluded from development because the parcels are more than one-half mile from a major road.

Maps 8.6 and 8.7 illustrate the implications that the two alternatives have for the study area. As shown in map 8.4(c), most of the central part of the study area is more than five hundred feet from a water feature. Map 8.6 reveals that the vacant parcels in this part of the map are more suitable for residential development under the Conservation alternative than they are under the Development alternative. This reflects the fact that these parcels are further away from water features and the Conservation alternative has a low Distance to Water factor weight (table 8.9). This means that the two other factors are much more important in determining the suitability score, which is of course the intent of the Conservation alternative.

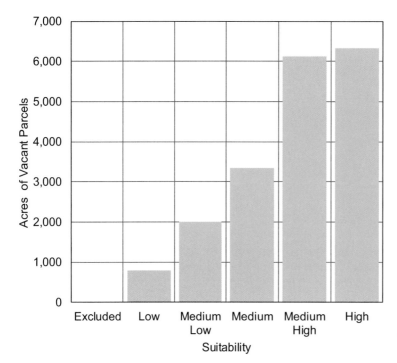

(a) Residential Suitability—
Conservation Alternative

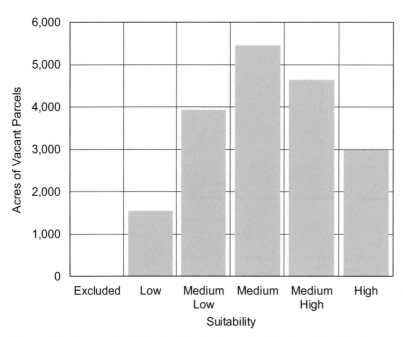

(b) Residential Suitability—
Development Alternative

Figure 8.10 Acres of Suitable Residential Parcels: Weighted Combination
Alternatives.

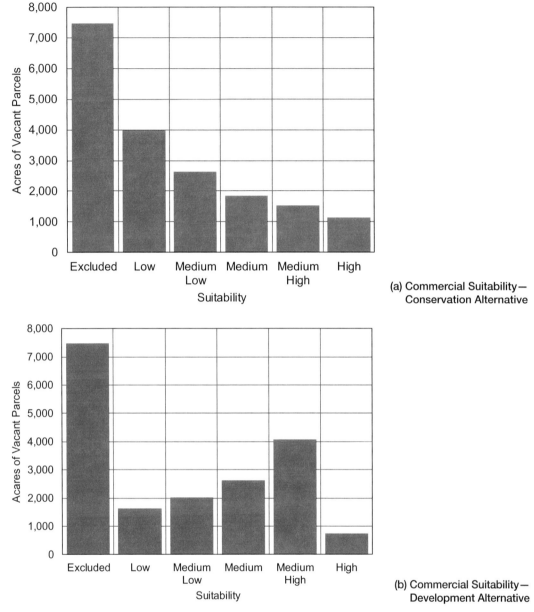

(a) Commercial Suitability—
Conservation Alternative

(b) Commercial Suitability—
Development Alternative

Figure 8.11 Acres of Suitable Commercial Parcels: Weighted Combination Alternatives.

Map 8.7 reveals that slightly more of the vacant parcels in the study area are more suitable for commercial development under the Development alternative than they are under the Conservation alternative, which again makes sense given the objectives of the two alternatives. While these observations can be quantified, it is important to consider whether they make sense in terms of the assumptions expressed in the alternatives.

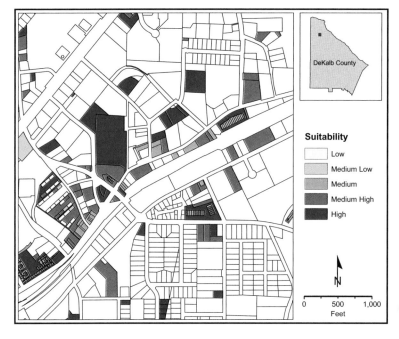

(a) Residential Suitability—
    Conservation Alternative

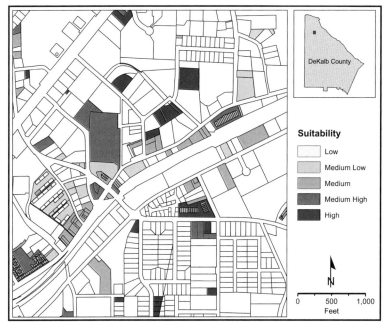

(b) Residential Suitability—
    Development Alternative

Map 8.6 Suitable Residential Parcels: Weighted Combination Alternatives.

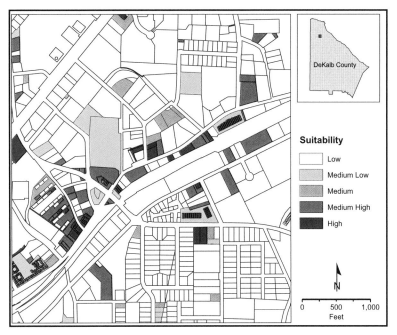

(a) Commercial Suitability—
Conservation Alternative

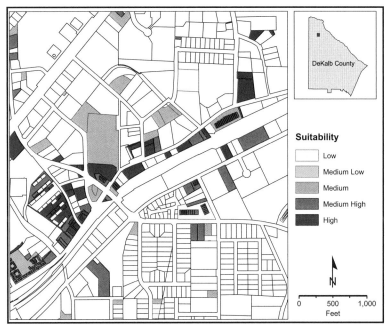

(b) Commercial Suitability—
Development Alternative

Map 8.7 Suitable Commercial Parcels: Weighted Combination Alternatives.

Sensitivity Analysis It is often helpful in evaluating two or more alternatives to consider how sensitive the results are to the alternatives' underlying assumptions. Table 8.12 reports the vacant parcel areas in six suitability classes for residential and commercial development under the Conservation and Development alternatives. The table shows that only 18.4 percent of the land (highlighted in the table) is rated differently for residential development by the two alternatives. Less than 1 percent of the land is ranked differently for commercial development by the two alternatives. These results suggest that the suitability of the vacant parcels is generally not dependent on the alternative assumptions because most of parcels are assumed to be suitable or unsuitable under either set of assumptions.

Map 8.8 provides more information on the implications of adopting one alternative or the other. The figure maps five relationships between the suitability scores for the Conservation and Development alternatives. The "High-High" parcels are rated High or Medium High for both alternatives. The "Low-Low" parcels are rated Low or Medium Low for both alternatives. The "Medium" parcels are rated Medium for one of the alternatives. The "High-Low" parcels are rated High or Medium High for the Conservation alternative and Low or Medium Low for the Development alternative. The "Low-High" parcels are rated Low or Medium Low for the Conservation alternative and High or Medium High for the Development alternative.

Map 8.8(a) reveals no differences between the commercial suitability of vacant parcels for the two alternatives. All the parcels are rated High or Medium on both alternatives. Map 8.8(b) reveals substantial differences between the alternatives. A large number of parcels are mapped as "High-Low," that is, High or Medium High for the Conservation alternative and Low or Medium Low for the Development alternative. This reflects the fact these parcels are more than one thousand feet from a water feature. The Development alternative assumes that these parcels are poorly suitable for residential development and the Conservation alternative assumes they are highly suitable. Together table 8.12 and map 8.8 indicate that while two alternatives may generally yield similar suitability scores, they may have substantially different implications for particular areas.

TABLE 8.12 **Sensitivity Analysis: Weighted Combination Method**

Land Use	Suitability		Vacant Parcels	
	Conservation Alternative	Development Alternative	Area (Acres)	Percent
(1)	(2)	(3)	(4)	(5)
Residential	Excluded	Excluded	0	0.0
	Low/Medium Low	Low/Medium Low	2,594	13.9
	Low/Medium Low	High/Medium High	2,623	14.1
	Medium	Medium	3,515	18.9
	High/Medium High	Low/Medium Low	795	4.3
	High/Medium High	High/Medium High	9,077	48.8
	Total	**Total**	**18,604**	**100.0**
Commercial	Excluded	Excluded	7,489	40.3
	Low/Medium Low	Low/Medium Low	5,139	27.6
	Low/Medium Low	High/Medium High	162	0.9
	Medium	Medium	3,488	18.8
	High/Medium High	Low/Medium Low	0	0.0
	High/Medium High	High/Medium High	2,325	12.5
	Total	**Total**	**18,604**	**100.0**

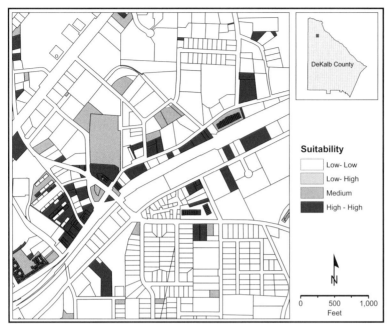

(a) Combined Suitability—
Commercial Alternatives

(b) Combined Suitability—
Residential Alternatives

Map 8.8 Suitable Parcels: Combined Weighted Combination Alternatives.

Computing Standardized Suitability Scores

It is sometimes useful to convert the ordinal or weighted combination suitability scores into standard scores that are consistent for more than one land use or suitability alternative. This allows planners to determine whether a location is more suitable for one land use or another, even though their suitability scores are not consistent.

Inconsistent suitability scores can be converted to standard scores by applying equation 8.6:

$$ss_i = \left(\frac{s_i - s_{min}}{s_{max} - s_{min}} \right) \times 100, \qquad\qquad 8.6$$

where ss_i = standardized suitability score, s_i = raw suitability score, s_{min} = minimum suitability score, and s_{max} = maximum suitability score.

The standardized scores range from 0 for the lowest suitability score to 100 for the highest suitability score.

For example, the weighted combination scores for residential development in table 8.10 range from 10 to 30; the commercial suitability scores range from 10 to 25. Table 8.13 uses equation 8.6 to convert these two sets of scores into a consistent set of scores.

The suitability scores for residential development in column one of table 8.13 range from 10 to 30. The standardized score for the first residential suitability score (10) in column two is computed as follows:

$$ss_1 = \left(\frac{s_1 - s_{min}}{s_{max} - s_{min}} \right) \times 100 = \left(\frac{10 - 10}{30 - 10} \right) \times 100 = \left(\frac{0}{20} \right) \times 100 = 0.0.$$

| TABLE 8.13 | Standardized Suitability Scores: Weighted Combination Method |

Residential Suitability		Commercial Suitability	
Suitability Score	Standardized Score	Suitability Score	Standardized Score
(1)	(2)	(3)	(4)
10	0.0	10	0.0
11	5.0	11	6.7
12	10.0	12	13.3
13	15.0	13	20.0
14	20.0	14	26.7
—	—	—	—
21	55.0	21	73.3
22	60.0	22	80.0
23	65.0	23	86.7
24	70.0	24	93.3
25	75.0	25	100.0
26	80.0	—	—
27	85.0	—	—
28	90.0	—	—
29	95.0	—	—
30	100.0	—	—

The standardized score for the last residential suitability score (30) is:

$$ss_{21} = \left(\frac{s_{21} - s_{min}}{s_{max} - s_{min}} \right) \times 100 = \left(\frac{30 - 10}{30 - 10} \right) \times 100 = \left(\frac{20}{20} \right) \times 100 = 100.0.$$

The standardized score for the second residential suitability score (11) is:

$$ss_2 = \left(\frac{s_2 - s_{min}}{s_{max} - s_{min}} \right) \times 100 \left(\frac{11 - 10}{30 - 10} \right) \times 100 = \left(\frac{1}{20} \right) \times 100 = 5.0.$$

The suitability scores for commercial development in column three of table 8.13 range from 10 to 25. The standardized score for the first residential suitability score (ten) in column four is computed as follows:

$$ss_1 = \left(\frac{s_1 - s_{min}}{s_{max} - s_{min}} \right) \times 100 = \left(\frac{10 - 10}{25 - 10} \right) \times 100 = \left(\frac{0}{15} \right) \times 100 = 0.0.$$

The standardized score for the second commercial suitability score (11) is computed as follows:

$$ss_2 = \left(\frac{s_2 - s_{min}}{s_{max} - s_{min}} \right) \times 100 \left(\frac{11 - 10}{25 - 10} \right) \times 100 = \left(\frac{1}{15} \right) \times 100 = 6.7.$$

As table 8.13 indicates the standardized commercial suitability scores are higher than the standardized residential suitability scores when the raw suitability scores are identical. This reflects the fact that the commercial suitability scores are compared to a smaller range of scores than the residential suitability scores.

Other Land Suitability Analysis Methods

The preceding discussion used three land suitability analysis methods to evaluate the suitability of vacant land parcels for accommodating new development. Accommodating the projected growth for well-defined spatial units such as land parcels is a popular land suitability analysis application. However, other land suitability analysis methods can be used to consider other spatial units and other policy issues. Two particularly important methods are fuzzy overlay methods and land supply analysis.

Fuzzy Overlay Methods

Fuzzy overlay methods are useful when the suitability factors are continuous variables without well-defined break points.[12] Many natural features have of these characteristics. For example, slope, elevation, and distance are measured continuously from low to high and natural land cover often ranges from forest to grassland with borders containing a mixture of trees, brush, and grass.

As described in chapter 7, traditional set theory assumes that a value can be clearly assumed to belong to a set or do not belong to a set. Fuzzy overlay methods use "fuzzy logic" to specify the extent to which a value belongs (or does not belong) to a set. It uses a "fuzzy membership" number to represent the extent to which a value belongs to a set. A value of 1 represents full membership, 0 represents non-membership, and values between 0 and 1 represent degrees of partial membership.

Assume, for example, that an environmental planner wants to identify the most suitable location for reintroducing a native plant species that grows best on low slopes at high elevations. She can use a DEM to create raster map layers representing the slope and height of one-hundred-meter cells in the study area. She can then use GIS mathematical functions to assign fuzzy membership values to the height and slope layers. The membership values for slopes will range from 1 for the lowest slopes to 0 for the highest slopes. The membership values for elevations will range from 1 for the highest elevations to 0 for the lowest elevations. She can then combine the fuzzy membership layers to create an output layer recording the overall membership likelihood of locations that satisfy logical or mathematical operations she specifies. The membership values in the output layer will range from 0 to 1. Locations having higher membership values in the composite layers will generally be more suitable in the output layer. An excellent discussion of fuzzy overlay methods is provided in Mitchell (2012, 129–167).

Land Supply Monitoring

The methods considered in this chapter are widely used to analyze the potential for greenfield development (or conservation) in areas where an abundance of undeveloped land is available. In contrast, **land supply analysis** is used in urbanized areas to estimate the supply and development capacity of the area's vacant, partially utilized, and underutilized land parcels. Land supply analysis uses the binary selection method and **sensitivity analysis** to consider the implications of applying different criteria to select locations that can accommodate new uses or reuses. Criteria that can be used to identify suitable parcels include their size, value, condition, current land use, location, zoning, and other growth management or development incentives policies. Some applications select locations that would be suitable for development if the current zoning were changed.

As land records transactions and building permit systems become more computerized, a near-real-time awareness of locations that are currently available for development may be possible, facilitating the monitoring of progress toward growth management or sustainability goals. Web-based site selection applications have also become available that facilitate the selection of properties that are suitable and available for development.[13] These systems typically contain extensive information on the characteristics of structures and sites as well as locational and regulatory information. These sites are often funded by local or regional governments as part of their economic development efforts to facilitate the identification and development of available sites and help implement local comprehensive and redevelopment plans. Excellent discussions of land supply analysis are provided in Moudon and Hubner (2000) and Knaap (2001).

Opportunities for Public Involvement

The preceding analysis demonstrates that land suitability analysis is not simply an objective process that should be the exclusive domain of technical experts. It can be much richer and more inclusive than this because it provides an analytic framework that uses publicly available information to explicitly incorporate normative judgments about the suitability of different locations for different uses. This framework can make the public's role in decision making more meaningful and more representative of the values of different segments of the public.

Judgment plays a role from the beginning of the land suitability analysis process. It is required to choose the issue to be considered (e.g., identifying areas to be protected or to be developed), the area to be analyzed (e.g., a neighborhood, city, county, or multicounty region), and the suitability factors to be considered (e.g., the features' physical or social characteristics, their location relative to which other features, and the public policies to be considered). As the preceding analysis demonstrates, even seemingly technical decisions such as selecting the data to be used, choosing the suitability method to be applied, and applying a method for defining class breaks inevitably shape the analysis results.

Judgment is involved most clearly in ranking the factor types in the ordinal combination method and weighting the factors in the weighted combination method. These choices can be used to prepare a range of alternatives that reflect different policy perspectives, the positions of stakeholders, and the concerns of interested individuals. GIS-based land suitability analysis allows the alternatives to be prepared quickly and easily and makes the implications of considering different suitability factors and applying different class ranks and factor weights readily apparent in maps, tables, and graphs. The tools used to develop the alternatives can then be used to compare, combine, evaluate, and change them, providing an excellent opportunity for public conversation and collaboration. This provides an opportunity for changing planning from a closed process of supposedly technical analysis to an open process that allows a community to work together, selecting areas they wish to protect and develop.[14]

Notes

[1] The discussion in this chapter draws heavily on Mitchell (2012, 31–128) and Hopkins (1977). Mitchell's discussion provides very useful guidance on using GIS to implement vector and raster land suitability analysis processes that varies somewhat from the methods described in this chapter.

[2] The other kinds of questions GIS can answer are: "What is at . . . ?"; "What has changed since . . . ?"; "What spatial patterns exist . . . ?"; and "What if . . . ?"(Rhind 1989).

[3] See, e.g., Carr and Zwick (2007, 42–72).

[4] Steinitz et al. (1976) and Collins et al. (2001) provide interesting histories of land suitability analysis.

[5] Given a "Parcel" layer containing the boundaries of an area's parcels and a "Flood" layer containing the flood zone boundaries, the parcels that have more than 50 percent of their area in a flood zone can be identified as follows: (1) add a new "Total_Area" field to the Parcel layer that stores the current area in each parcel, (2) create a new "Parcel_Flood_Union" layer that is the Union of the Flood and Parcel layers, (3) select the polygons in the Parcel_Flood layer that are inside the Flood layer and save them as a new "Parcel_Inside" layer, (4) add a new "Inside_Area" field to the Parcel_Inside layer that stores the current area of the Parcel_Inside polygons, (5) add a new "Percent_Inside" field to the Parcel_Inside layer that is equal to ([Inside_Area]/[Total_Area]) * 100, and (6) select the polygons in the Parcel_Inside layer that have Percent_Inside values greater than 50 and store them in a new "Parcel_50_Percent_Inside" layer.

[6] The weighted combination method is a spatial version of the **Multi-Criteria Evaluation** (**MCE**) methods planners have used for many years. Carr and Zwick (2007) provide an example of land suitability analysis using the Analytic Hierarchy Process (AHP).

[7] The discussion in this section draws heavily on Hopkins (1977, 389).

[8] Excellent introductions to these methods are provided in Hopkins (1977), Malczewski (2006), and Carr and Zwick (2007, 45–72).

[9] Other limitations can be addressed by implementing the rules of combination method described in Hopkins (1977).

[10] The quantiles method, a third widely used classification method, is not considered because it groups the residential suitability scores into four classes.

[11] The suitability scores were not distributed evenly across their ranges so the classes were created with the Natural Breaks method.

[12] The following discussion draws heavily on Mitchell (2012, 129–167), which provides an excellent discussion of fuzzy overlay methods.

[13] See, e.g., http://www.zoomprospector.com/ and http://spokane.zoomprospector.com/.

[14] See, e.g., Klosterman (2007, 202–209).

Using Planning Support Methods 9

Chapter 1 suggested that planners should use simple models and methods to help private citizens, stakeholders, and public officials learn about their community's past and present and think about what its future may—and should—be. This chapter draws on the analyses in chapters 2 through 8 to identify six guidelines for supporting community-based planning: (1) use graph and charts, (2) use maps, (3) combine different information, (4) compare to projections by other organization, (5) document assumptions, and (6) use scenarios.[1]

Use Graphs and Charts

Planners routinely present quantitative information to people who have very little familiarity with—or interest in—mathematics or statistics. Graphs and charts are particularly useful for presenting information clearly and understandably to a wider, non-quantitative audience.

For example, table 9.1 and figure 9.1 both report substantial increases in DeKalb County's projected population over the age of seventy. The table is useful for recording the precise values for the county's observed and projected population by age and sex. However, the population pyramid is much more effective for demonstrating the projected increase in the county's older population to a nontechnical audience.

Types of Graphs and Charts

Line and Point-and-Line Graphs **Line graphs** and point-and-line graphs are generally used to represent the variation in variables such as population or employment over time. In these cases, the horizontal axis records points in time and the vertical axis records the values at different points in time. **Point-and-line graphs** that use points to identify particular values and lines to connect the points are appropriate for identifying precise values at different points in time. Line graphs that do not identify individual data points are appropriate for showing general trends when the values at particular points in time are not important.

Bar Chart A **bar chart** is a chart with vertical or horizontal rectangular bars whose lengths are proportional to the values they represent. Bar charts are used to represent categories of data that do not make up a whole. For example, a bar chart may represent a city's population at different dates. A bar chart in very useful for displaying discrete data, unlike histograms that use categories to display continuous data.

Stacked Bar Chart A **stacked bar chart** is a bar chart in which the rectangular bars are divided into sections. The size of each section represents the cumulative effect of the attributes for each bar. For example, the sections of a stacked bar chart can be used to represent the proportion of a city's population in different racial categories when the height of each bar represents the city's total population.

| TABLE 9.1 | | Observed and Projected Population by Age and Sex, Reduced Net Out-Migration Projection: DeKalb County, 2010 and 2040 | | | | | |

Cohort	Age Interval	Male Population			Female Population		
		2010	2040	Percent Change	2010	2040	Percent Change
n		$(PM_n^{2010})^a$	$(PM_n^{2040})^b$		$(PF_n^{2010})^a$	$(PF_n^{2040})^b$	
(1)	(2)	(3)	(4)	(5)	(6)	(7)	(8)
1	0–4	25,856	29,335	13.5	24,551	27,715	12.9
2	5–9	23,110	27,495	19.0	22,180	25,959	17.0
3	10–14	21,915	26,277	19.9	20,882	24,680	18.2
4	15–19	22,737	27,042	18.9	21,906	26,003	18.7
5	20–24	25,555	31,765	24.3	26,144	31,593	20.8
6	25–29	29,422	34,908	18.6	31,514	34,823	10.5
7	30–34	28,134	33,800	20.1	29,579	33,320	12.6
8	35–39	27,346	30,036	9.8	28,245	29,988	6.2
9	40–44	25,043	27,055	8.0	26,511	27,553	3.9
10	45–49	23,810	24,230	1.8	26,424	25,065	–5.1
11	50–54	21,027	20,572	–2.2	24,984	22,274	–10.8
12	55–59	17,928	20,104	12.1	22,198	23,065	3.9
13	60–64	14,530	17,900	23.2	18,134	20,750	14.4
14	65–69	9,311	16,379	75.9	11,793	19,305	63.7
15	70–74	6,107	13,466	120.5	8,254	17,000	106.0
16	75–79	4,182	11,277	169.7	6,552	15,539	137.2
17	80–84	3,046	7,755	154.6	5,287	12,264	132.0
18	85+	2,296	5,751	150.5	5,400	11,471	112.4
Total	—	331,355	405,149	22.3	360,538	428,365	18.8

Source:
[a] US Bureau of the Census (2013c).
[b] Table 5.23.

Pie Charts A **pie chart** is a circular chart divided into sectors like slices of a pie. The size of the sections represents the proportion that a value contributes to the total for all values. Pie charts like figure 9.4 are useful for comparing information on different parts of a population at one point in time for characteristics such as age, race, location, or employment. For example, a pie chart can use segments of a circle to represent the portion of a population in different racial or ethnic groups. Pie charts are widely used in the media, but it is often hard to interpret the size of different sectors or compare data between different pie charts. As a result, bar graphs and histograms are generally preferable to pie charts.

Histogram A **histogram** is a chart with vertical or horizontal rectangular bars whose lengths are proportional to the values they represent. Unlike bar graphs, the bars of histograms represent the frequency of continuous variables that are grouped into discrete intervals (or bins). The total area of the histogram represents the total of all values; the area of each bar represents the portion of the observations in each interval. For example, a histogram may represent the total population of a city with bars that record the population in different parts of the city. **Population pyramids** like figure 9.1 are horizontal histograms that use the horizontal axis to record the male and female population in different age groups recorded on the vertical axis.

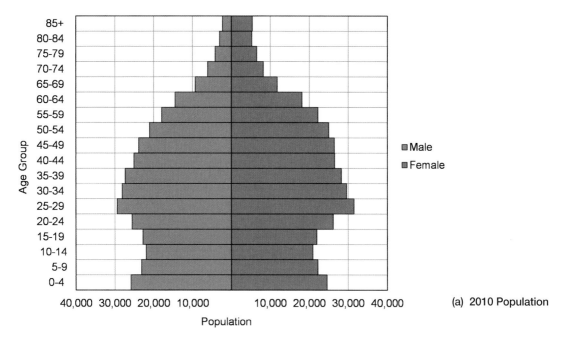

(a) 2010 Population

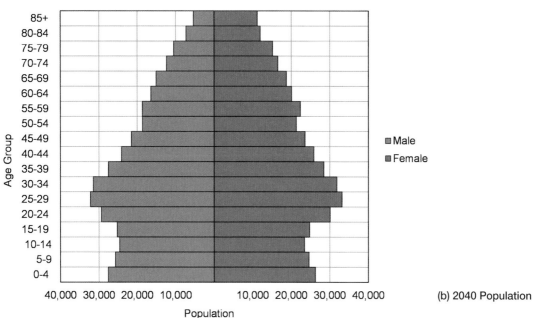

(b) 2040 Population

Figure 9.1 Observed and Projected Population by Age and Sex, Reduced Net Out-Migration Projection: DeKalb County, 2010 and 2040.

Source: Tables 5.1 and 5.2.

Using Graphs and Charts

Graphs and charts are useful for several analytic and reporting tasks: (1) revealing data errors and anomalies, (2) identifying fit periods, (3) reporting and evaluating projections, and (4) evaluating projection assumptions.

Revealing Data Errors and Anomalies Data plots are extremely useful for revealing data errors and anomalies. Even the most careful analysts can make mistakes in collecting and analyzing data that may themselves contain errors or uncertainties. While data anomalies and errors can be reported in tables, they are much more readily recognized when plotted.

For example, figure 9.2 reveals somewhat surprisingly that Decatur's population (the bar graph) declined from 1960 to 1990 and grew slightly since 1990, even though DeKalb County's population (the point-and-line graph) grew steadily from 1900 to 2010. This suggests that the Decatur population data may be incorrect or that something unusual happened to the city's population over the last fifty years.

As chapter 2 pointed out, Decatur's past population trend reflects substantial changes in the average household size of the city's residents. Table 9.2 indicates that the city's population declined between 1970 and 1990 because the average household size declined by over 40 percent and the number of households grew by only 10 percent. As chapter 2 also pointed out, Decatur has only thirty-one acres of vacant land. This suggests that any growth in Decatur's population will result from residential redevelopment or from changes in the average household size of its residents.

Identifying Fit Period Graphs are also extremely useful for determining the most appropriate time interval (or fit period) for projecting future values. For example, figure 9.2 reveals that Decatur's population declined between 1970 and 1990 and grew from 1990 to 2010. This suggests that 1990–2010 provides the best fit period for projecting Decatur's future population, if this growth is assumed to continue.

Figure 9.2 is less clear in suggesting the best fit period for projecting DeKalb County's population. The county's population grew steadily between 1970 and

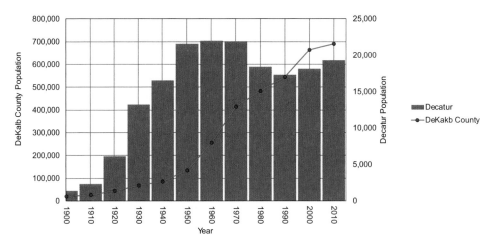

Figure 9.2 Decatur and DeKalb County Population, 1900–2010

Source: Minnesota Population Center (n.d.).

1990, more rapidly between 1990 and 2000, and much less rapidly between 2000 and 2010. As a result, four different time periods could be used to project the county's future population: 1970–2010, 1980–2010, 1990–2010, and 2000–2010. There is no way to know which trend will continue, so it is advisable to prepare projections for all four fit periods, as has been done in chapter 3.

Reporting and Evaluating Projections Graphs of past trends with one or more projections are particularly useful for determining whether the projected growth patterns are reasonable extensions of past trends. Projections for unprecedented growth or decline are often warning signs that something may be wrong with the analysis. Even if nothing is technically wrong with the projections, large deviations from historic patterns and expected trends are often met with suspicion by public officials and stakeholders.

For example, figure 9.3 displays four projections for DeKalb County's 2040 population, given the 1990–2010 fit period. The projections are all mathematically correct but project substantially different values, as figure 9.3 makes clear. The parabolic

TABLE 9.2   **Population and Housing Statistics: Decatur, 1970–2010**

Variable	Year				
	1970	1980	1990	2000	2010
(1)	(2)	(3)	(4)	(5)	(6)
Total population	21,942	18,404	17,336	18,147	19,335
Group quarters population	919	703	856	989	691
Household population	21,023	17,701	16,480	17,158	18,644
Number of households	7,479	7,893	8,230	8,051	8,599
Average household size	2.81	2.24	2.00	2.13	2.17

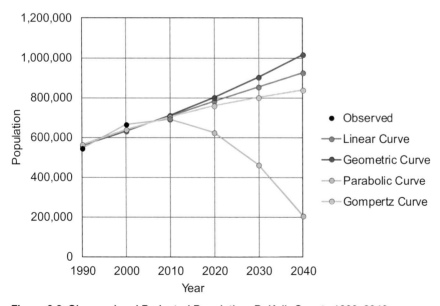

Figure 9.3 Observed and Projected Population: DeKalb County 1990–2040

Source: Figures 3.9 and 3.10.

projection fits the observed values perfectly but is obviously unlikely in projecting a dramatic decline in the county's population. The figure reveals that the other projections are more reasonable.

Charts are useful for describing the breakdown of a population by race, ethnicity, location, or employment. For example, the pie charts in figure 9.4 represent the observed population of Decatur's block groups in 2010 and their projected population in 2040. The charts graphically display the dramatic growth of block groups 226.1, 226.2, and 227.3 over the thirty-year projection period.

Evaluating Projection Assumptions Graphs are also helpful for considering the assumptions that underlie the projected values. For example, figure 9.5 records DeKalb County's projected births for two fertility assumptions. The Constant Fertility Rates projection assumes that the county's age-specific fertility rates between 2005 and 2010 continue without change. The Projected Fertility Rates projection assumes the county's future fertility rates will parallel the fertility assumptions incorporated in the Census Bureau's national population projections. The projected births for the Constant Rates projection grow steadily as the county's population grows, which seems reasonable. The projected births for the Projected Fertility Rates projection decline by more than 1,500 between the 2005–2010 projection period and

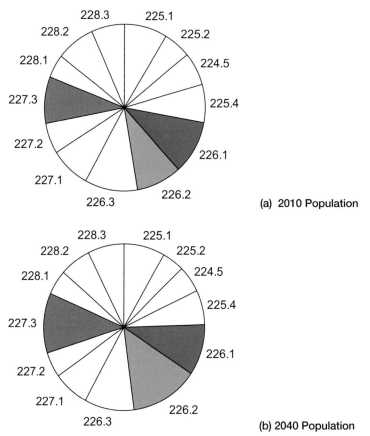

(a) 2010 Population

(b) 2040 Population

Figure 9.4 Observed and Adjusted-Share-of-Change Population Projections: Decatur Block Groups, 2010 and 2040.

Source: Table 4.8.

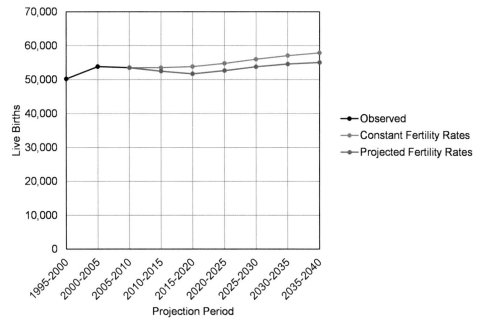

Figure 9.5 Projected Live Births: DeKalb County, 2005–2040

Source: Figure 5.12.

the 2015–2020 projection period and do not match the earlier rates for twenty years. This is unlikely for a steadily growing population. As a result, figure 9.5 calls the Projected Fertility Rates projection into question.

Use Simple Graphs and Charts

Modern spreadsheet and graphics programs provide an array of different ways to present quantitative information. While this diversity makes it tempting to present information in unusual ways, it is generally preferable to use simple and straightforward graphs that will be readily recognized and easily understood by the public.

For example, figure 9.6(a) uses a bar graph and figure 9.6(b) uses a 3-D cylinder graph to report Decatur's population from 1970 to 2010. The population values are much easier to read in the bar graph than they are in the 3-D cylinder graph. This example illustrates Edward Tufte's suggestion that data graphs should draw the viewer's attention to the substance of the data, not to something else (Tufte 1983, 91–105). For Tufte, this meant that data graphs should maximize the share of a graph's ink that is used to present information and eliminate distracting graphic elements. This suggests that, whenever possible, planners should use simple graphs and charts that display information clearly and understandably to nontechnical audiences.

Planners should also be careful that their graphs do not mislead their audiences. For example, figures 9.6(a) and 9.7 both use bar graphs to display Decatur's population from 1970 to 2010. The bar graph in figure 9.6(a) plots population values from 0 to 25,000. The bar graph in figure 9.7 plots population values from 12,000 to 24,000. The two graphs display the same information correctly but the bar graph in figure 9.7 overemphasizes the decline in Decatur's population between 1970 and 1990. The city's population declined by roughly 12 percent over this period, but

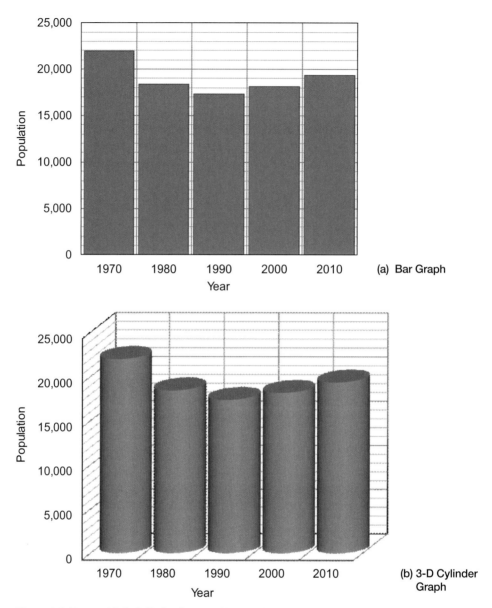

Figure 9.6 Bar and 3-D Cylinder Graphs: Decatur Population, 1970–2010.

Source: Figure 9.2.

figure 9.7 suggests the decline was much larger than that. This may mislead readers who are less experienced in dealing with quantitative data analysis or who spend less time interpreting the information in the graph. Figure 9.6(a), which plots values from 0 represents the city's past population trend more clearly.

The academic literature contains a wealth of guidelines for presenting quantitative information persuasively and clearly. Excellent advice is provided in Edward Tufte's classic *The Visual Display of Quantitative Information* (Tufte 1983) and *Envisioning Information* (Tufte 1990). Dowell Myer's excellent *Analysis with Local Census Data: Portraits of Change* (Myers 1992) also provides useful suggestions for presenting quantitative data clearly and understandably.

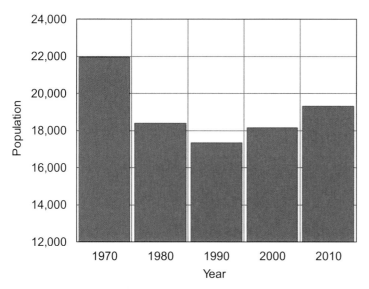

Figure 9.7 Truncated Bar Graph: Decatur Population, 1970–2010

Source: Figure 9.2.

Use Maps

All the natural and human-made phenomena, activities, policies, and actions that are of concern to planners and the public take place in one or more locations. As a result, maps, like graphs and charts, are extremely valuable for displaying information clearly and understandably to the public. Maps are particularly useful for: (1) orienting data users, (2) displaying analysis information, (3) displaying analysis results, and (4) evaluating analysis results.

Orienting Data Users

Maps are useful first for orienting data users to the location considered in the analysis. Members of the public will generally be familiar with their communities, but all data users may not be. As a result, it is always advisable to provide one or more maps that orient data users.

For example, map 9.1 locates DeKalb County directly adjacent to the city of Atlanta in the center of the twenty-eight-county Atlanta region. The inset map locates the Atlanta region within the state of Georgia. As maps 9.1 and 9.4 illustrate, inset maps are particularly useful for locating smaller analysis areas within a larger city, county, or state.

Displaying Analysis Information

Maps are also useful for displaying the information that was used in the analysis. For example, map 9.2 shows Decatur's land uses and the streams that were used to identify land parcels and structures that are subject to different flood risks. The map includes a legend, scale bar, inset map, and north arrow and uses the American Planning Association's Land Based Classification Standards' standard land use colors (American Planning Association n.d.). The inset map shows Decatur's location in DeKalb County.

Similarly, map 9.3 maps the slope and distance to water factors that were used to determine the suitability of DeKalb County's vacant parcels for accommodating

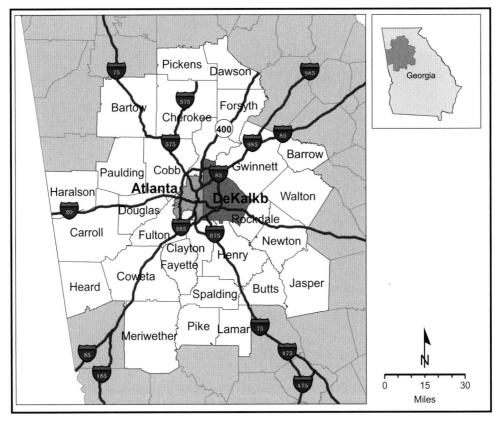

Map 9.1 DeKalb County and the Atlanta Region, 2010

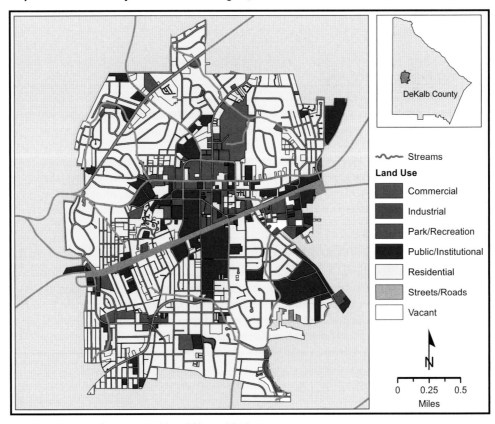

Map 9.2 Decatur Streams and Land Uses, 2014

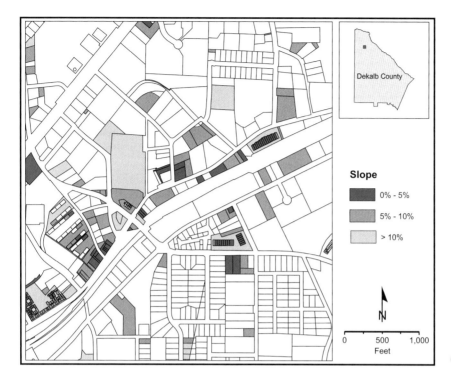

(a) Slope

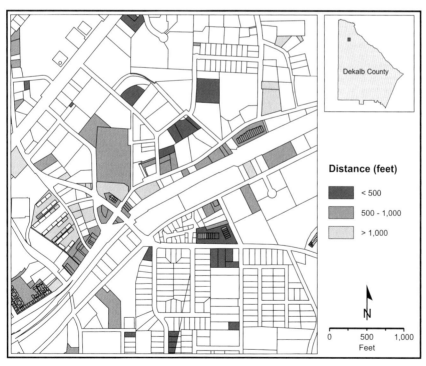

(b) Distance to Water

Map 9.3 DeKalb County Suitability Factors, 2014.

new residential and commercial development. The maps clearly locate the slope and distance to water suitability layers and associated land parcels for a portion of the county.

Displaying Analysis Results

Maps are also extremely useful for displaying analysis results to lay audiences. For example, map 9.4 clearly identifies land parcels, buildings, and roads that are in three different flood zones. The figure makes the implications of different flood events clear to people and organizations located near one of the city's water features.

In a similar way, map 9.5 displays the results of an ordinal combination suitability analysis for the study area in map 9.3. Several of the vacant parcels on the left-hand side of map 9.5 are highly suitable for commercial development and less suitable for residential development. This reflects the information in map 9.3 that these parcels have lower slopes, which makes them highly suitable for commercial development and less suitable for residential development. Conversely, several vacant parcels at the top of map 9.5 are more suitable for residential development than they are for commercial development. This again reflects the information in map 9.3 that they are close to a water feature, making them highly suitable for residential development and less suitable for commercial development.

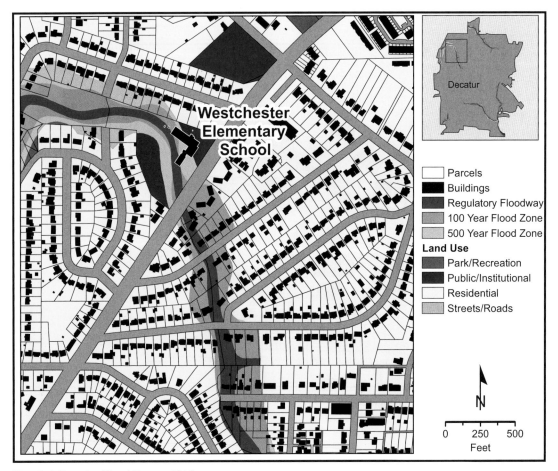

Map 9.4 Decatur Flood Zones, 2014

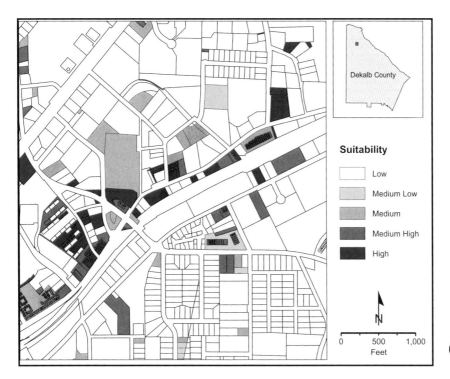

(a) Commercial
Suitability

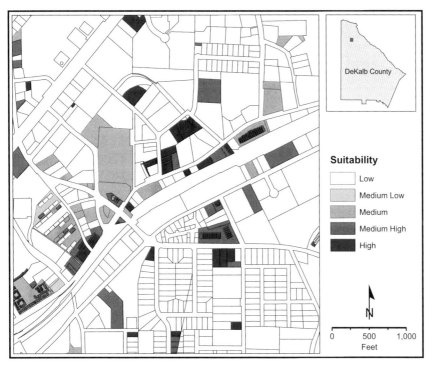

(b) Residential
Suitability

Map 9.5 DeKalb County Ordinal Combination Commercial and Residential Suitability.

Map 9.6 reports the observed population (in gray) and adjusted-share-of-change 2040 projections (in black) for Decatur's block groups. The figure allows community members to easily relate the projected population values to familiar locations in the city and their associated land uses. Thus, for example, it reveals that the largest growth will occur in Block Group 226.2, which is near the city center and contains most of the city's commercial development. The pie charts in figure 9.4 provide another way to easily interpret the projected population values for Decatur's block groups.

Use Appropriate Map Options

As was true for graphs and charts, the academic literature provides a wealth of guidelines for producing attractive and informative maps. Resources that are particularly useful for planners include Kent and Klosterman (2000), Monmonier (1996), Brewer (2005), Kryger and Wood (2005), and Lewis (2015, 162–181).

Document Assumptions

The projection methods considered in this book implicitly assume that past trends will continue or change in assumed ways. This process is revealed most clearly in graphical trend projection methods that use visual observation and manual techniques to

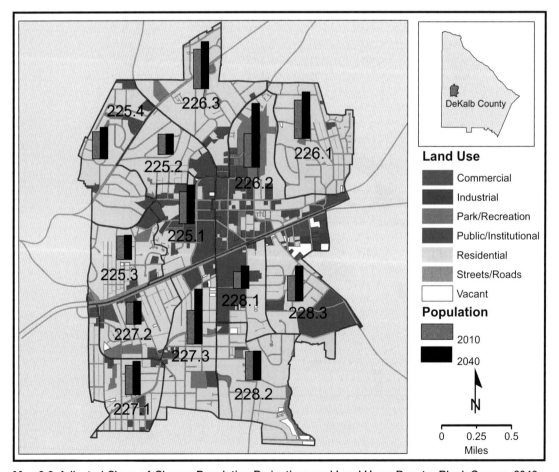

Map 9.6 Adjusted-Share-of-Change Population Projections and Land Uses: Decatur Block Groups, 2040

identify past trends and extend them to project future values. However, the quantitative procedures for identifying and extending past trends that underlie the other methods are only more explicit and replicable forms of these procedures.

Unfortunately, there is no assurance that past trends will continue. Close agreement with past may result from mere coincidence or from the aggregate effect of underlying causal processes that may not continue. Growth trends for rapidly growing or declining regions cannot continue forever and many regions experience long-term cyclical patterns of growth and decline that cannot be predicted from past trends. Most importantly, blindly continuing past trends ignores the potential for public action that attempts to change the future.

Professional judgment is required to consider an unknown future, and it must be documented. As the analyses in the preceding chapters demonstrate, a large number and variety of computationally correct projections can easily be prepared by applying different methods, using different data, and making different assumptions about what the future will be. Analytic methods alone cannot determine what the future will be. They are only tools for tracing the implications of the assumptions that are chosen independently from—and prior to—the method that implements them. When the core assumptions are valid, the choice of methodology is secondary or obvious. And when the core assumptions are wrong, the projection method rarely makes a difference (Ascher 1981).

Explicitly stating the assumptions that underlie a projection will not produce more accurate forecasts for an unknowable future. However, it will make the decisions that underlie a forecast more understandable to intelligent nonspecialists and to experts. Revealing a projection's underlying assumptions also will not eliminate debate about alternative forecasts; it is likely to increase it. Yet the resulting discussion will be more informed than the uncritical acceptance of projections prepared by unfathomable black boxes, described by unnecessarily complex language (**bafflegab**).

Nonspecialists cannot be expected to understand the detailed computations that underlie a projection. However, they should be able to understand the general logic of the projection process and, most importantly, the assumptions that underlie a projected future. As a result, the assumptions embodied in the projections should be revealed in clearly understood narrative and/or numerical statements.

Assumptions that should be documented include: (1) *ceteris paribus* "all other things being equal" assumptions (e.g., that there will be no wars, natural disasters, or extreme economic conditions), (2) the historical data or growth trends that underlie the projected values, and (3) the sources for all the data used to prepare the projections. In addition, projections that assume that past trends will continue must recognize that they do not account for efforts to plan and implement public policies and actions that will shape future trends.

Most importantly, all the assumptions that underlie a projection should be described clearly and understandably. For example, table 9.6 projects the need for Decatur's new housing units in 2040. The projections are based on the following assumptions in tables 9.5 and 9.6: (1) the **trimmed average trend projections** for Decatur's 2040 population, (2) the city's group quarters population will not change, (3) the city's **vacancy rate** will remain constant, (4) the projected average household size will assume one of the values in table 9.5, and (5) 5 percent of the city's existing housing stock will be removed over the thirty-year projection period. Modifications to any of these assumptions will change the projected demand for new housing units. Given appropriate software tools, planners should try modifying these assumptions to determine how sensitive the future values are to different projection assumptions.

Compare to Projections by Other Organizations

It is often helpful to compare locally produced population and employment projections to the projections prepared by other organizations such as officially sanctioned regional, state, or national forecasts. For example, the Atlanta Regional Commission (ARC) has developed a series of population and employment forecasts for its twenty-county region to the year 2040 (Atlanta Regional Commission 2016). The forecasts were prepared with an integrated system of three sophisticated models. The Regional Economic Models Inc.'s (REMI) model produced a regional forecast that was disaggregated by the Production Exchange Consumption Allocation System, which fed local socioeconomic data to an activity-based travel model that modeled household-level and person-level travel choices. The forecasts support the transportation project prioritization and land-use planning processes that are the basis for the Atlanta region's plan. ARC projects a 1.5 percent growth in the region's population between 2015 and 2040, which is lower than the region's historic annual growth of 3 percent.

Table 9.3 compares the Atlanta Regional Commission's 2040 population forecast for Decatur to the four trimmed average trend projections for the 1990–2010 fit period developed in chapter 3. ARC projects a 33.9 percent increase in the city's population, which is substantially larger than the four trend projections and highly unlikely in a community that is almost entirely built out. ARC's projections for DeKalb County's 2040 population in table 9.4 is slightly lower than the 1970–2010 trend projection. This suggests that 880,000 is a reasonable projection for the county's 2040 population.

Combine Different Information

The previous chapters used several demographic, economic, and spatial analysis methods to examine Decatur's and DeKalb County's past, present, and possible futures. The chapters considered each perspective in isolation, but a much better understanding of an area can be gained by combining the information generated by the different methods. Particularly useful ways of combining this information include: (1) combining trend with cohort-component projection information, (2) combining trend with share projection information, (3) combining trend projection with housing information, and (4) combining population and employment projections with land suitability analysis information.

TABLE 9.3	Atlanta Regional Commission and Trend Population Projections: Decatur, 2040

Projection	Population		Percent Change
	2010 $(P^{2010})^a$	2040 $(P^{2040})^b$	
(1)	(2)	(3)	(4)
Atlanta Regional Commission	19,335	25,895	33.9
Linear curve	19,335	22,271	15.2
Geometric curve	19,335	22,707	17.4
Parabolic curve	19,335	25,161	30.1
Gompertz curve	19,335	21,793	12.7
Trimmed average	19,335	22,489	16.3

Source:
[a] US Bureau of the Census (2013c).
[b] Atlanta Regional Commission (2016); Table 3.5.

| TABLE 9.4 | **Atlanta Regional Commission and Trend Population Projections: DeKalb County, 2040** | | | |

Projection	Population		Percent Change
	2010 (P^{2010})[a]	2040 (P^{2040})[b]	
(1)	(2)	(3)	(4)
Atlanta Regional Commission	691,893	874,424	26.4
1970–2010 Trend	691,893	880,330	27.2
2000–2010 Trend	691,893	768,797	11.1

Source:
[a] US Bureau of the Census (2013c).
[b] Atlanta Regional Commission (2016); Table 3.7.

Combining Trend and Cohort Component Projection Information

The trend projection methods in chapter 3 project an area's population by continuing past trends into the future. The cohort component projection methods in chapter 5 project an area's population by age and sex by applying projected mortality, fertility, and migration rates to its current population. The two projection methods can be combined to consider the impacts that projected trends will have on the community's future age structure.

For example, the trend projection analysis suggested that 833,000 is a reasonable continuation of DeKalb County's 1970–2010 trend and a 769,000 is reasonable continuation of the county's 2000–2010 trend. The cohort component method can be used to generate very similar projections. It projects a 2040 population of 833,514 if the observed survival and fertility rates between 2005 and 2010 continue without change and the 2005–2010 net out-migration is reduced by 65 percent. The corresponding projection for DeKalb County' 2040 population by age and sex is reported in table 9.1 and figure 9.1. The cohort component method projects a 2040 population of 771,901 if the observed survival and fertility rates continue without change and the net out-migration increases by 15 percent.

The cohort component projection casts doubt on the 2000–2010 trend projection. It is highly unlikely that net out-migration will increase in a growing county, that is, that the county's population will grow in the future even though more people move out of the county than did between 2005 and 2010. It is much more reasonable to assume that the 2005–2010 trend reflects the unusual effect of the Great Recession which temporarily reduced house buying and migration across the United States and cannot be assumed to continue.

Combining Trend with Share Projection Information

The trend projection methods described in chapter 3 and the share projection methods described in chapter 4 can be used to consider the implications that projections for a large area such as a county or city will have on its subareas.

For example, the trend projection analysis suggested that 22,500 is a reasonable projection for Decatur's 2040 population. The corresponding adjusted-share-of-change projections for the city's thirteen block groups are displayed in the pie chart in figure 9.4 and the map in map 9.6. Other trend and share projection methods provide a range of projections for the population in Decatur's block groups. The projected demands must be matched with a detailed analysis of the block groups' development potential, zoning restrictions, and other public policies to determine what future development patterns will—and should—be.

Combining Population Projections with Housing Information

The trend projection methods in chapter 3 and the cohort component projection method in chapter 5 can be used with the housing unit method described in chapter 2 to project the number of housing units required to accommodate an area's projected population.[2]

The **housing unit method** is expressed in equation 9.1:

$$P^t = (HH^t \times AHS^t) + GQ^t, \qquad 9.1$$

where P^t = study area population in year t, HH^t = number of households (occupied housing units) in year t, AHS^t = average household size in year t, and GQ^t = group quarter population in year t.

A **housing unit** is a house, apartment, mobile home or trailer, a group of rooms, or a single room that is occupied as separate living quarters, or if vacant, is intended for occupancy as separate living quarters. The **household population** includes everyone who uses a housing unit as their usual place of residence. It is equal to the total population minus the **group quarters population,** which includes people residing in college dormitories, military quarters, nursing facilities, and correctional institutions. The **average household size** is the average number of persons per household, that is, the household population divided by the number of households.

Equation 9.2 rearranges the terms in equation 9.1 to project the number of households in year t (HH^t):

$$HH^t = \frac{(P^t - GQ^t)}{AHS^t}. \qquad 9.2$$

Table 9.5 uses equation 9.2 to project the number of households in Decatur in 2040, given the trimmed average projection for Decatur's 2040 population in column one. The 2040 group quarters population in column two is projected from the 1990–2010 trend in table 9.2. Three assumptions for the city's average household size in 2040 are reported in column three: 2.17 is the current average household size; 2.45 is the trend projection for the observed values for 1990–2010; and 2.31 is midway between the city's current value and the trend projection. The household projections in column four are computed by applying equation 9.2. For example, given the projected total and group quarters populations and the current average household size:

$$HH^{2040} = \frac{(P^{2040} - GQ^{2040})}{AHS^{2040}} = \frac{(22,489 - 746)}{2.17} = 10,020.$$

The number of households is equal to the number of occupied housing units. As a result, a housing unit projection must adjust a household projection to include

---

TABLE 9.5   **Projected Housing Units: Decatur, 2040**

Total Population 2040 (P^{2040})[a]	Group Quarters Population 2040 GQ^{2040}	Average Household Size 2040 AHS^{2040}	Households 2040 HH^{2010}	Vacancy Rate	Housing Units 2040 HU^{2040}
(1)	(2)	(3)	(4)	(5)	(6)
22,489	746	2.17	10,025	7.9	10,885
22,489	746	2.31	9,417	7.9	10,225
22,489	746	2.45	8,879	7.9	9,641

Source:
[a] Table 3.8.

vacant housing units. The adjustment divides the projected number of households by one minus the projected vacancy rate. For example, the projected number of housing units for the current average household size in column six of table 9.5 is equal to $10,025/(1 - 0.079)$ or 10,885.

Table 9.6 computes the number of new housing units required to accommodate the projected population demand for the three average household size assumptions.[3] The number of existing housing units that remain in 2040 (column four) is equal to the number of existing housing units (column two) minus the projected number of housing units that will be abandoned, lost to fires, converted to other uses, or otherwise removed from the housing stock (column three). The required number of new housing units in column six is equal to the projected demand (column five) minus the number of remaining units (column four).

The projected demand for new housing units ranges from 2,017 (more than 20 percent of the existing housing stock) for the current average household size to 773 (8 percent of the existing housing stock) for the projected average household size. This range dramatically illustrates the impact that the average household size assumptions have on the projected demand for new housing units.

Combining Projections with Land Suitability Information

The population and employment projection methods described in chapters 3 through 6 are widely used to project the demand for transportation systems, schools, hospitals and other population- and employment-related facilities and activities. They can also be combined with the land suitability analysis methods described in chapter 8 to guide the process of determining where different land uses should be located.

Table 9.7 computes the additional residential land that is needed to accommodate two projections for DeKalb County's 2040 population. DeKalb County's current residential density is equal to its current residential area (77,380 acres) divided by its current population (691,893) or 0.112 acres per person. Assuming the residential density is constant, the projected demand for residential land in 2040 is equal to the projected population multiplied by the average number of residential acres per person. The projections for additional residential land are 8,624 acres for the 2000–2010 trend and 21,373 for the 1970–2010 trend.

Table 9.8 records the suitability of DeKalb County's vacant parcels for two suitability scenarios described in chapter 8. It reports that DeKalb County does not

| TABLE 9.6 | Projected Demand for New Housing Units: Decatur, 2040 |

Average Household Size 2040	Existing Housing Units 2010	Housing Units Lost 2010–2040	Housing Units Remaining 2040	Projected Housing Units 2040	New Housing Units 2010–2040
$(AHS^{2040})^a$	$(HU^{2010})^a$	$HU_{Lost}^{2010,2040}$	$HU_{Remaining}^{2040}$	HU^{2040}	$HU_{New}^{2010,2040}$
(1)	(2)	(3)	(4)	(5)	(6)
2.17	9,335	467	8,868	10,885	2,017
2.31	9,335	467	8,868	10,225	1,357
2.45	9,335	467	8,868	9,641	773

Source:
[a] US Bureau of the Census (2013c)..

TABLE 9.7   Projected Residential Land Demand: DeKalb County, 2040

Projection	Population		Residential Area (Acres)		
	2010[a]	2040	2010[b]	2040	Additional
(1)	(2)	(3)	(4)	(5)	(6)
1970–2010 Trend	691,893	883,000	77,380	98,753	21,373
2000–2010 Trend	691,893	769,000	77,380	86,004	8,624

Source:
[a] US Bureau of the Census (2013a).
[b] Table 2.7.

TABLE 9.8   Weighted Combination Vacant Parcel Residential Suitability: DeKalb County, 2010

Suitability	Vacant Parcels			
	Conservation Alternative		Development Alternative	
	Area (Acres)	Percent	Area (Acres)	Percent
(1)	(2)	(3)	(4)	(5)
Excluded	0	0.0	0	0.0
Low	809	4.3	1,279	6.9
Medium low	4,605	24.8	2,110	11.3
Medium	5,695	30.6	3,515	18.9
Medium high	5,260	28.3	6,894	37.1
High	2,236	12.0	4,806	25.8
Total	18,604	100.0	18,604	100.0

Source:
Table 8.11.

have enough suitable land to accommodate the projected residential demand for the 1970–2010 trend in table 9.7.[4] This indicates that the county's future residential density will be substantially higher than it is currently if the 1970–2010 trend continues.

Land use plans and policies are generally prepared and implemented for cities, not counties. As a result, the demand projections and suitability analysis for DeKalb County are only illustrative of the procedures that would be conducted by cities and other subcounty areas. Procedures for allocating the projected housing demand to different residential uses and projecting the demand for residential support facilities are described in Berke et al. (2006, 401–420).

Chapters 3–5 prepared several projections for Decatur's 2040 population that were converted into equivalent demands for new housing units in tables 9.5 and 9.6. The economic analysis methods in chapter 6 identified four economic sectors that accounted for nearly 60 percent of Decatur's employment in 2013. Table 9.9 reports Decatur's projected 2040 employment in these sectors and in the remaining sectors combined.

As reported in chapter 2, Decatur has very little vacant land that can accommodate the projected housing and employment-related land use demands. As a result, the land supply methods described in chapter 8 must be used to determine whether infill and redevelopment will be sufficient to accommodate the projected demands.

TABLE 9.9			Observed and Projected Employment for Selected Employment Sectors: Decatur, 2013 and 2020		

NAICS Code	NAICS Sector	Employment		Percent Change
		2013 $(E_i^{2013})^a$	2040 $(E_i^{2040})^b$	
(1)	(2)	(3)	(4)	(5)
54	Professional Services	1,318	1,561	18.4
61	Educational Services	1,658	2,487	50.0
62	Health Care	2,579	1,532	−40.6
72	Accommodation	1,642	2,633	60.4
—	Other Sectors	5,061	8,038	58.8
—	**Total**	**12,257**	**16,251**	**32.6**

Source:
[a] US Bureau of the Census (2013c).
[b] Atlanta Regional Commission (2016).

Procedures for locating the projected demand for employment-related land uses are described in Berke at al. (2006, 347–381).

Support Other Applications

The analysis and projection methods described in this book are very useful in their own right. However, they are also useful for supporting two applications of particular interest to planners: land use planning and allocating projected land use, population, and employment growth to small areas. The role the planning support methods described in this book play in these applications is described briefly below.

Land Use Planning

Urban land use planning has long been a central concern of planners. The importance of land use planning in planning education and practice and education is reflected in five books on urban planning methods produced over nearly fifty years by the planning faculty at the University of North Carolina at Chapel Hill from Chapin (1957) through Berke et al. (2006). Berke et al (2006, 291–313) describe a five-step process that can be used to prepare area-wide policy plans, community-wide design plans, and small-area land use plans.

The first task develops principles and standards for locating development districts, for land uses or land-use mixes, for transportation and other community facilities, and for the spatial relationship between these uses. The second task uses the design principles and standards developed in the first task to prepare maps showing the relative suitability of different locations for locating land policy districts, land use design components, or community facilities. The third task estimates the amount of land needed to accommodate the projected demand for land policy districts, land use activities, or community facilities. The fourth task analyzes the capacity of the suitable locations for accommodating the land policy districts or land use design components. The final task develops design alternatives for the location and size of future development and redevelopment areas, activity centers, land use design components, land use sectors, community facilities, and open space.

The geographic information systems described in chapter 7 are essential for collecting and analyzing the spatial information used in all stages of this process. The suitability analysis methods described in chapter 8 are used in the second task to prepare the suitability maps that are the foundation for later tasks. And the trend and share projection methods described in chapters 3 and 4 are used in the third task to project the demand for land policy districts, land use activities, and community facilities.

Allocating Growth to Small Areas

Planners are often required to allocate area-wide land use, population, and employment growth to smaller areas such as census enumeration areas, political jurisdictions, school districts, traffic analysis zones. The What if? planning support system (*What if?*) uses several of the methods described in this book to provide these small-area projections.[5] As its name suggests, *What if?* draws on the lessons of chapter 1 and does not attempt to predict an unknowable future. Instead, it is an explicitly policy-oriented planning support tool that can be used to determine *what* would happen *if* clearly defined policy choices are made and assumptions concerning the future prove to be correct. Policy choices that can be considered in the model include the staged expansion of public infrastructure and the implementation of land use plans, zoning ordinances, and open space protection programs. Assumptions about the future that can be considered in the model include future population and employment trends, assumed household characteristics, and future development densities.

What if? is a simple, rule-based model that does not attempt to duplicate the complex spatial interaction and market clearing processes that shape the urban fabric. Instead, it incorporates a set of explicit decision rules that project the future by balancing the supply of, and demand for, land suitable for different uses at different locations. Alternative visions for an area's future can be explored by defining different suitability, growth, and allocation projections. The results generated by considering these projections provide concrete and understandable expressions of the likely results of the projections' underlying policy choices and assumptions. For instance, the model might demonstrate that there is insufficient land simultaneously to accommodate high growth, low residential densities, and strict agricultural protection policies, forcing a community to choose between highly desirable, but inconsistent, policy goals.

What if? uses the powerful geographic information systems technology described in chapter 7 to analyze the available spatial data for any area. *The What if?* Suitability component uses the weighted combination land suitability analysis method described in chapter 8 to determine the relative suitability of different locations for accommodating future land use demands. The suitability analyses can consider developers' preferences for particular kinds of locations, community preferences for preserving particular land uses, and public policies that limit the amount of developable land.

The *What if?* Demand component converts the five main types of land use demand—residential, industrial, commercial, preservation, and locally oriented uses— into the equivalent land use demands. The residential, industrial, and commercial demands are computed by converting area-wide trend projections prepared with the trend projection methods described in chapter 3 and the share projection methods described in chapter 4 into equivalent land use demands. The model allows users to specify the amount of land that should be preserved for environmental uses such as agriculture, forestry, and open space protection. It also allows the user to specify how much land will be required, per capita, to satisfy locally oriented land uses such as local parks and recreational areas and local retail.

The *What if?* Allocation component projects the future by allocating the projected land use demands—specified by a demand projection—to different locations based on their relative suitability—defined by a suitability analysis. The projections can be prepared for up to five periods, allowing the system to incorporate a staged development process in which future development patterns are based on the previous development patterns and the staged expansion of infrastructure such as the construction of major roads or the extension of sewer or water service.

What if? allocates the projected demand for all land uses in each projection year and converts the projected land uses into the equivalent small-area population and employment projections. The process stops when all the demand has been allocated for all the projection years or the model runs out of land because there is not enough suitable land to satisfy the projected demand. If this happens, the user must modify the model assumptions to increase the supply of suitable land or reduce the demand for land. The projected land uses, population and employment for user-defined subareas are reported in easily understood maps, graphs, and tables.

Use Scenarios

There are two ways to use projections when deciding what to do. The traditional approach selects a preferred projection and uses that projection when making decisions. A better approach considers several different projections and makes decisions informed by an understanding of a range of futures that might happen in combination with different actions. These multiple futures are called scenarios.

Scenarios are much more than projections that extend past trends to identify a single future.[6] Instead, they describe a process of change and provide a range of plausible and divergent stories outlining how the future may unfold and how it might be different from the past. Scenarios are used in the private sector to help organizations survive and prosper in the face of future contingencies and competition. In the public sector, scenario planning allows community members to learn about their community, identify what the community can control, recognize what it cannot control, and consider the implications of choices they make today. Scenarios recognize the uncertainty and complexity of foresight and provide a way to think carefully about the future without trying to predict what it will be.

In current practice, planners often select a single preferred future from a small set of alternative futures that generally include high/medium/low options—a desired future with two bookends—or two options, one of which is a baseline "business as usual" option that serves a foil for the preferred option. Assuming they can control the future, planners then discard the less desirable options and attempt to make the desired future happen.

A better approach creates a range of scenarios about future circumstances that communities cannot control. On this approach the different scenarios do not bracket a preferred option and cannot be described simply as high or low. Instead they provide a collection of plausible futures, some of which may include new, sometimes even uncomfortable, possibilities.

History suggests that planners should not be too constrained in considering what the future may be. The amazingly rapid development of Chinese cities is a warning that planners should not think that things will continue as they are. The solution to horse manure in urban streets was the automobile, whose introduction created new problems of air pollution, traffic congestion, and urban sprawl. A possible switch to self-driving electric cars powered by wind and solar energy sources will bring unanticipated benefits and costs.

Scenarios cannot be used independently from public efforts to plan and act. Scenarios that are initially imagined as resulting from circumstances beyond a community's control should be combined with the expected effects of responses to the anticipated futures. Climate-change scenarios must consider efforts to control emissions and adapt to new weather conditions. Migration from areas that are anticipated to experience water shortages should incorporate public efforts to recycle and reuse water. An anticipated return of industrial employment from places it has left will depend on investments in infrastructure and education and the emergence of new industrial products through research and development.

In practice, the interactions between uncontrolled and controlled will be a two-step process. That is, planners should first create scenarios that incorporate aspects of change beyond the community's control. They can then use these scenarios to consider decisions and actions that address the anticipated implications of the external scenario. The planning process should consider both possible futures and public actions in response to them. As Isserman (1984, 209) suggests, "[I]t should be impossible for a planner to think about what *will be* without thinking about what *can* and *ought* to be."

Scenario planning will employ many of the analysis and projection methods described in this book but with different data inputs and assumptions. For example, rather than basing future fertility rates on the recent past, scenarios might consider what would happen if recently declining rates increased with the introduction of new immigrant populations. Other scenarios might consider what would happen if regions impacted by sea level rise or excessive heat become less attractive than they have been, reversing past migration trends. This may require replacing the historical data for the study area with data from other regions that underwent the possible changes or creating realistic synthetic data based on hypothesized changes in technology or institutions.

Planning scenarios should be described in stories that provide clear and convincing images of how a community can move from the where it is now to a possible future. These stories will be useful for ensuring that the expected future is plausible and attainable. More importantly, compelling, evidence-based stories can create a strong sense of a place that invites the public to think about where they are, where they have been, and where they want to go. Done well, scenario planning that uses the methods described in this book can encourage the people in a community to cowrite a compelling story about their past and present that motivates them to work together to create a better future.

Notes

[1] Armstrong (2001b) provides a comprehensive list of 139 principles for formulating a forecasting problem, obtaining required information, selecting and applying methods, evaluating methods, and using forecasts.

[2] Smith (1986) provides an extensive evaluation of the housing unit method.

[3] The following discussion draws on Berke et al. (2006, 406–408).

[4] Table 2.1 indicates that DeKalb County has 30,626 acres of vacant land, more than enough to accommodate the projected demand in table 9.7. This is substantially different from the value reported in table 9.8 because the tables use two measures of "vacant land." The vacant land in table 9.8 includes only the area of parcels which are primarily vacant, i.e., have centroids in vacant land. The vacant land in table 2.1 also includes the vacant portions of other parcels which are primarily devoted to other uses. As a result, the quantities in table 9.8 are a better measure of the amount of land that is available to accommodate future land use demands.

[5] Further information on the What if? planning support system is available at Klosterman (1999, 2007), Pettit et al. (2013), and http://www.whatifinc.biz/index.php.

[6] The following discussion draws on Klosterman (2013, 164) and Isserman (2007). Further information on scenario planning is available from Hopkins and Zapata (), Myers and Kitsuse (2000), and Xiang and Clarke (2003).

Appendix A

US Census Geography

The US Bureau of the Census (or Census Bureau) has three closely related responsibilities.[1] It provides accurate population counts for each state in years ending in zero for allocating representatives to the House of Representatives based on the state population. It conducts an array of enumerations, sample **surveys**, and other statistical programs to collect data on the people, businesses, and organizations in the United States. Equally importantly, it tabulates, publishes, and disseminates the results of these efforts to public and private data users. These responsibilities require the Census Bureau to assign each person, household, housing unit, business establishment, and other reporting entity to a location and assign that location to the appropriate tabulation entities. As a result, geography is an important component of the Census Bureau's programs for collecting, organizing, and disseminating information.

The Bureau of the Census classifies all geographic entities into two broad categories: (1) **legal and administrative entities** and (2) **statistical entities**. Legal and administrative entities are created by state and federal legislation, court decisions, legal actions, and similar official actions. Data for these entities are required by federal, state, and local governments to conduct elections and manage a wide variety of programs. The size and boundaries of legal and administrative units are established by law and often follow physical features and property lines that may not be easily identified.

Statistical entities are established by the Bureau of the Census to reflect the practice, customs, and needs of data users. These entities are established when the geographic coverage of legal and administrative entities is incomplete, inadequate, or inconsistent over time. Their boundaries are easily recognized linear features such as roads, railroads, and rivers that help data users relate information from other data sources to the appropriate statistical entities.

The Bureau of the Census provides data for a wide range of geographic entities. The most widely used entities are described below. The relationship among these entities is outlined in figure A.1. Statistical entities are identified in figure A.1 by bold fonts; legal and administrative entities are displayed in normal fonts. The discussion and figure do not include American Indian, Alaska Native, and Native Hawaiian Areas.

As a rule, more data and more detailed data are available for larger census entities than they are for smaller areas. In addition, data for smaller areas are more likely to be suppressed to prevent reporting data for individuals and firms than they are for larger areas. The data for smaller areas are also more likely to contain data errors than the data for larger areas.

Legal and Administrative Entities

The Nation

Data for the United States includes the fifty states and the District of Columbia. They do not include Puerto Rico and the Island Areas (American Samoa, the Commonwealth of the Northern Mariana Islands, Guam, and the US Virgin Islands).

States

In addition to the fifty states, the Bureau of the Census treats the District of Columbia, Puerto Rico, and the Island Areas as the statistical equivalents of states.

Counties

Counties are the primary legal divisions of most states. In Louisiana, these divisions are known as parishes. Equivalent subareas are defined in Alaska, which has no counties, and in four states (Maryland, Missouri, Nevada, and Virginia) where one or more incorporated places are independent of any county organization. The Bureau of the Census treats the *municipios* in Puerto Rico as equivalent to counties.

Minor Civil Divisions

Minor Civil Divisions (**MCDs**) are the primary governmental or administrative divisions of a county in twenty-nine states and the county equivalents

261

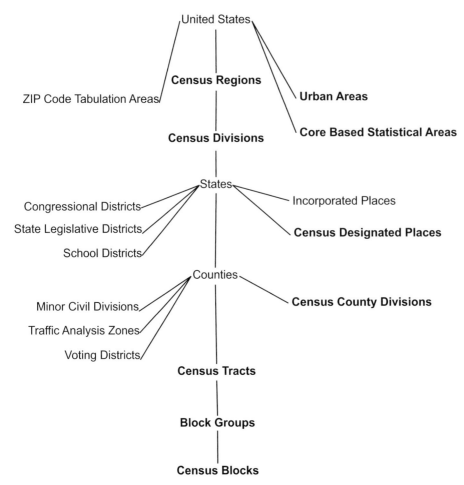

Figure A.1 US Census Geographic Entities

Source: US Bureau of the Census (2010).

in Puerto Rico and the Island Areas. MCDs have many designations, including townships, towns, parishes, reservations, and barrios.

Incorporated Places

Incorporated places are legally established by a state to provide governmental functions for a concentration of people, as opposed to MCDs, that are generally created to provide services to an area without regard to its population. Incorporated places include cities, towns, villages, and boroughs.

Election Districts

The Bureau of the Census reports data for the 435 congressional districts from which people are elected to the US House of Representatives and the state legislative districts from which members are elected to state legislatures.

School Districts

The Bureau of the Census tabulates data for elementary school districts, secondary school districts, and unified districts that provide education to children in their service areas.

Traffic Analysis Zones

Traffic analysis zones or **transportation analysis zones** (**TAZs**) are spatial entities that are commonly used in transportation planning models. The size of a TAZ varies from very large zones in rural areas to census blocks (see below) in urban areas.

Statistical Entities
Census Regions

Census Regions include four groupings of the fifty states and the District of Columbia: Northeast, Midwest, South, and West.

Census Divisions

Census Divisions are nine groupings of states and the District of Columbia that subdivide the four census regions.

ZIP Code Tabulation Areas

ZIP Code Tabulation Areas (**ZCTAs**) are approximate spatial equivalents to the US Postal Service five-digit ZIP Code service areas. ZCTAs generally assign a census block to the most frequently occurring ZIP Code for the addresses in that block. The ZCTA boundaries may overlap county and state boundaries.

Urban and Rural Population

The **urban population** includes the population residing in urbanized areas and urban clusters. An **urbanized area** (**UA**) is a densely developed territory containing fifty thousand or more people. An **urban cluster** (**UC**) is a densely developed territory with at least two thousand five hundred people but fewer than fifty thousand people. UAs and UCs incorporate areas of high population density and urban land use that represent urban footprint. The **rural population** includes all the population located outside UAs and UCs.

Core Based Statistical Areas

Core Based Statistical Areas (**CBSAs**) consist of one or more counties or equivalent entities containing at least one urbanized area or urban cluster of at least ten thousand population, plus adjacent counties having a high degree of social and economic integration with the core, reflected by commuting ties with the core county or counties. CBSAs include: (1) **Metropolitan Statistical Areas** that contain at least one urbanized area with a population of at least fifty thousand and (2) **Micropolitan Statistical Areas** that contain at least one urban cluster with a population of at least ten thousand but less than fifty thousand.

Census County Divisions

Census County Divisions (**CCDs**) are subcounty statistical areas delineated by the Bureau of the Census in twenty states. CCDs have no legal or government function and are defined where there are no legally established MCDs or the MCDs are not suitable for the Bureau's purposes.

Census Designated Places

Census Designated Places (**CDPs**) are the statistical counterparts of incorporated places that are designated by the Bureau of the Census to provide data for settled concentrations of population that are identifiable by name but are not legally incorporated by a state.

Census Tracts

Census tracts are small, relatively permanent statistical subdivisions of a county or equivalent entity. Census tracts generally have a population size between 1,200 and 8,000 people, with an optimum size of 4,000 people. When possible, census tract boundaries are maintained over long time periods to allow cross-census comparisons, but they are occasionally split due to population growth or merged because of substantial population decline.

Block Groups

Block groups (**BGs**) are statistical divisions of census tracts that generally contain between six hundred and three thousand people. BGs never cross state, county, or census tract boundaries but may cross the boundaries of other geographic entities.

Census Blocks

Census blocks are statistical areas bounded by visible features, such as streets, roads, streams, and railroad tracks, and by nonvisible boundaries such as the boundaries of political and administrative entities. Census blocks are generally small, for example, a city block bounded on all sides by streets, but may encompass hundreds of square miles in rural areas. Census blocks are the smallest census geographic unit; cover the entire territory of the United States, Puerto Rico, and the Island Areas; and nest within all the other census geographic entities.

Geographic Codes

The **American National Standards Institute codes (ANSI codes)** are standardized numeric or alphabetic codes issued by the ANSI to provide uniform geographic identifiers for all federal government agencies. ANSI issues two types of geographic codes: (1) the **Federal Information Processing Series** (**FIPS**) codes that were previously managed by the National Institute of Standards and Technology and (2) the **Geographic Names**

Information System identifiers, established by the US Geological Survey.

The FIPS codes were known as the **Federal Information Processing Standards** codes before 2005. The Bureau of the Census continues to maintain FIPS codes for its geographic entities, albeit with a revised meaning for the FIPS acronym because the codes are no longer considered a standard. FIPS codes usually are unique within the next highest level of census geography in which a nesting relationship exists. For example, FIPS state, congressional district, and CBSA codes are unique within the United States; FIPS county, place, county subdivision, and sub-MCD codes are unique within a state.

Note

[1] The following discussion draws on US Bureau of the Census (n.d.d).

Appendix B

American Community Survey

The Census Bureau's most important and longest-standing obligation is providing accurate population counts for each State in every year ending in zero.[1] This requirement is contained in the US Constitution, which requires that representation in the House of Representatives be allocated based on each state's population. As a result, a population **census** has been conducted in the United States for every decade since 1790. This census was conducted in 2010 and would be conducted every decade as long as the United States exists.

The 2000 census and previous censuses included "short form" and "long form" surveys. The short form **surveys** included basic questions about the respondent's age, sex, race, Hispanic origin, household relationship, and owner/renter status. The long form survey was collected from a sample of roughly one sixth of the US households. It included the basic questions on the short form and detailed questions about the respondent's social, economic, and housing characteristics. The questions on the long form supplied the data needed for a wide range of government programs related to education, employment, housing and community development, public health care, and assistance programs for low-income families and children. Roughly $300 billion in federal funds are distributed every year in whole, or in part, based on these data.

In 2010, the Census Bureau replaced the long form sample with the **American Community Survey (ACS)**. The ACS includes nearly all the topics that were included in the 2000 long form survey. However, it utilizes a rolling sample and continuous measurements instead of the "point-in-time" approach used in the decennial census. The ACS data also are updated annually, unlike the decennial census data that are released once a decade.

The ACS data provide more current information on the characteristics of America's population than was provided by the decennial censuses. The 2010 census data were collected over several months in 2010 and provided estimates for a specific date, April 1, 2010. In contrast, the ACS data are collected nearly every day throughout the calendar year. The ACS data are based on a sample of three million addresses drawn from the Census Bureau's master address file. This sample is large enough to provide reliable estimates for areas with populations larger than sixty-five thousand from one year's worth of responses. Larger samples from surveys conducted over several years must be accumulated for areas with smaller populations. Three-year estimates were provided for areas with populations of twenty thousand or more from 2005–2007 to 2011–2013. Five-year estimates are provided for all units of census geography down to block groups.

Using Single-Year and Multiyear Estimates

It is important to recognize that the ACS estimates are period estimates that should be interpreted as the average for the entire period. For example, the 2015 single-year estimates are the averages for responses collected for the entire calendar year from January 1, 2015, to December 31, 2015. The five-year 2011–2015 estimates include survey responses for sixty months from January 1, 2011, through December 31, 2015. The single-year and multiyear estimates are calculated from survey responses collected over the entire period, not from the averages of responses collected in individual months or years. That is, the ACS single-year and multiyear estimates are for a specified period, not point-in-time estimates for a particular date, like the decennial census. Similarly, the multiyear ACS estimates are averages for the entire period, not estimates for the period's midpoint.

Most areas have consistent population characteristics throughout the calendar year and their period estimates may not be very different from estimates obtained from point-in-time surveys. However, some areas experience substantial changes in estimates collected at different times of the year. Consider, for example, a small community on the Gulf of Mexico whose population is dominated by retirees in the winter months and by locals in the summer months. The community's age, income, and labor force participation rates will be substantially different for surveys conducted in the winter and summer months. An ACS single-year estimate for the entire year averages out these differences.

Single-year ACS estimates provide more current information about areas that have changing population and/or housing data because they are based on the most current information—data for the previous year. The multiyear ACS data are less current because they provide information from not only the previous year but also data that are two or four years old. The differences between the single-year and multiple year ACS estimates may be small for stable areas but can be substantial for areas that grew or declined substantially over a given period. The multiyear estimates are more reliable than single-year estimates because they are based on larger samples. The larger samples may not be necessary for most applications but can be required for analyzing the population in the sex, age, and racial cohorts for smaller areas.

The ACS estimates are used in three primary ways. They can be used to understand an area's population to, for example, locate schools or hospitals, assess the need for public services, or project the demand for transportation and infrastructure enhancements. The ACS estimates can also be used to compare different areas at the same time, for example, to compare poverty rates in several counties. And they can be used to study changes in an area's population over time, for example, to compare a city's current population structure to what it was five or ten years previously.

Understanding an Area's Characteristics

The five-year ACS estimates are the only option for analyzing areas with populations less than twenty thousand. One-year estimates are provided for areas with populations over sixty-five thousand. The key trade-off to be made in deciding to use single-year or multiyear estimates is between currency and precision. The single-year estimates are generally preferred because they are more relevant to current conditions. However, the single-year data may be highly uncertain for small populations. Fortunately, the Census Bureau provides sample error measures that can be used to determine whether a single-year estimate is acceptable or five-year estimates should be used. The statistics provided to assess the sample variability of the ACS estimates are discussed below.

Making Comparisons

Planners often want to compare the characteristics of one area to those of another area. Thus, for example, they may want to rank the age and income of different communities or compare a study area to the larger areas in which it is located. It is important to use the same type of estimates for making these comparisons to ensure that the data are comparable. That is, single-year estimates should only be compared to single-year estimates; three-year estimates should only be compared to other three-year

estimates; and five-year estimates should only be compared to five-year estimates.[2]

Some comparisons may involve a large area or group and one of its subsets, for example, to compare the estimate for a county to the estimate for the state in which it is located or to compare an estimate for an area's black female population to the corresponding estimate for all females in an area. In these cases, the two estimates are not independent because the estimate for the larger area or population is partially dependent on the estimate for the subarea or subpopulation. If the user is confident that the two estimates are independent, it is acceptable to ignore the partial dependence. However, if the two estimates are positively correlated, a finding of statistical significance will be correct but a finding of a lack of statistically significant differences will not be valid.

Assessing Changes

Single-year estimates are generally preferred for assessing changes in a population over time. This is particularly true because the ACS data for consecutive multiyear estimates will include data for overlapping time periods. Consider for example, a comparison between the 2012–2014 three-year ACS data and the 2013–2015 three-year ACS data. The 2012–2014 estimate includes survey responses data for January 1, 2012, through to December 31, 2014; the 2013–2015 estimate contains responses for January 1, 2013, to December 31, 2015. Both estimates contain data for January 1, 2013, through December 31, 2014, which make up half of the data in each sample. This makes it extremely difficult to isolate the changes in the area's population between 2012 and 2015.

The ACS multiyear estimates provide a smoothing of short-term population trends that may provide a better understanding of changes in an area's population over time. Conversely, the single-year estimates may provide an early indicator of changes in an area's population trend that will only be revealed when the updated multiyear estimates become available. Comparisons across time periods should always use estimates for comparable time periods (e.g., single-year estimates should not be compared to multiyear estimates) because they measure the characteristics of the population in two different ways.

Differences between ACS and Decennial Census Data

There are many similarities and differences between the 2000 decennial long form questionnaire and the ACS survey. They are both based on information collected from a sample of the population, unlike the decennial census that attempts to count everyone living in the United States. The questions asked in the 2000 census and the ACS surveys also are identical, or nearly identical. However, as pointed out previously, the ACS provides estimates for a specified period whereas the census estimates provide a snapshot of the US population on a specific date. Other important differences between the two concern the residence rules and reference periods they employ.

Residence Rule Differences

The survey results from the 2000 census and previous censuses are based on the respondent's "usual residence," that is, where they lived or stayed most of the time, as reported on April 1 of the census year. The ACS asks for the respondent's "current residence," that is, where they are currently living, as long as their stay at that address will exceed two months. The differences between the ACS and decennial census data resulting from the different residence rules are small for most areas. However, some populations may have significantly different responses to questions about their usual residence and their current residence. For example, migrant workers move during the crop season and large resort areas can have substantial numbers of short-term residents. Differences in the locations of these populations during the year can lead to substantial differences in the population estimates for some areas.

Reference Period Differences

Most of the decennial census information is centered on the reference date of April 1 of the census year. However, some of the census questions have their own reference points, such as "last week," which refers to the week before the response is completed. The census data are not collected by contacting everyone in the United States on April 1 of the census year. Instead, they are collected from mail-out, mail-back responses returned in March and April and from nonrespondents who are contacted in May and June. As a result, a "last week" response can refer to any week between March and June of the census year. The ACS surveys are conducted throughout the year, so responses tied to time periods such as "last week" can refer to nearly every week in the year. These differences may be substantial for people with seasonal or part-time employment.

Measures of Sampling Error

The ACS estimates include some errors because they are based on a sample of a population, rather than the entire US population. This uncertainty, the **sample error**, is the difference between the estimates derived from a sample survey and the values that would have been obtained if the entire

population were included in the survey. Because the ACS estimates are based on sample surveys, information about the errors associated with the estimates must be considered when analyzing a single estimate or comparing estimates for different areas, population groups, or time periods. Four measures can be used to express the sample errors in the ACS estimates: standard errors, margins of error, confidence intervals, and coefficients of variation.

Standard Errors

The **standard error (SE)** provides a quantitative measure of the extent to which an estimate derived from a sample survey can be expected to deviate from the value which would have been obtained if the entire population were surveyed. The SE is the foundational error measure from which all the other sample errors are derived. In general, the larger the sample size, the smaller the SE of the estimates produced from the sample. This relationship is the reason ACS estimates for less populous areas are only published for multiple years of data. The larger samples obtained from aggregating data from more than one year reduce the errors of the sample estimates.

Margins of Error

The **margin of error (MOE)** describes the precision of an estimate at a given confidence level. The confidence level associated with the MOE reflects the likelihood that the sample estimate is within a specified distance (the MOE) of the true population value. The MOE provided for all the ACS estimates use a 90-percent confidence interval. That is, users can assume at a 90-percent confidence level that the true population value lies within the MOE distance from the published ACS estimate.

For example, table B.1 reports a portion of the ACS five-year population estimates for Decatur's population for the period between January 1, 2008, and December 31, 2012. Column two reports that the five-year estimate for Decatur's total population is 19,443. This estimate is consistent with the city's 2010 decennial census count, 19,335, for April 1, 2010, near the midpoint of the five-year period. Column four reports that 43.5 percent of Decatur's population over this period is estimated to be male and 56.5 percent is estimated to be female.

Column three reports the MOEs for the population estimates in column two. For example, it reports a MOE of ±27 for Decatur's total population. This value is computed for a 90 percent confidence interval, which means users can be certain with a 90 percent confidence level that the true population average is within a range 27 larger or smaller than 19,443.

| TABLE B.1 | American Community Survey 2008–2012 Five-Year Population Estimates: Decatur, 2008–2012 |

Subject	Estimate	Margin of Error	Percent	Percent Margin of Error
(1)	(2)	(3)	(4)	(5)
Total population	19,443	±27	—	—
Male	8,458	±434	43.5	±2.2
Female	10,985	±433	56.5	±2.2

Source: US Bureau of the Census (2014b).

It is worth pointing out that the margins of error are much larger for the male and female population estimates than they are for Decatur's total population. It may seem that these margin of errors should be smaller because the population values are smaller. However, the opposite is true: the errors are larger because a smaller sample has been collected so the true population value is less certain.

Computing Margins of Error for Other Confidence Levels The Census Bureau's MOE assume a 90 percent confidence level. The published MOEs can be converted to corresponding values for 95 and 99 percent confidence levels by applying equations B.1 and B.2:

$$\text{MOE}_{95} = \left(\frac{1.960}{1.645}\right)\text{MOE}_{\text{ACS}}, \qquad \text{B.1}$$

$$\text{MOE}_{99} = \left(\frac{2.576}{1.645}\right)\text{MOE}_{\text{ACS}}, \qquad \text{B.2}$$

where MOE_{ACS} = positive published ACS margin of error, MOE_{95} = positive 95 percent margin of error, and MOE_{99} = positive 99 percent margin of error.

For example, given equation B.1 and the data in table B.1, the MOE at a 95 percent confidence level for Decatur's average population for the period between 2008 and 2012 is computed as follows:

$$\text{MOE}_{95} = \frac{1.960}{1.645}(27) = 32.$$

That is, users can assume with a 95 percent confidence that Decatur's average population over this five-year period is within a range 32 larger or smaller than 19,443.

Computing the Standard Error from the Margin of Error The SE at a given confidence level can be computed by dividing the published MOE by 1.645. For example, given the data in table B.1, the SE for Decatur's average population for the 2008–2012 period at a 90 percent confidence level can be computed as follows:

$$\text{SE} = \frac{27}{1.645} = 16.$$

Computing Margins of Error for Derived Estimates US Bureau of the Census (2009, A-14–A-17) provides detailed guidelines for computing MOEs for derived estimates such as aggregated counts, derived ratios, and the product of two estimates.

Confidence Intervals

The **confidence interval (CI)** is the range of values that is expected to contain the true population value at an assumed confidence level. The CI at the 90 percent confidence level can be computed by adding and subtracting the SE from the published estimate. For example, the lower bound for Decatur's estimated population between 2008 and 2012 is equal to the estimated value (19,443) minus the MOE (27) or 19,416. The upper bound is equal to the estimated value (19,443) plus the MOE (27) or 19,470. That is, the 90 percent CI for Decatur's average population for the five-year period between 2008 and 2010 is 19,416–19,470. The CI is useful when graphing estimates to show the extent of the sampling error in the estimates and for comparing the precision of different estimates.[3]

Coefficients of Variation

The **coefficient of variation (CV)** measures the relative amount of sampling error that is associated with a sample estimate. The CV is equal to the SE divided by the estimate and usually is expressed as a percentage. The CV is a function of the sample size and the size of the population of interest. In general, the CV decreases as the estimation period increases or the sample size increases. A small CV indicates that the sample error is small relative to the estimate and the user can be more confident that the estimate is close to the true population value.

The CV is computed by applying equation B.3:

$$CV = \frac{SE}{\hat{X}} \times 100, \qquad \text{B.3}$$

where CV = coefficient of variation, SE = computed standard error, and $\hat{X}$ = sample estimate.

For example, given the data in table B.1, the CV for Decatur's estimated population in Decatur over the 2008–2012 period is equal to

$$CV = \frac{27}{19,443} \times 100 = 0.1\%.$$

That is, the errors related to Decatur's estimated population for the 2008–2012 period is an insignificant 0.1 percent. In contrast, the sampling error for Decatur's male population is 3.1 percent.

Notes

[1] The discussion in this Appendix draws extensively on US Bureau of the Census (2009). Additional information on the American Community Survey is available at http://www.census.gov/programs-surveys/acs/guidance.html.

[2] US Bureau of the Census (2009, A-19-A-20) provides a procedure for computing the standard errors for overlapping estimation periods.

[3] Caution must be used to insure that the calculated interval is logical. For example, a small population estimate may have a calculated lower bound that is less than zero. In this case, the lower bound should be set to zero because a negative population does not make sense.

Appendix C

US Data Sources

A wealth of demographic, vital statistics, economic, and spatial data are available online from a wide range of federal and state agencies. Only the most important sources of the data required to apply the analysis and projection methods described in this book will be considered here.

Demographic Data Sources

Current Population Information

The US Bureau of the Census (or Census Bureau), located in the Department of Commerce, is the leading source of statistical information about the nation's population. Information is provided by the Census Bureau for population characteristics such as age, sex, race, Hispanic origin, migration, ancestry, and language use; for population groups such as children, veterans, and the foreign-born; and for particular topics including people's health, education, employment, and income.

The Census Bureau's population data come primarily from the Decennial Census of Population and Housing, from annual surveys such as the **American Community Survey (ACS)** and the **Current Population Survey (CPS)**, and from the periodic Survey of Income and Program Participation.

Decennial Census of Population and Housing Historically, the primary source of information on the US population was the **Decennial Census of Population and Housing**, often referred to as the **Decennial Census**. The Decennial Census provides accurate population for states, counties, places, and other units of census geography in years ending in zero. The data are collected in part to meet the requirement contained in the US Constitution that representation in the United States House of Representatives be allocated based on each state's population at the beginning of every decade. In recent decades, the reference day for the Decennial Census—the date for which the information pertains—has been April 1. After the information is collected, it is coded, compiled, cross-checked, and eventually released in stages from one and a half to three years after the reference day.

The 2000 census and other previous censuses included a "short form" and a "long form." The short form included basic questions about age, sex, race, Hispanic origin, household relationship, and owner/renter status. The long form was sent to a sample of households and included the basic questions on the short form and detailed questions about respondents' social, economic, and housing characteristics. As a result, the Decennial Census is the best source of population data on and before 2000.

The long form was eliminated with the 2010 Decennial Census, and only the short-form information is currently being collected. In 2010, the Census Bureau replaced the long-form sample with the ACS described in appendix B. The ACS data are updated annually, unlike the Decennial Census long-form data that were collected and released every ten years. The ACS data also are collected nearly every day throughout the entire calendar year instead of on one reference day. As a result, the ACS provides more current information on the characteristics of America's population than was provided by the Decennial Censuses. Information about the ACS is available at https://www.census.gov/programs-surveys/acs/.

Current Population Survey The **Current Population Survey (CPS)**, sponsored jointly by the US Census Bureau and the US Bureau of Labor Statistics (BLS), is the primary source of labor force statistics for the US population. The CPS is a survey of about sixty thousand occupied households. The survey is conducted monthly during the week that includes the nineteenth of the month and collects information on respondents' activities during the prior week. In addition to basic labor force questions including employment status, pay rates, and disability status, the CPS often includes supplemental questions on subjects such as annual work activity and income, veteran status, school enrollment, and job tenure. Information about the CPS is available at https://www.census.gov/programs-surveys/cps.html.

Survey of Income and Program Participation The Census Bureau's **Survey of Income and Program Participation (SIPP)** collects data on various types of income, **labor force** participation, social program participation and eligibility, and general demographic characteristics of individuals and households in the United States. The SIPP data are used to evaluate the effectiveness of federal, state, and local transfer programs and analyze the impacts of modifications to those programs. Information about the SIPP is available at https://www.census.gov/sipp/.

Historical Population Information

Historical demographic data for the United States, states, counties, and subcounty areas are readily available online. The primary sources for these data are the National Historical Geographic Information System, the Census Bureau publication archives, and the Census Bureau's historical state and county population report.

National Historical Geographic Information System The National Historical Geographic Information System (NHGIS, https://www.nhgis.org/) provides free online access to summary statistics and GIS boundary files from US censuses and other national surveys from 1790 to the present. The NHGIS does not provide tools for data analysis, mapping, or reporting. Instead, it provides data files that can be used in spreadsheet programs such as Microsoft Excel, in statistical software such as SPSS, and in GIS packages such as Esri's ArcGIS. The NHGIS is managed by the Minnesota Population Center at the University of Minnesota and funded by grants from the National Science Foundation and the Eunice Kennedy Shriver National Institute for Child Health and Human Development.

The NHGIS provides summary data tables as comma separated value (.csv) files from the US Decennial Censuses, the ACS, and many other national data sources. These data include: (1) Decennial Census data for states and counties from 1790 to the present, for census tracts starting from 1910, and for census enumeration areas down to blocks from 1970 onward; (2) ACS data for five-year periods starting from 2005 to 2009, for three-year periods from 2008–2010 to 2011–2013, and for one-year periods from 2010; and (3) County Business Patterns data from 1970.

The NHGIS also provides **time series** tables that combine comparable statistics (e.g., the number of occupied housing units) measured at multiple times (e.g., for all census years from 1970 to 2010) at selected geographic levels (e.g., for states or counties) in one downloadable bundle. It also provides sixty-five time series tables that provide 2000 and 2010 Decennial Census data standardized to 2010 census boundaries. These data make it possible to measure demographic changes for all census enumeration areas down to block groups, even when the census boundaries changed between 2000 and 2010. The data cover a broad range of topics including population counts by age, sex, race, Hispanic or Latino origin, household and group quarters population, and average household size. Further information on using the NHGIS data is provided at https://www.nhgis.org/.

Census Bureau Publication Archives The Census Bureau provides scanned copies of all the Decennial Census printed reports from 1790 to the present on line. The census reports are provided in Adobe portable document format (.pdf). The reports are available at https://www.census.gov/prod/www/decennial.html.[1] Copies of selected reports from the Economic Census and other censuses are available at https://www.census.gov/prod/www/index.html. Census publications released after June 23, 2014 are available at http://census.gov/library/publications.html.

State and County Population Report A special Census Bureau report, *Population of States and Counties of the United States: 1790–1990* (Forstall 1996), contains the population of states and their counties reported by the twenty-one decennial US censuses conducted from 1790 to 1990.

Population Estimates and Projections

The Census Bureau's **Population Estimates Program (PEP)** produces population estimates for the United States, states, counties, minor civil divisions, places, and for the Commonwealth of Puerto Rico and its *municipios*. Demographic components of population change (births, deaths, and migration) and housing unit estimates also are produced for the nation, states, and counties. The estimates are used to allocate federal funds to state, county, and local governments. With each new release of annual estimates, the entire time series of estimates is revised for all years back to the last Decennial Census. The previously published estimates are superseded and archived.

The PEP provides monthly population estimates by age, sex, race, and Hispanic origin for the United States. It produces annual population estimates by age, sex, race, and Hispanic origin and housing-unit estimates for states and counties. Total population estimates are also prepared annually for incorporated places and minor subdivisions in some states. The reference date for all estimates is July 1, unless otherwise specified. The most current population estimates and information on the PEP are available at https://www.census.gov/popest/data/index.html/.

The Census Bureau's **Population Projections Program** produces projections of the United States' **resident population** by age, sex, race, Hispanic origin, and nativity. The 2014 National Projections project the nation's population from July 1, 2014 to July 1, 2060. The projections are produced using the cohort-component method with assumptions and models of future births, deaths, and net international migration. The Census Bureau releases new national projections periodically. It does not provide population projections for states and smaller areas and has no plans to produce them.[2] The most current projections for the US population are available at https://www.census.gov/population/projections/data/national/.

American FactFinder

Population data for the United States, Puerto Rico, and the Island Areas can be easily obtained from the Census Bureau's **American FactFinder** website (https://factfinder.census.gov/faces/nav/jsf/pages/index.xhtml). The American FactFinder site combines data from the following sources: (1) the ACS, (2) the Decennial Census of Population and Housing, (3) the PEP, and (4) the Puerto Rico Community Survey. Information can be downloaded from the American FactFinder as comma-delimited value (.csv) files that can be imported into Microsoft Excel and other spreadsheet and database programs. Customized reports also can be printed and downloaded in .pdf, Microsoft Excel (.xlsx), or Rich Text Format (.rtf) formats. Large data tables also can be downloaded via the file transfer format.

Additional datasets are gradually being added to the American FactFinder interface. For example, data from the quinquennial Economic Censuses are now available, as is information from annual surveys of manufactures, entrepreneurs, and governments. Detailed guidance on using the American FactFinder is provided at http://factfinder.census.gov/faces/nav/jsf/pages/index.xhtml.

Vital Statistics Data Sources

The Center for Disease Control and Prevention, National Center for Health Statistics' (NCHS) Wide-Ranging Online Data for Epidemiologic Research (WONDER) is the primary source for vital statistics data in the United States. The WONDER system (National Center for Health Statistics n.d.a) provides data, reference materials, reports, and guidelines on health-related topics including information about deaths, births, census population data, and many other health-related topics. The data are provided in formats that can be used with word processors, spreadsheet programs, and GIS software and readily summarized and analyzed with dynamically calculated statistics, charts, and maps.

Life Tables

National Center for Health Statistics (n.d.b) provides life tables for the United States population by sex and one-year age cohorts for: (1) the total population, (2) the white and black populations, (3) the Hispanic population, and (4) the white and black non-Hispanic populations. The life tables are available as PDF documents and as Excel spreadsheets. National Center for Health Statistics (n.d.c) provides state life tables by sex and one-year age cohorts for the total population and the white and black populations. The life tables are available as printed reports and as Excel spreadsheets.

Fertility Information

National Center for Health Statistics (n.d.e) provides counts of live births occurring within the United States to US residents and nonresidents. The counts can be obtained by state, county, child's gender and weight, mother's race, mother's age, and many other variables.

Mortality Data

National Center for Health Statistics (n.d.d) provides mortality and population counts for all US counties from 1968 to the present. Counts and rates of death can be obtained by underlying cause of death, state, county, age, race, sex, and year. National Center for Health Statistics (n.d.f) provides detailed mortality data for all US counties. The number of deaths, crude death rates or age-adjusted death rates, and 95 percent confidence intervals and standard errors for death rates can be obtained by place of residence, age group (single year-of age, five-year age groups, ten-year age groups, and infant age groups), race, Hispanic ethnicity, gender, year, and cause of death.

Economic Data Sources

Three federal agencies, the US Bureau of the Census, the Bureau of Economic Analysis (BEA), and the Bureau of Labor Statistics (BLS), provide most of the economic data for local and regional analysis in the United States. Data from these sources are used because the agencies are experienced and widely

trusted and because the data they produce are available consistently across many geographic scales.

US Census Bureau

In addition to the Decennial Census of Population
and Housing and the ACS, the US Bureau of the
Census provides a wealth of useful economic data.

The **Survey of Business Owners** (http:// www
.census.gov/econ/sbo) provides information regarding some 1.75 million businesses and their owners. The Census Bureau also conducts the **County
Business Patterns (CBP)** program (http://www
.census.gov/econ/cbp). The CBP provides annual
data on the number of establishments and employees
during the week of March 12, first quarter payroll,
and annual payroll data for states and counties.

Zip Code Business Patterns (ZBP) provides the same data for fifty thousand five-digit
ZIP code areas nationwide. The CBP/ZBP com-
piles these data from the administrative records of
the Internal Revenue Service, the Social Security
Administration, and the BLS, providing highly
reliable measures of the country's economy. The
CBP/ZBP data are available from the American
FactFinder website (https://factfinder.census.gov
/faces/nav/jsf/pages/index.xhtml).

The CBP and the ZBP data cover most of the
country's economic activity but do not provide data
on self-employed individuals, employees of private
households, railroad employees, agricultural production employees, and most government employees. The CBP has been produced as a consistent,
annual series since 1964; data for ZIP code areas
have been available since 1994. The CBP/ZBP data
can be accessed online from 1998. Printed reports
were published annually through 2004 and at irregular intervals since 1946.

Bureau of Economic Analysis

Like the Census Bureau, the **Bureau of Economic
Analysis (BEA)** is housed in the Department of
Commerce. The BEA produces the US national
economic accounts, including data on domestic
production, personal income, consumer spending,
imports and exports, and industry transactions
(http://bea.gov). The BEA also estimates much of
this information for smaller geographies, including
states, metropolitan areas, and counties.

Bureau of Labor Statistics

The **Bureau of Labor Statistics (BLS)** is located
in the Department of Labor and provides labor
and production information (http://www.bls

.gov). BLS data series useful for local and regional
economic analysis include measures of employment
and unemployment, prices and inflation, wages and
other compensation, and productivity.

Each state operates a designated Labor Market
Information (LMI) agency. The LMI agencies
collect state and local data that are provided to
the BLS, which aggregates the data to release as
national-level statistics. The LMI agencies often
provide data to state agencies and directly to the
public.

Other Economic Data Sources

Other federal data providers include the Economic
Research Service of the Department of Agriculture
(http://www.ers.usda.gov/), the Bureau of
Transportation Statistics within the Department of
Transportation (http://www.rita.dot.gov/bts/),
and the HUD User interface of the Department of
Housing and Urban Development (huduser.gov).

Every state operates an agency that collects,
estimates, and disseminates economic data at the
state and substate level. A substantial portion of this
information overlaps with the data available from
federal agencies since many federal data programs
rely on information gathered by the state agencies.
However, the state agency may offer other kinds of
information, more recently updated data, or information organized into different spatial units such
as substate planning regions or workforce investment regions. Economic data may also be available
from regional and local units of government, private
firms, and a wide range of nonprofit and quasi-public
organizations.

Most economic data are collected by the
place of business (i.e., the physical location of a
firm or establishment). However, some economic
data reported for individuals pertain to residential locations—where people live—rather than the
places where people work. For example, the data
provided in the US Decennial Census of Population
and Housing and the ACS supply a host of information on the number of workers, their occupations
and income, and other variables, organized by the
counties, places, and census tracts where the workers reside. Comparing data compiled by place of
work and by place of residence is often a useful way
to examine commuting patterns and to construct
other measures of interest.

Spatial Data Sources
US Government Data Portal

The US Government's data portal (https://www .data.gov/) provides free, online access to over 192,000 data sets, tools, and resources for conducting research, developing web and mobile applications, and designing data visualizations.

National Historical Geographic Information System

The National Historical Geographic Information System (NHGIS, https://www.nhgis.org/), managed by the Minnesota Population Center at the University of Minnesota, provides free online access to the following census boundaries as Esri shapefiles: (1) states counties and states since 1790, (2) census tracts since 1910, (3) metropolitan areas since 1950; (4) places and county subdivisions since 1980; and (5) all standard census reporting areas since 1990.

TIGER/Line Files

The Census Bureau provides census geography as TIGER/line files (https://www.census.gov/geo /maps-data/data/tiger.html). TIGER is an abbreviation for the Topologically Integrated Geographic Encoding and Referencing format used to store the census boundaries and related spatial features (Klosterman 1991). The Census Bureau releases the TIGER data in several formats, including Esri shapefiles and geodatabases. Many state offices provide census geography files available for download, sometimes with limited demographic data. These files may be helpful for local use when the state agency has converted the data to the official state coordinate systems.

Open Street Map

Open Street Map (https://www.openstreetmap .org/) is a global effort to map the world's streets and transportation features. It is particularly useful in countries where spatial data are not freely available.

USGS National Map

The US Geological Service (USGS) National Map (https://viewer.nationalmap.gov/launch/) provides tools for viewing and downloading an extensive array of spatial data. Data available on the USGS National Map include: (1) current and historical topographic maps, (2) the National Boundary Dataset, (3) elevation data, (4) hydrography and watersheds, (5) one-foot and one-meter remote sensed imagery, (6) the National Land Cover Database, and (7) the National Structures Dataset.

NRCS Geospatial Data Gateway

The US Department of Agriculture, Natural Resources Conservation Service (NRCS) Geospatial Data Gateway (https://datagateway.nrcs.usda .gov/) makes a variety of data available in one location, organized by state and county. The available data varies for different locations but may include (1) cadastral boundaries, (2) geology, (3) hydrography and hydrologic units, (4) land use/land cover, (5) orthography, and (6) soils.

NRCS Web Soil Survey

The NRCS Web Soil Survey (http://websoilsurvey .sc.egov.usda.gov/App/HomePage.htm) is the preferred source for soils data in the United States. The Web Soil Survey (WSS) provides tools and guidance that help users identify and download the data they require. The complete soils dataset is quite complex and difficult to use, making the WSS a valuable resource for most users.

National Land Cover Database

The USGS makes land cover data available for the coterminous United States through the Multi Resolution Land Characteristics Consortium (http://www.mrlc.gov/nlcd2011.php). The land cover data are derived from Landsat remote sensing imagery classified into a sixteen-class land cover classification system. Thirty-meter resolution data are available for several years. Other data are provided, including: (1) land cover change data, (2) percent impervious surface, and (3) percent tree canopy.

USGS Earth Explorer

The USGS Earth Explorer (https://earthexplorer .usgs.gov/) indexes and distributes a wide variety of aerial and satellite imagery for the entire globe. These data are particularly useful for data-poor areas or where there are data access restrictions.

Notes

[1] A few documents or portions of documents are unavailable due to the loss of or damage to archival paper copies.

[2] Many states produce their own projections of population and demographic components for state and substate geographies; these should be used cautiously since neither the methods nor the resulting estimates are consistent across states.

Glossary

accuracy The magnitude of the difference between the projected and actual value of a variable. Accuracy can be measured in either absolute or percentage terms.

adjusted-share-of-change method A projection method that projects an area's population by adding the area's adjusted projected population growth to its current population.

adjusted-share employment projection methods Adjustments to the standard share employment projection methods that attempt to account for an area's economic advantages or disadvantages.

age-specific rate The number of demographic events (e.g., births or deaths) for a specific at-risk population, for example, the number of births to women between the ages of twenty and twenty-five.

age-specific fertility rate The probability that the women in an age cohort will give birth in a time period.

aggregate consistency An aggregation of projected or estimated values that are consistent across the relevant levels of sectoral or geographic aggregation.

American Community Survey (ACS) An annual survey of the American population conducted by the US Bureau of the Census that collects information previously contained on the Bureau's *decennial census* longform surveys.

American FactFinder A website maintained by the US Bureau of the Census that provides easy access to the Census Bureau's data for the United States, Puerto Rico, and the Island Areas.

American National Standards Institute (ANSI) codes A standardized set of numeric and alphabetic codes issued by the American National Standards Institute (ANSI) to ensure uniform identification of geographic entities through all US federal government agencies. These standards replace the Federal Information Processing Standards (FIPS) codes previously issued by the National Institute of Standards and Technology (NIST).

area A map *feature* that bounds a region at a given scale, such as a country on a world map or a census tract on a city map.

areal interpolation A process that uses the spatial and attribute data for one set of polygons to estimate the values for a second set of polygons, for example, estimating the population of school districts from population data for block groups.

area-weighted interpolation An *areal interpolation method* that uses the area of the overlapping source and target zones to convert data from the *source zones* to the *target zones*.

arithmetic mean The sum of a collection of numbers divided by the number of numbers in the collection.

aspect A position facing in a direction.

asymptote An upper or lower limit that the values of an *asymptotic curve* approach but never equal.

asymptotic curve A curve whose values approach but never equal an upper or lower limit, the *asymptote*. The Gompertz curve is an asymptotic curve.

at-risk population The group of people to whom a demographic event (e.g., birth or death) could potentially occur.

attribute data Nonspatial information for *features* such as their magnitude, classification, or ownership. Attribute data are usually stored in a *database* and displayed in a *map legend*.

attribute query A *query* that selects *features* based on their *attributes*. Attribute queries can normally be answered by *querying* a *database*.

attribute table A *database* containing information about a set of *features*, usually arranged so that each row represents a *feature* and each column represents a feature *attribute*.

average See *arithmetic mean*.

average household size The average number of persons per *household*, that is, the *household population* divided by the number of households.

average projection The average of a number of individual projections. The averages can include any number of projections, different trend curves, both trend and share projections, or different fit periods.

bafflegab Professional jargon that confuses more than it clarifies. Writing that sounds impressive while saying nothing. The term was coined in 1952 by Milton A. Smith, who defined it as "multiloquence characterized by a consummate interfusion of circumlocution and other familiar manifestation of abstruse expatiation."

bar chart A chart with vertical or horizontal rectangular bars whose lengths are proportional to the values they represent. Bar charts are used to represent categories of data that do not make up a whole such as a city's population at different dates.

base year The year corresponding to the earliest data used to make a *projection*.

bias A systematic error in which deviations between the projected and actual values tend to be in one direction, that is, on average the projections are overestimates or underestimates.

binary selection method A *land suitability analysis* method that uses binary (yes/no) rules to divide the suitability *factor types* and locations into two groups: suitable and unsuitable..

block group A *statistical entity* that is a subdivision of a *census tract* and a combination of *census blocks*. A block group is the smallest geographic unit for which the US Bureau of the Census tabulates sample data. About seven hundred people reside in a block group.

Boolean operator A logical operator used in the formulation of a *Boolean expression*. Boolean operators include (1) AND, which specifies a combination of conditions (A and B must be true); (2) OR, which specifies a list of alternative conditions (A or B must be true); (3) NOT, which negates a condition (A but not B must be true); and (4) XOR (exclusive or), which specifies mutually exclusive conditions (A or B may be true, but A and B cannot both be true).

Boolean query An *attribute* or *spatial query* that uses *Boolean operators* to represent relationships between features. The operations are named after George Boole (1815–1864), the French mathematician who developed this approach.

bottom-up adjustment An *aggregate consistency* adjustment that preserves the values for the lower level of aggregation and replaces the top-level value by the sum of the lower-level values.

buffer A *polygon* enclosing a point, line, or area at a specified distance.

buffering A *GIS* operation that creates two zones around a point, line, or polygon: the buffer zone that is within a specified distance of a feature, and a second zone that is outside the specified distance.

Bureau of Economic Analysis (BEA) An agency housed in the US Department of Commerce that produces the US national economic accounts, including data on domestic production, personal income, consumer spending, imports and exports, and industry transactions.

Bureau of Labor Statistics (BLS) An agency located in the US Department of Labor that provides labor and production information.

cadaster A map of land ownership that provides a record of land ownership to facilitate land taxation.

Cartesian coordinate system A two-dimensional measurement system that locates features on a plane based on their distance from an origin (0,0) along two perpendicular axes. The Cartesian coordinate system is named for the French mathematician and philosopher René Descartes (1596–1650).

cartography The art and science of preparing maps.

cartographic data *Information* on where *features* are located and how they can be displayed on a map. For example, the cartographic data needed to display a line include its beginning and ending points, its type (solid, dotted, or dashed), and color.

categorical data *Data* that are measured on a *nominal* or *ordinal scale* that allow observations to be assigned to categories but do not allow arithmetic operations such as computing means and ratios.

category A group of similar things.

census The collection of *data* about the entire population of a country or region at a specific point in time, as opposed to a *survey* that uses *data* for a sample of the *population* to *estimate* values for the entire population.

census block The smallest *statistical entity* that the US Bureau of the Census uses to tabulate decennial *census* data. Many census blocks correspond to city blocks bounded by streets but blocks in rural areas may include several square miles and have boundaries that are not streets. The next larger census geographic unit is the *block group*.

Census County Division (CCD) A subcounty statistical areas delineated by the Bureau of the Census in twenty states where there are no legally established *Minor Civil Divisions (MCDs)* or the MCDs are not suitable for the Bureau's purposes.

Census Designated Place (CDP) The statistical counterparts of *incorporated places* that are designated by the Bureau of the Census to provide data for settled concentrations of population that are identifiable by name but are not legally incorporated by a state.

census division Nine groupings of states and the District of Columbia that subdivide the four Census Regions.

census region Four groupings of the fifty states and the District of Columbia: Northeast, Midwest, South, and West.

census tract A relatively permanent statistical subdivision of a *county* that the US Bureau of the Census uses to tabulate census and survey data. Census tract boundaries normally follow visible features but may follow governmental unit boundaries and other non-visible features in some instances. Census tracts typically contain between 2,500 and 8,000 inhabitants. The next larger census geographic unit is a county; the next smallest unit is a *block group*.

child-to-woman ratio The number of children born to the women in a *cohort* over a time period divided by the female population in the cohort.

choropleth map A *map* that divides an area into general purpose zones that can be used to map several variables. For example, a choropleth parcel map shows the parcel boundaries that can be used to display the parcels' land uses, zoning, assessed values, and many other variables. See *dasymetric map*.

clip A *GIS* operation that combines two *polygon* layers to create a new layer that includes only the parts of the input layer that lie within the clip layer.

coefficient of variation A measure of relative variation that is computed by dividing the *standard deviation* for a set of values by the *mean* value.

cohort A group of people with the same demographic characteristics, for example, are within the same age interval and have the same sex, race, and ethnicity.

cohort change ratio The population in a *cohort* in a year divided by the population in a previous year.

cohort survival rate The probability that the members of a cohort will survive and move into the next higher age cohort.

cohort-component method A population projection method that projects the fertility, mortality, and migration components of population change for the age, sex, and race cohorts of a population.

cohort-survival method See *Hamilton-Perry method*.

cohort life table A *life table* based on the mortality rates experienced by an actual population, for example, the US population born in 1900 during their entire lifetime. See *Current life table*.

Competitive Shift See *Local Shift*.

components of population change The demographic events that determine population change: fertility, mortality, and migration.

composite projection Projections that use a combination of methods that have been found to perform better for areas with particular characteristics.

computed value The *dependent value* for a curve when the *independent variable* is equal to a value *X*.

computer-aided design or computer-aided drafting (CAD) Computer systems for preparing drawings and maps, typically for engineering and architectural applications. An alternate definition is CADD, computer-aided design and drafting.

confidence interval An estimated range of values that is likely to include an unknown value.

Consolidated Metropolitan Statistical Area An obsolete term previously used to identify urbanized counties in the United States. See *Metropolitan Statistical Area* and *Micropolitan Statistical Area*.

constant-share method A projection method that assumes an area's share of the population or employment a larger *context area* in which it is located is constant over time.

constrictive population pyramid A *population pyramid* that has fewer people in the younger age groups than in the older age groups.

context area A large area that contains the *study area*.

continuous data Data uninterrupted in space, time, or sequence such as slopes or distances.

continuous features *Features* that represent seamless geographic phenomena. Continuous features include values that vary continuously across space such as temperature and elevation and phenomena with indistinct boundaries such as natural vegetation areas.

coordinate system A reference framework used to define locations in either two or three dimensions. See *Cartesian coordinate system* and *Geographic coordinate system*.

Core Based Statistical Area (CBSA) A geographic area defined by the US Office of Management and Budget consisting of a county that contains one or more urbanized area with a population of ten thousand and adjacent counties that have a high degree of economic and social integration with the core area. See *Metropolitan statistical area* and *Micropolitan statistical area*.

county The primary legal divisions of most states.

coarse scale A map that covers a large area containing large features and shows little detail (e.g., a national map). See *Fine scale*, *Large scale*, *Map scale*, and *Small scale*.

County Business Patterns (CBP) A US Bureau of the Census program that provides annual data on the number of establishments and employees during the week of March 12, first quarter payroll, and annual payroll data for states and counties. See *ZIP Code Business Patterns (ZBP)*.

coverage A topological vector data format used in *Esri*'s *GIS* products.

criteria *Attributes* used to evaluate alternatives in *multicriteria evaluation*.

current life table A *life table* based on the mortality rates experienced by the different components of a population over a short time period, for example, all women residing in the United States between 2009 and 2011. See *Cohort life table*.

Current Population Survey (CPS) The primary source of labor force statistics for the US population. The CPS is a monthly survey of about sixty thousand households sponsored jointly by the US Census Bureau and the US Bureau of Labor Statistics (BLS).

curve extrapolation   The process of extending past trends to project future values.

curve fitting   The process of finding the mathematical expression that best describes a set of data measured over time.

dasymetric interpolation   A spatial interpolation method that uses axillary information to improve the *target zone* estimates produced by the *area-weighted interpolation*. See *dasymetric map*. See *choropleth map*.

dasymetric map   Maps that display statistical data in meaningful spatial zones. Dasymetric maps can be preferable to *choropleth maps* that show data by enumeration zones, because the dasymetric zones more accurately represent the underlying data distributions.

database   An organized collection of related *data* that is normally stored in one or more tables consisting of *records* (or rows) and *fields* (or columns).

database management system (DBMS)   A computer system for creating and maintaining *databases*. Database management systems provide tools for adding, storing, changing, deleting, and retrieving *data*.

decennial census   See *Decennial Census of Population and Housing*.

Decennial Census of Population and Housing   A complete count of the US population conducted by the US Bureau of the Census in years ending in zero.

dependent variable   The variable in an equation whose values are assumed to be determined by changes in the *independent variable*. The dependent variable is normally denoted by Y and plotted along the vertical axis of a graph.

Differential Shift   See *Local Shift*.

digital elevation model (DEM)   A digital model that uses a grid of regularly spaced elevation values to represent the Earth's surface.

discrete feature   A *feature* that is unique and identifiable. Discrete features include (1) features with clearly defined boundaries such as buildings, (2) locations such as addresses, (3) linear features such as roads, and (4) administrative units such as *census tracts* and land *parcels*.

employment data   Place-of-work information on the people who work in an area and live inside or outside the area.

enterprise   One or more than one location performing the same or different types of economic activities. An enterprise may operate a single *establishment* or multiple *establishments*.

equal interval classification   A method of grouping values that divides a set of attribute values into groups that contain an equal range of values. For example, if three classes are used to group features whose values range from 1 to 300, the equal interval method will use class ranges of 1–100, 101–200, and 201–300. See *manual classification, natural breaks classification, quantiles classification*, and *standard deviation classification*.

Esri Inc. (Environmental Systems Research Institute Inc.)   A privately held GIS company founded by Jack and Laura Dangermond in 1969. Esri is located in Redlands, California, and the developer of *ArcGIS*, the world's most widely used GIS system.

establishment   A place where business is conducted, services are provided, or industrial operations are performed. Establishments include factories, stores, hotels, movie theaters, mines, farms, airline terminals, sales offices, warehouses, and central administrative offices.

estimate   A calculation of the current or past value for a variable. See *Projection (mathematical)*.

EXCLUSIVE (or XOR) query   A *Boolean query* that selects all of members of overlapping sets that satisfy one condition or a second condition but not both.

expansive population pyramid   A *population pyramid* that has many more people in younger age groups than in older age groups.

extrapolation   The process of using mathematical or graphical procedures to determine values that fall beyond the last known value for a series of values.

factor types   The set of possible values for a *suitability factor*. For example, there could be five different slope types.

factor weight   See *Importance weight*.

feature   An entity or activity that can be located on a *map* at a given map scale, for example, a parcel on a property map.

feature attribute table   An *attribute table* that is created and maintained by a *GIS* system to provide a direct connection to a *cartographic database*. The connection is maintained by a unique feature code that is stored in the cartographic database and the feature attribute table.

feature code   A unique identifying code assigned to the *features* in a GIS *database*.

Federal Information Processing Series (FIPS) codes   A standardized system of numeric and alphabetic coding issued by the US National Institute of Standards and Technology (NIST) to ensure uniform identification of geographic entities through all US federal government agencies. The FIPS codes were known as the Federal Information Processing Standards codes before 2005. The

Bureau of the Census continues to maintain FIPS codes for its geographic entities, albeit with a revised meaning for the FIPS acronym.

field   A column in a *data table* that stores the values of a single *attribute* for all the records (rows) in the table, for example, the assessed value of all parcels in a tax assessor's *database.*

fine scale   A map that covers a small area containing small features and shows a lot of detail (e.g., a neighborhood map). See m*ap scale,* c*oarse scale,* l*arge scale,* and s*mall scale.*

FIPS codes   See *Federal Information Processing Series codes.*

firm   See *enterprise.*

fit period   The time interval used to fit a trend curve, that is, the period between the *base year* and the *launch year.*

floating point   A number with a decimal point. The term is derived from the fact that there is no fixed number of digits before and after the decimal point; that is, the decimal point can float.

forecast   The *projection* that is selected as the most likely to accurately predict the future value of a variable.

functional economic area   An idealized region that encompasses all the activities, flows, and influences relevant to the economy in question.

fuzzy overlay method   A *land suitability analysis* method that that uses "fuzzy logic" to specify the extent to which a value belongs (or does not belong) to a set.

generational life table   See *Cohort life table.*

Gini coefficient   A measure of the inequality among values of a frequency distribution (for example, levels of income or wealth).

geodatabase   An *object*-based data model developed by Esri Inc.

geographic coordinate system   A reference system that uses *latitude* and *longitude* to define locations on the Earth's surface.

geographic information system (GIS)   A computer system for capturing, storing querying, analyzing, and displaying spatially related data. GIS can be used to view and manage information about spatially located entities and activities, analyze spatial relationships, and model spatial processes.

geometric curve   A curve that describe phenomena that grow by a *constant rate.*

geometric mean   The nth root of the product of a set of values.

Geographic Names Information System   A system of standardized numeric and alphabetic codes

maintained by the US Geological Survey that provide uniform geographic identifiers for all federal agencies.

georeferenced raster   Cell-based systems that are made up of cells of a known size that are tied to locations in space and stored as a matrix of numeric values. Examples include the USGS *digital elevation models (DEMs).*

georelational data model   A *vector data model* developed by Esri, Inc. that stores *spatial data* and *attribute data* in separate but related data tables.

GIS   See *Geographic information system.*

Gompertz curve   An *asymptotic curve* that recognizes that a region's population or employment will eventually approach an upper or lower growth limit or *asymptote.*

goodness of fit statistic   A measure of how well a set of computed curve estimates correspond to the observed data values.

gross migration   The number of people who move into or out of an area. See *Net migration.*

group quarters population   All the people living in group quarters, that is, not living in *households.* The group quarters population includes people residing in college dormitories, military quarters, nursing facilities, and correctional institutions.

growth increment   The change in the *dependent value* of an equation for a unit change in the *independent variable X.*

growth rate   The change in the *dependent value* of an equation divided by the initial value of the dependent variable Yt.

Hamilton-Perry method   A simplified cohort-component method that projects the aging of a population over time without directly considering the three components of population change.

histogram   A chart with vertical or horizontal rectangular bars whose lengths are proportional to the values they represent. The bars of histograms represent the frequency of continuous variables that are grouped into discrete intervals (or bins).

household   All the people who occupy a *housing unit* as their usual place of residence. The number of households is equal to the number of *housing units* minus the number of vacant housing units.

household population   Everyone who uses a housing unit as their usual place of residence. It is equal to the total population minus the *group quarters population.*

household size   The total number of people living in a *housing unit.*

housing unit   A house, an apartment, a mobile home or trailer, a group of rooms, or a single room occupied as separate living quarters, or if vacant, intended for occupancy as separate living quarters. The occupants of separate living quarters live separately from other individuals and have direct access to their residence from outside the building or through a common hall.

housing unit method   A method that estimates an area's population by multiplying the number of occupied *housing units* (or *households*) by the *average household size* and adding the *group quarters population*.

IDENTITY (or NOT) query   A *Boolean query* that selects the members of overlapping sets that satisfy one condition but not a second condition.

image   A visual representation of something: such as a likeness of an object produced on a photographic material or an electric devise such as a computer screen.

incorporated place.   A legally established government entity created by a state to provide governmental functions for a concentration of people.

independent variable   One or more variables in an equation whose values are assumed to determine changes in the *dependent variable*. The independent variable is normally represented by X and plotted along the horizontal axis of a graph.

industry   A collection of business units (usually *establishments* but sometimes *firms*) that are grouped together because they produce similar products or use similar production methods or technologies.

Industry Mix   The local employment change that would occur if the local industry's employment change matched the rate of change for the industry in the reference region relative to the change in the entire reference region's economy.

in-migrant   A person who moves into a place from another place in the same country. An immigrant is a person who moves into a country from a different country.

integer   A whole number (not a fractional number) that can be positive, negative, or zero.

interpreted data   Data that are based on expert judgment. An example is the suitability scores that use expert judgment to combine user-defined suitability ratings and weights.

intersect   A GIS overlay operation that combines two polygon layers to create a new polygon layer that includes only areas that are common to the two input layers.

INTERSECT (or AND) query   A *Boolean query* that selects all of members of overlapping sets that satisfy one condition and a second condition.

interpreted data   Values based on expert judgment.

interval scale   A measurement system that has equal intervals between values but not a meaningful 0 value. An example is temperature measured on a Fahrenheit or Celsius scale. Interval-scale data can be added, subtracted, or to compute averages but cannot be multiplied or divided. See *nominal scale*, *ordinal scale*, and *ratio scale*.

labor force   The sum of the full- or part-time employed and the unemployed population at least 16 years old that is actively seeking employment.

land suitability analysis   The process of determining the suitability of one or more pieces of land for one or more uses.

land supply analysis   An analysis used in urbanized areas to estimate the supply and development capacity of the area's vacant, partially utilized, and underutilized land parcels.

large scale   This term has two somewhat contradictory meanings. For noncartographers, a large-scale map covers a large area and contains large features with little detail. For cartographers, a large-scale map has a large *representative fraction* (e.g., 1:1,000) and covers a small area with a high level of detail. This book avoids the term and uses *coarse scale* to refer to maps that cover a large area and show large features with little detail (e.g., a national map) and *fine scale* to refer to maps that cover small areas and show small features with a lot of detail (e.g., a neighborhood map).

latitude   Angular measurements north and south of the equator that are represented on a globe as lines parallel to the equator that circle the globe.

launch year   The year corresponding to the most recent data used to make a *projection*.

layer   A collection of features that are related to a common *theme* and have the same geometric type (e.g., points, lines, or areas).

legal and administrative entities   Government units created by state and federal legislation, court decisions, legal actions, and similar official actions.

least squares criterion   A decision rule specifying that the curve which best fits a given set of observations must minimize the sum of squared deviations between the observed values for the dependent variable Y and the curve estimates Y_C.

life table   A summary statistical table that records the mortality and survivorship rates and life expectations of a population. See *Cohort life table* and *Current life table*.

line   A feature geometry representing map features that have length but no width or are too narrow to be visible at a given map scale.

line-in-polygon overlay   An *overlay* operation that combines a line layer and a polygon layer to create line segments that are split at the polygon boundaries and contain the attribute values from the polygon within which they are located.

line clip overlay An *overlay* that combines a *line* layer (the input layer) and a *polygon* layer (the clip layer) to create an output layer consisting of the portion of the line layer that is inside the polygon layer.

line erase overlay An overlay that combines a *line* layer (the input layer) and a *polygon* layer (the erase layer) to create an output layer made up of the portion of the line layer that are outside the polygon layer.

line graph Graphs that generally use lines to represent the variation in variables such as population or employment over time. In these cases, the horizontal axis records points in time and the vertical axis records the values at different points in time.

linear curve The simplest and most widely used trend curve. The curve has a constant slope and plots as a straight line.

linear referencing system A spatial reference system that identifies locations along a network by measuring the distance from a defined point in each direction along a defined path along the network.

local economy An area's economy that includes local businesses, public or not-for-profit organizations, in-commuters, consumers who buy goods and services from local firms, and the area's residents who work in the place, commute to jobs outside the place, draw pensions from nonlocal sources, or receive transfer payments from the government.

location quotient A measure of an area's specialization in an attribute relative to a reference area's specialization in the attribute. For example, an employment location quotient is the local share of the employment in an industry divided by the industry's employment share in the reference area.

Local Shift The local employment change that reflects differences between an industry's growth or decline in the study area and its growth or decline in the reference area.

longitude Angular measurements east or west of the Greenwich meridian passing near London that are represented on a globe by circles intersecting the north and south poles.

Lorenz curve Graphs that are generally used to represent the inequality of a population's income or wealth distribution.

MSA See *Metropolitan Statistical Area.*

manual classification A method of grouping values that allows the user to manually define the upper and lower limits for each class. See *equal interval classification,* n*atural breaks classification,*

q*uantiles classification,* and s*tandard deviation classification.*

map A two- or three-dimensional model of all or part of the Earth or other objects that can be reproduced on paper or displayed on a computer screen.

map legend A visual explanation of the symbols (points, lines, and areas) used on a map. It typically includes a sample of each symbol and a short description of what the symbol means.

map overlay method A *land suitability analysis* method that overlays two or more map layers to identify the most suitable locations. The method was popularized by Ian McHarg.

map scale. The ratio of a distance on a map to the corresponding distance on the ground. See c*oarse scale* and f*ine scale.*

MAPE See *Mean absolute percentage error.*

margin of error (MOE) An accuracy measure that describes the precision of an estimate at a given confidence level.

mean See *arithmetic mean* and *geometric mean.*

mean absolute error (MAE) The average *residual error,* ignoring signs.

mean absolute percentage error (MAPE) The average of the sum of the percentage errors for a data set, ignoring signs.

mean error The average *residual error,* recognizing positive and negative errors.

mean percentage error (MPE) The average percentage error, considering signs.

median The value of the middle item in a series of items arranged in order of magnitude. The median for an even number of items is the average of the two middle values. See *mean.*

Metropolitan Statistical Area (MSA) A geographic area defined by the US Office of Management and Budget consisting of a county that contains one or more *urbanized area* with a population of fifty thousand and adjacent counties that have a high degree of economic and social integration with the core area.

Micropolitan Statistical Area A geographic area defined by the US Office of Management and Budget consisting of a county that contains one or more *urban clusters* with a population between ten thousand and fifty thousand and adjacent counties that have a high degree of economic and social integration with the core area.

Minor Civil Division (MCD) The primary governmental or administrative divisions of a county in twenty-nine states and the county equivalents in Puerto Rico and the Island Areas. MCDs have many designations including: townships, towns, parishes, reservations, and barrios.

mortality count method A *cohort component* method that computes survival rates directly from mortality and population data specific to the study area.

multi-criteria evaluation (MCE) A procedure for combining several, potentially conflicting, criteria that incorporate user-specified preference weights and trade-off functions.

NAICS See *North American Industry Classification System.*

National Historical Geographic Information System (NHGIS) A website managed by the Minnesota Population Center at the University of Minnesota that provides free online access to summary statistics and GIS boundary files from US censuses and other national surveys from 1790 to the present.

National Shift See *Reference Shift.*

natural breaks classification A method of grouping values that creates groups that minimizes the difference between the members of each group and maximizes the difference between groups. See *equal interval classification, manual classification, quantiles classification,* and *standard deviation classification.*

natural increase The excess of births over deaths. Natural decrease is the excess of deaths over births.

network A collection of interconnected lines that represent paths of movement. Examples include roads, rivers, and utility distribution systems.

nominal scale A measurement system that assigns labels to features that do not express order or magnitude and do not allow arithmetic operations such as addition, subtraction, or computing means and ratios. An example is the *ZIP codes* assigned to mail delivery areas in the United States. See *interval scale, ordinal scale,* and *ratio scale.*

nominal value See *nominal scale.*

non-georeferenced raster An *image* that is not tied to locations in space and does not store attribute information on the objects in the image or topological information on how the objects are related to each other.

nonspatial attribute table An *attribute table* that does not have a direct connection to a cartographic database. Nonspatial attribute tables are not created or maintained by a GIS system. Examples include dBase tables, Excel tables, and character delimited text files.

nonspatial query See *Attribute query.*

nonspatial data *Data* without inherently spatial qualities that do not have a direct connection to a cartographic database.

North American Industry Classification System (NAICS) A system for classifying business establishments that was developed jointly by the US Economic Classification Policy Committee (ECPC), Statistics Canada, and Mexico's Instituto Nacional de Estadística y Geografía to allow for a high level of comparability in business statistics among the North American countries. NAICS replaced the *Standard Industrial Classification (SIC)* codes in 1997.

numeric data *Data* that are measured on an *interval scale* or a *ratio scale* that allows arithmetic operations such as computing means and ratios.

numeric value See *numeric data.*

observation interval The time interval between the dates in the *observation period.*

observation period The period between the first and last years for which data are available.

observed value The measured value for the *dependent variable* when the *independent variable* is equal to *X.*

occupational data Place-of-residence information on the people who live in an area and work inside or outside the area.

one-hundred-year floodplain Areas that can be expected to flood at least once during a one-hundred-year period. The US Federal Emergency Management Agency (FEMA) distributes maps for the one-hundred-year floodplains in the United States.

ordinal combination method A *land suitability analysis* method that ranks the *factor types* on their suitability for that land use, assuming the suitability factors are equally important.

ordinal scale A measurement system in which features have been assigned a meaningful order but where the intervals between features are not necessarily equal. An example is the order in which contestants complete an athletic competition. Ordinal-scale data cannot be added, subtracted, multiplied, divided, or used to compute averages. See *interval scale, nominal scale,* and *ratio scale.*

ordinal number See *ordinal scale.*

out-migrant A person who moves from a place to another place inside the same country. An emigrant is a person who moves out of an area and into another country.

overlay A GIS operation that combines the geometries and attributes of two or more input layers to create an output layer.

parabolic curve   A curve that has a constantly changing slope and one bend. Given a sufficient range of values, the parabolic curve is positively inclined (or "upward sloping") in one section and negatively inclined (or "downward sloping") in another section.

parcel   A piece or unit of land that conveys land ownership and other rights.

parameter   A constant that defines the relationship between the *independent* and *dependent* variables in an equation. For example, a linear curve is defined by the values of two parameters: *a* and *b*.

period life table   See *Current life table.*

pie chart   A circular chart divided into sectors like slices of a pie. The size of the sections represents the proportion that a value contributes to the total for all values.

place of residence   The location where a person lives and sleeps most of the time.

planning support system (PSS)   A computer system that provides specialized analysis and visualization tools that support the planning process.

plant   See *establishment.*

point   A feature geometry representing features that are too small to be visible at a given map scale (for example buildings on a county map) or have no width or length such as parcel centroids.

point-and-line graph   Graphs that generally use points and lines to represent the variation in variables such as population or employment over time. In these cases, the horizontal axis records points in time and the vertical axis records the values at different points in time. Point-and-line graphs use points to identify particular values and lines to connect the points are appropriate for identifying precise values at different points in time. See *line graph.*

point clip overlay   An *overlay* of a *point layer* (the input layer) and a *polygon layer* (the clip layer) that selects all the points inside the clip layer.

point erase overlay   An *overlay* for a *point layer* (the input layer) and a *polygon layer* (the erase layer) that selects points are outside the erase layer.

point spatial join   A GIS operation that joins the records from an input layer that satisfy a specified spatial relationship with the join layer to create an output layer containing the attributes for both layers.

point-in-polygon overlay   An *overlay* that combines a point layer and a polygon layer to create a point layer that contains the attributes for the polygon layer in which the points are located.

polygon   Two-dimensional shapes that represent *areas* in a *vector GIS*. Polygons consist of line segments that enclose an area.

polygon erase overlay   An *overlay* that creates an output layer containing only the features in the input layer that lie outside the boundaries of the erase layer.

polygon spatial join   A GIS operation that joins the part of an input layer that shares locations with the join layer to create an output layer containing the shared locations and the associated attributes from the input and join layers.

polygon-on-polygon overlay   An *overlay* that combines two or more polygon layers to create a new polygon layer that contains portions of the input layers and their attributes.

population   All the people, male and female, child and adult that live in a geographic area. See *Resident population.*

population pyramid   Horizontally stacked *histograms* that plot a population's age groups on the vertical axis and the number of people or the percentage of the population in each age group on the horizontal axis. By convention, the male population is plotted to the left of the vertical axis and the female population is plotted to the right.

Population Estimates Program (PEP)   A program managed but the US Bureau of the Census that produces *population estimates* for the United States, states, counties, minor civil divisions, and places and for the Commonwealth of Puerto Rico and its municipios.

Population Projections Program   A program managed by the US Bureau of the Census that produces projections of the United States' resident population by age, sex, race, Hispanic origin, and nativity.

postal code   A nongeographic system of alphanumeric codes associated with mail delivery areas that increase the efficiency of sorting and delivering mail. An example is the *ZIP code* system used in the United States.

projected coordinate system   A reference system that locates features on a two-dimensional plane. It locates features on the earth's surface on a two-dimensional plane based on their distance from an origin (0,0) along two axes, a horizontal X-axis representing east–west and a vertical Y-axis representing north–south.

projection (geography)   A mathematical model that represents the Earth's irregular spherical surface on a flat, two-dimensional plane.

projection (mathematics)   The numerical outcome of a set of assumptions regarding the future values of a variable. See *Estimate.*

projection period   The time interval between the last observation year (the *launch year*) and the last projection year (the *target year*).

projection interval The time increments for which projections are made, for example, one-year or five-year projection increments.

proportion A statistic that records the portion of a whole that a quantity represents. It is computed by dividing quantities that measure the same variable. For example, dividing the number of people in an age interval by the total population shows the proportion population in that age interval.

PSS See *Planning support system.*

quantiles classification A method of grouping values that places the roughly same number of values in each category. For example, if five classes are used to group twenty values, the quantiles method will place five values in each class. See *equal intervals classification, manual classification, natural breaks classification, and standard deviation classification.*

quantities The number or value of the features being considered. Examples include the number people for *census tracts*, the size of land parcels, and the average daily traffic for road segments

query A question posed to a digital data set that selects features that satisfy specified conditions. See *attribute query* and *spatial query.*

rank Show relative values that put features into an order from high to low. Ranks can be represented by labels, for example, low, moderate, and high, or by ordinal numbers, for example, from 1 (low) to 9 (high).

Raster data model A *spatial data model* that defines space as an array of equally sized cells arranged in rows and columns. Each cell contains an attribute value and location coordinates. Groups of cells that share the same value represent the same type of geographic feature.

ratio The relationship between two numbers indicating how many times the first number contains the second. For example, a ratio of 3:1 indicates that there are three units of the first item for every unit of the second item.

ratio scale A measurement system that has equal intervals between values and a true 0. An example is temperature measured on a Kelvin scale where the 0 value is the temperature and atomic motion ceases. Ratio-scale data allow the full range of arithmetic operations. *See interval scale, nominal scale, and ordinal scale.*

record A set of related data fields, generally a row in a *database* that contain all the *attribute* values for a single *feature*, for example, a parcel in a tax assessor's database.

Reference Shift The local employment change that would occur if the local industry's growth or decline matched the reference area's rate of change.

regional economy A region's economic system and activities that serve the region's population or provide goods and services to consumers, businesses, and organizations from outside the region.

representative fraction The ratio between a distance or area on a map and the corresponding distance or area on the Earth's surface, expressed as a fraction or ratio. For example, for a map with a scale of 1:1,000 one inch on the map represents 1,000 inches or 83.3 feet on the Earth's surface.

resident population The people who usually live in a specific area. For example, the resident population of the United States includes everyone living in the country and excludes the US Armed Forces overseas and civilian US citizens whose usual place of residence is outside the United States.

residual error The difference between an actual value and an estimated value.

residual net migration estimation method A cohort-component method that estimates migration as the residual difference between a cohort's observed and its expected populations, given natural processes of birth, death, and aging. See *vital statistics net migration estimation method* and *survival rates net migration estimation method.*

Residual Shift See *Local Shift.*

rural population Includes all the population living outside of *urbanized areas* and *urban clusters.*

sample error The difference between the *estimates* derived from a sample *survey* and the values that would have been obtained if the entire population were included in the survey.

scenario A description of an imagined situation or sequence of events such as an outline of a possible sequence of future events or an outline of an intended course of action.

sector A group of *industries* that produce similar products or use similar production methods or technologies.

selective top-down adjustment An *aggregate consistency* adjustment that preserves the value for the top level of aggregation and for one or more lower-level values.

self-defeating A *forecast* that proves to be false because of actions resulting from the forecast. For example, by forecasting a potential disaster, a person or organization can ensure that steps are taken to ensure that it does not occur.

self-fulfilling A *forecast* that affects what happens to ensure that it becomes true. For example, an area that is projected to decline may do so because people act on the assumption that it will decline.

sensitivity analysis An analysis that introduces variations in the *independent variables* of a *model* to determine the effect these changes will have on the *dependent variable*.

sex ratio The number of males per hundred females in a population.

shapefile A data format developed by the *Esri Inc.* for storing nontopological *vector data*.

share-of-change method A projection method that assumes an area's future growth or decline is proportional to its change over the base period.

shift-share analysis method An analysis method that examines the changes in an area's economic composition that occurred during a specified period, relative to a reference region, which is very often the nation. Do not confuse with the *shift-share projection method*.

shift-share projection method A projection method that modifies the *constant-share method* by adding a "shift" term that accounts for the observed difference between the employment changes in the study area and their changes in the reference area. Do not confuse with the *shift-share analysis method*.

share-trend method A projection method that uses information on the trend in the study area's share of the context area's population or employment to project the study area's future.

slope The incline, or steepness, of a surface. Slope can be measured in angular degrees from horizontal (0 degrees) to vertical (90 degrees), or percent slope (the rise divided by the run, multiplied by 100). For example, a 3 percent slope will change three feet vertically for every 100 feet or horizontal change.

small scale This term has two somewhat contradictory meanings. For noncartographers a small-scale map covers a small area and shows small features with a lot of detail. For cartographers, a small-scale map has a small representative fraction (e.g., 1:1,000,000) and covers a large area with little detail. This book avoids the term and instead uses *coarse scale* to refer to maps that cover a large area showing large features with little detail (e.g., a national map) and *fine scale* to refer to maps that cover small areas showing small features with a high degree of detail (e.g., a neighborhood map).

source zone One or more sets of areas that are used to assign values to an overlaying and incompatible set of geographic areas (the *target zones*).

stationary or nearly stationary population pyramid A *population pyramid* that has roughly equal population numbers for most age groups. The graph looks more like a column than a pyramid, except at the very top, where the size of older age groups naturally diminishes.

stationary population A hypothetical population in which one hundred thousand persons are born and die each year with the proportion dying in each age interval corresponding to the *life table* probability of dying values.

spatial analysis The process of examining the locations, attributes, and spatial relationships between *features* located in space.

spatial data Data about the location and shapes of geographic *features* and the spatial relationships between them.

spatial feature See *Feature.*

spatial query A *query* that selects *features* based on their location or spatial relationship to other features. Spatial queries can normally be answered by examining a map.

special population A group of persons who reside in an area because of administrative or legislative actions. Examples include college students, prison inmates, and military personnel and their dependents.

SQL See *Standard (or structured) query language.*

S-shaped curve A curve such as the *Gompertz curve* that starts growing slowly with an increasing rate, then grows rapidly with an increasing rate, and eventually grows slowly with a decreasing rate as it approaches an *asymptote*.

stacked bar chart A *bar chart* in which the rectangular bars are divided into sections. The size of each section represents the cumulative effect of the attributes for each bar.

standard deviation A statistical measure of the spread of values from their *mean*, calculated as the square root of the sum of the squared deviations from the mean value, divided by the number of elements minus one. The standard deviation for a distribution is the square root of the variance.

standard error (SE) A statistic that provides a quantitative measure of the extent to which an *estimate* derived from a sample survey can be expected to deviate from the value which would have been obtained if the entire population were surveyed.

standard deviation classification method A classification method of grouping values that creates classes based on the number *of standard deviations* the values are from the *mean* value for all the values. See e*qual interval classification, manual classification, natural breaks classification*, and q*uantiles classification,*

Standard Industrial Classification (SIC) An industrial classification system that was replaced by the *North American Industrial Classification System* classification system in 1997.

statistical entity Geographic units established by the US Bureau of the Census when the geographic coverage of *legal and administrative entities* is incomplete, inadequate, or inconsistent over time.

Structured (or Standard) Query Language (SQL) A standard set of terminology and operations for querying and manipulating data..**suitability factors.** Characteristics of the land that can be used to determine the relative suitability of different locations for alternative uses. Suitability factors can include natural features such as slope and soil type and man-made features such as accessibility to roads and amenities.

suitability rating A numerical value indicating the relative *suitability* of locations for a *factor type* (for example the value for a given slope).

suitability score A numerical value indicating a location's overall suitability for a land use when all the *suitability factors* are considered.

sum of squared residual error Sum of the squared difference between a set of observed and estimated curve values.

surface A feature geometry representing continuous features that cover an area and include measurements along a third dimension. Elevations, air pollution, and noise levels are all surfaces.

Survey of Income and Program Participation (SIPP) A survey conducted by the US Bureau of the Census that collects data on income, labor force participation, social program participation and eligibility, and general demographic characteristics of individuals and households in the United States.

survey The collection of data for a population sample that are used to *estimate* the values for the entire population. See *census*.

Survey of Business Owners A survey conducted by the US Bureau of the Census that provides information regarding 1.75 million businesses and their owners.

survival rates net migration estimation method A *residual net migration estimation method* that uses the recorded number births and deaths in the study area to compute the expected population. See *vital statistics net migration estimation method*.

target year The last year for which a variable is projected.

target zone One or more sets of areas that are assigned values from an overlaying and incompatible set of geographic areas (the *source zones*).

theme A collection of related geographic *features* such as streets, parcels, or rivers, along with their attributes. All features in a theme share the same coordinate system, are located within a common geographic extent, and have the same attributes.

theory A hypothesis that has received substantial support and assumed to have predictive value. The term is often misused to mean "complicated and obscure arguments."

time series An ordered sequence of values measured at equally spaced time intervals.

top-down adjustment An *aggregate consistency* adjustment that retains the value for the top level of aggregation and adjusts the bottom-level values to match the top-level value.

top-down projection A projection for the disaggregate components of a collection of items based on the projected value for their aggregate. For example, the projected population for subcounty areas derived from a county projection.

topography The study and mapping of land surfaces, including the terrain of the Earth's surface and the position of natural and constructed features located on the Earth's surface. See *topology*.

topological data Data describing spatial relationships, that is, where things are located relative to other things. See *topology*.

topology The branch of geometry that deals with the properties of a figure that remain unchanged when the figure is bent, stretched, or otherwise distorted. Topology is used in GIS to express the spatial relationships between map features. Examples include (1) adjacency ("What is next to what?"), (2) connectivity ("What is connected to what?"); and (3) containment ("What is within what?"). See *topography*.

total fertility rate The average number of children that a group of children would have during their lifetimes if their fertility behavior conformed to a set of age-specific birth rates.

traffic (or transportation) analysis zone (TAZ) A small geographic area defined for transportation modeling and planning purposes.

triangulated irregular network (TIN) A *data model* that represents a three-dimensional surface with a set of contiguous nonoverlapping triangles. The vertices of each triangle are sample data points with X, Y, and Z values. The points are connected by lines to form triangles that represent a surface.

trimmed average projection An *average projection* that excludes questionable projections. For example, the trimmed average may exclude the highest and lowest projections or questionable projections.

tract See *Census tract*.

union An *overlay* that combines two *polygon layers* to create a new polygon layer that contains all the locations in the input layers.

UNION (or OR) query A *Boolean query* that selects all of members of overlapping sets that satisfy either: (1) one condition, (2) a second condition, or (3) both conditions.

urban cluster   A *statistical entity* delineated by the US Bureau of the Census consisting of densely settled *census tracts*, *blocks*, and adjacent densely settled territory that together contains at least 2,500 people.

urban population   The population residing in *urbanized areas* and *urban clusters*.

urbanized area   A *statistical entity* delineated by the US Bureau of the Census consisting of densely settled *census tracts*, *blocks*, and adjacent densely settled territory that together contains at least 50,000 people.

vacancy rate   The proportion of *housing units* that is vacant and available for sale.

variable   A characteristic, number, or quantity that takes different values.

vector data model   A *spatial data* model that represents geographic features as points, lines, and polygons. Each point is represented as a single coordinate pair; line and polygon features are represented as ordered lists of coordinate pairs.

Venn diagram   A diagram that uses overlapping circles to represent logical relationships between two or more classes or sets of entities.

vital statistics   Data obtained from the registration of vital events such as births, deaths, marriages, divorces, and abortions.

vital statistic net migration estimation method   A *residual net migration method* that uses fertility and survival rates to estimate the number of births and deaths that are used to compute the expected population. See *survival rates net migration estimation method*.

weight   The importance value or preference assigned to criteria in multicriteria evaluation.

weighted combination method   A *land suitability analysis* method that uses *factor weights* to reflect the user's judgement regarding the relative importance of different *suitability factors*.

X-Axis   The horizontal line on a two-dimensional graph that runs right and left of the origin (0,0). Values to the right of the origin are positive; values to the left of the origin are negative.

Y intercept   The value of the *dependent value*, Y, of a curve when the independent variable, X, is 0.

ZIP codes   A system of numeric codes assigned to mail delivery areas by the US Postal Service to more efficiently sort and deliver mail.

ZIP Code Business Patterns (ZBP)   Provides annual data on the number of establishments and employees during the week of March 12, first quarter payroll, and annual payroll data for states and counties for fifty thousand five-digit ZIP code areas across the United States. See *County Business Patterns (CBP)*.

ZIP Code Tabulation Area (ZCTA)   Statistical entities that are approximate spatial equivalents to the US Postal Service five-digit ZIP Code service areas.

Note

The definitions in this glossary were obtained from several sources including Armstrong (1985), LeGates (2005), and Esri's GIS Glossary (http://wiki.gis.com/wiki/index.php/GIS_Glossary).

References

Alonso, William. 1968. "Predicting best with imperfect data." *Journal of the American Institute of Planners* 34 (4):248–255.

American Planning Association. n.d. *Land-based classification standards.* Available from https://www.planning.org/lbcs/.

Anonymous. 2016. *Population pyramids of the world from 1950 to 2100.* Available from https://populationpyramid.net/.

Arias, Elizabeth. 2014. United States life tables, 2010. In *National vital statistics reports,* volume 63, number 7. Washington DC: National Center for Health Statistics.

Armstrong, J. Scott. 2001a. "Evaluating forecasting methods." In *Principles of forecasting: A handbook for researchers and forecasters,* edited by J. Scott Armstrong, 443–472. Boston: Kluwer Academic.

———. 2001b. "Standards and practices for forecasting." In *Principles of forecasting: A handbook for researchers and forecasters,* edited by J. Scott Armstrong, 679–732. Boston: Kluwer Academic.

Armstrong, J. Scott, and Fred Collopy. 1992. "Error measures for generalizing about forecasting methods: Empirical comparisons." *International Journal of Forecasting* 8:69–80.

Armstrong, J. Scott. 1985. *Long-range forecasting: From crystal ball to computer,* 2nd edition. New York, NY: John Wiley.

———. n.d. *Forecasting principles.* Available from http://forecastingprinciples.com/.

Arnstein, Sherry R. 1969. "A ladder of citizen participation." *Journal of the American Institute of Planning* 35 (4):387–394.

Ascher, William. 1978. *Forecasting: An appraisal for policy-makers and planners.* Baltimore: John Hopkins University Press.

———. 1981. "The forecasting potential of complex models." *Policy Sciences* 13 (3):247–267.

Atkinson, Peter M., and Christopher D. Lloyd. 2009. "Geostatistics and spatial interpolation." In *The SAGE handbook of spatial analysis,* edited by A. Stewart Fotheringham and Peter A. Rogerson. Thousand Oaks, CA: SAGE.

Atlanta Regional Commission. 2016. *Forecasts.* Available from http://www.atlantaregional.com/info-center/forecasts.

Avin, Uri P. 2007. "Using scenarios to make urban plans." In *Engaging the future: Forecasts, scenarios, plans and projects,* edited by Lewis D. Hopkins and Marisa A Zapata. Cambridge, MA: Lincoln Institute of Land Policy.

Batty, Michael. 2013. *The new science of cities.* Cambridge: MIT Press.

Berke, Phillip R., David R. Godschalk, Edward J. Kaiser, and Daniel A. Rodriguez. 2006. *Urban land use planning,* 5th edition. Urbana: University of Illinois Press.

Box, George E. P., and Norman R. Draper. 1987. *Empirical model-building and response surfaces.* New York: John Wiley.

Brail, Richard K. 2008. *Planning support systems for cities and regions.* Cambridge, MA: Lincoln Institute of Land Policy.

Brail, Richard K., and Richard E. Klosterman. 2001. *Planning support systems: Integrating geographic information systems, models and visualization tools.* Redlands, CA: ESRI Press.

Brewer, Cynthia A. 2005. *Design better maps: A guide for GIS users.* Redlands, CA: ESRI Press.

Carr, Margaret H., and Paul D. Zwick. 2007. *Smart land-use analysis: The LUCIS Model.* Redlands, CA: ESRI Press.

Center for Family and Demographic Research, Bowling Green University. n.d. *Creating population pyramids using Microsoft Excel.* Available from https://www.bgsu.edu/content/dam/BGSU/college-of-arts-and-sciences/center-for-family-and-demographic-research/documents/Help%20Resources%20and%20Tools/Statistical%20Analysis/Excel-Basics-Creating-Population-Pyramids-Using-Microsoft-Excel.pdf.

Centers for Disease Control and Prevention. 2016a. *CDC Wonder.* Centers for Disease Control and Prevention. Available from https://wonder.cdc.gov/ucd-icd10.html.

———. 2016b. *National Center for Health Statistics: Life tables.* Centers for Disease Control and Prevention. Available from https://www.cdc.gov/nchs/products/life_tables.htm.

———. 2016c. *National Center for Health Statistics: National vital statistics system.* Centers for Disease Control and Prevention. Available from https://www.cdc.gov/nchs/nvss/mortality/lewk4.htm.

———. 2016d. *Natality information.* Centers for Disease Control and Prevention. Available from http://wonder.cdc.gov/natality.html.

Chapin, F. Stuart. 1957. *Urban land use planning*. New York: Harper.

Clarke, Keith C. 2003. *Getting started with geographic information systems*, 4th edition. New York: Prentice Hall.

Colby, Sandra L., and Jennifer M. Ortman. 2015. "Projections for the size and composition of the U.S. population: 2014." In *Current population reports, P25-1143*. Washington, DC: US Bureau of the Census.

Collins, Michael G., Frederick R. Steiner, and Michael J. Rushman. 2001. "Land-use suitability in the United States: Historical development and promising technological achievements." *Environmental Management* 28 (5):611–621.

De Gooijer, Jan G., and Rob J. Hyndman. 2006. "25 years of time series forecasting." *International Journal of Forecasting* 22 (3):443–473.

DeMers, Michael N. 2001. *GIS modeling in rasters*. New York: John Wiley.

Dublin, L. I., and A. J. Lotka. 1930. "The present outlook for population growth." *American Sociological Society Publications* 24 (2):106–114.

Eicher, Cory L., and Cynthia A. Brewer. 2001. "Dasymetric mapping and areal interpolation: Implementation and evaluation." *Cartography and Geographic Information Sciences* 28 (2):125–138.

Forstall, Richard L. 1996. *Population of counties by decennial census: 1900 to 1990*. Available from https://www.census.gov/population/www/censusdata/PopulationofStatesandCountiesoftheUnitedStates1790-1990.pdf.

Franklin, Rachel S. 2014. "An examination of the geography of population composition and change in the United States, 2000–2010: Insights from geographical indices and a shift–share analysis." *Population, Space and Place* 20 (1):18–36.

Geertman, Stan, Joseph Ferreria Jr., Robert Goodspeed, and John Stillwell. 2015. *Planning support systems and smart cities*. Berlin: Springer-Verlag.

Geertman, Stan, and John Stillwell. 2003. *Planning support systems in practice*. Berlin: Springer Verlag.

———. 2009. *Planning support systems: Best practices and new methods*. New York: Springer.

Georgia Department of Labor. n.d. *Georgia Labor Market Explorer*. Available from https://explorer.dol.state.ga.us/vosnet/Default.aspx.

Goldstein, Harvey A. 1990. "A practitioner's guide to state and substate industry employment projections." *Economic Development Quarterly* 4 (3):260–275.

———. 2005. "Projecting state and area industry employment." Produced for the Employment and Training Administration Projections Workshop. Washington, DC: US Department of Labor.

Hamilton, C. Horace, and Josef Perry. 1962. "A short method for projecting population by age from one decennial census to another." *Social Forces* 41:163–170.

Harris, Britton. 1989. "Beyond geographic information systems: Computers and the planning professional." *Journal of the American Planning Association* 55:85–92.

Hopkins, Lewis D. 1977. "Methods for generating land suitability maps: A comparative evaluation." *Journal of the American Institute of Planners* 43:386–400.

Hopkins, Lewis D., and Marisa A. Zapata. 2007. *Engaging the future: Forecasts, scenarios, plans, and projects*. Cambridge, MA: Lincoln Institute of Land Policy.

Hunt, J. D. 2005. *Integrated land-use and transport models: An introduction*. Washington, DC: Transportation Research Board, Workshop 162.

Ihrke, David K. 2014. Reason for moving: 2012 to 2013. Washington DC: US Bureau of the Census, P20-574.

Isard, Walter, et al. 1960. *Methods of regional analysis: An introduction to regional science*. Cambridge: MIT Press.

Isserman, Andrew M. 1977. "The accuracy of population projections for sub-county areas." *Journal of the American Institute of Planners* 43 (3):247–259.

———. 1984. "Projection, forecast, and plan: On the future of population forecasting." *Journal of the American Planning Association* 50 (2):208–221.

———. 1985. "Economic-demographic modeling with endogenously determined birth and migration rates." *Environment and Planning A* 17 (1):25–45.

———. 1993. "The right people, the right rates: Making population estimates and projections with an interregional cohort-component model." *Journal of the American Planning Association* 59 (1):45–54.

———. 2007. "Forecasting to learn how the world can work." In *Engaging the future: Forecasts, scenarios, plans and projects*, edited by Lewis D. Hopkins and Marisa A. Zapata. Cambridge, MA: Lincoln Institute for Land Policy.

Isserman, Andrew M., and James Westervelt. 2006. "1.5 million missing numbers: Overcoming employment suppression in County Business

Patterns data." *International Regional Science Review* 29 (3):311–335.

Kent, Robert B., and Richard E. Klosterman. 2000. "GIS mapping: Pitfalls for planners." *Journal of the American Planning Association* 66 (2):189–198.

Klein, William R. 2000. "Building concensus." In *The practice of local government planning*, 3rd edition, edited by Charles J. Hoch, Linda C. Dalton, and Frank S. So. Washington, DC: International City/County Management Association.

Klosterman, Richard E. 1985. "Arguments for and against planning." *Town Planning Review* 56 (1):5–20.

———. 1987. "Politics of computer-aided planning." *Town Planning Review* 58 (4):441–452.

———. 1990. *Community analysis and planning techniques.* Savage, MD: Rowman & Littlefield.

———. 1991. *TIGER: A primer for planners*, Planning Advisory Service Report Number 436. Chicago: American Planning Association.

———. 1994. "An introduction to the literature on large-scale urban models." *Journal of the American Planning Association* 60 (1):41–44.

———. 1997. "Planning support systems: A new perspective on computer-aided planning." *Journal of Planning Education and Research* 17 (1):45–54.

———. 1999. "The What If? collaborative planning support system." *Environment and Planning B: Planning and Design* 26 (3):393–408.

———. 2007. "Deliberating about the future." In *Engaging in the future: Forecasts, scenarios, plans and projects*, edited by Lewis D. Hopkins and Marisa A. Zapata, 199–220. Cambridge, MA: Lincoln Institute for Land Policy.

———. 2012. "Commentary: Simple and complex models." *Environment and Planning B: Planning and Design* 39 (1):1–6.

———. 2013. "Lessons learned about planning: Forecasting, participation and technology." *Journal of the American Planning Association* 79 (2):161–169.

Knaap, Gerrit J. 2001. *Land market monitoring for smart urban growth.* Cambridge, MA: Lincoln Institute for Land Policy.

Krueckeberg, Donald A., and Arthur L. Silvers. 1974. *Urban planning analysis: Methods and models.* New York: Wiley.

Kryger, John, and Dennis Wood. 2005. *Making maps: A visual guide to map design for GIS.* New York: Guilford Press.

Lau, Sándor. n.d. *The art of walking backwards.* Available from http://www.horschamp.qc.ca/new_offscreen/walking_backwards.html.

LeGates, Richard T. 2005. *Think globally, act regionally: GIS and data visualization for social science and public policy research.* Redlands, CA: ESRI Press.

Lewis, Karen. 2015. *Graphic design for architects: A manual for visual communication.* New York: Routledge.

Lipovska, Barbora, and Roberta Stepenkova. 2016. *Research methods in urban and regional planning* (2 volumes). London: Koros Press.

Long, John F. 1995. "Complexity, accuracy, and utility of official population projections." *Mathematical Population Studies* 5:203–216.

Longley, Paul A., Michael F. Goodchild, David J. McGuire, and David W. Rhind. 2005. *Geographic information systems: Principles, techniques, management and application.* 2nd edition. New York: John Wiley.

Loveridge, Scott, and Anne C. Selting. 1998. "A review and comparison of shift-share identities." *International Regional Science Review* 21 (1):37–58.

Malczewski, Jacek. 2006. "GIS-based multi-criteria decision analysis: A survey of the literature." *International Journal of Geographical Information Systems* 20 (7):703–726.

Martin, Joyce A., Bradly E. Hamilton, Michelle J. K. Osterman, Sally C. Curtin, and T. J. Mathews. 2013. Births: Final data for 2012. Washington, DC: Centers for Disease Control and Prevention.

Mathews, T. J., and Brady E. Hamilton. 2005. Trend analysis of the sex ratio at birth in the United States. In *National vital statistics reports.* Washington, DC: US Centers for Disease Control and Prevention.

McHarg, Ian. 1969. *Design with nature.* Garden City, NY: Doubleday/Natural History Press.

Miller, Eric J., David S. Kriger, and John David Hunt. 1999. Integrated urban models for simulation of transit and land use policies: Guidelines for implementation and use. Transit Cooperative Research Program Report 48. Washington, DC: Transportation Research Board.

Miller, Ronald E., and Peter D. Blair. 2009. *Input-output analysis: Foundations and extensions.* Cambridge: Cambridge University Press.

Minnesota Population Center. n.d. National Historical Geographic Information System, Version 11.0. Minneapolis: University of Minnesota.

Mitchell, Andy. 1999. *The ESRI guide to GIS analysis, volume 1: Geographic patterns and relationships.* Redlands, CA: ESRI Press.

———. 2012. *The ESRI guide to GIS analysis, volume 3: Modeling suitability, movement, and interaction.* Redlands, CA: ESRI Press.

Molly, Raven, Christopher L. Smith, and Abigail K. Wozniak. 2011. "Internal migration in the United States." In *National Bureau of Economic Research*, Working Paper 17307. Cambridge, MA: National Bureau of Economic Research.

Monmonier, Mark. 1996. *How to lie with maps*, 2nd edition. Chicago: University of Chicago Press.

Moudon, Anne Vernez, and Michael Hubner. 2000. *Monitoring land supply with geographic information systems: Theory, practice, and parcel-based approaches.* New York: John Wiley.

Murphy, Sherry L., Xu Jiaquan, and Kenneth D. Kochane. 2013. Deaths: Final data for 2010. In *National vital statistics reports*, volume 61, number 1. Hyattsville, MD: National Center for Health Statistics.

Myers, Dowell. 1992. *Analysis with local census data: Portraits of change.* San Diego, CA: Academic Press.

Myers, Dowell, and Alicia Kitsuse. 2000. "Constructing the future in planning: A survey of theories and tools." *Journal of Planning Education and Research* 19 (3):221–231.

National Center for Health Statistics. n.d.a. *CDC Wonder.* Available from https://wonder.cdc.gov/.

———. n.d.b. *Life tables.* Available from https://www.cdc.gov/nchs/products/life_tables.htm.

———. n.d.c. *United States decennial life tables, 1999–2001: State life tables.* Available from https://www.cdc.gov/nchs/nvss/mortality/lewk4.htm.

———. n.d.d. *Compressed mortality files.* Available from https://www.cdc.gov/nchs/data_access/cmf.htm.

———. n.d.e. *Natality information.* Available from http://wonder.cdc.gov/natality.html.

———. n.d.f. *About underlying cause of death, 1999–2015.* Available from https://wonder.cdc.gov/ucd-icd10.html.

Pettit, Christopher J., Richard E. Klosterman, Marcos Nino-Ruiz, Ivo Widjaja, Patrizia Russo, Martin Tomko, Richard Sinnot, and Robert Stimson. 2013. "The online What if? planning support system." In *Planning support systems for sustainable urban development*, edited by Stan Geertman, Fred Toppen, and John Stillwell. Heidelberg: Springer-Verlag.

Pindyck, Robert S., and Daniel L. Rubinfeld. 1997. *Econometric models and economic forecasts*, 4th edition. New York: McGraw-Hill.

Rayer, Stefan. 2008. "Population forecasting errors: A primer for planners." *Journal of Planning Education and Research* 27: 417–430.

Rayer, Stefan, and Stanley K. Smith. 2010. "Factors affecting the accuracy of subcounty population forecasts." *Journal of Planning Education and Research* 30 (2):147–161.

Reibel, Michael, and Aditya Agrawal. 2007. "Areal interpolation of population counts using pre-classified land cover data." *Population Research and Policy Review* 26:619–633.

Renski, Henry, and Susan Strate. 2013. "Evaluating alternative migration estimation techniques for population estimates and projections." *Journal of Planning Education and Research* 33 (3):325–335.

Rhind, David W. 1989. "Why GIS?" *ArcNews* (Summer).

Rich, Robert. 2013. *The great recession of 2007–2009.* Available from http://www.federalreservehistory.org/Events/DetailView/58.

Robinson, Arthur H., Joel L. Morrison, Phillip C. Muehrcke, A. Jon Kimerling, and Stephen C. Guptill. 1995. *Elements of cartography*, 6th edition. New York: John Wiley.

Silva, Elisabete A., Patsy Healey, Neil Harris, and Pieter Van den Broeck, eds. 2015. *The Routledge handbook of planning research methods.* New York: Routledge.

Shryock, Henry S., Jacob S. Siegel, and Associates. 1975. *The methods and materials of demography*, third printing (rev.). Washington, DC: US Government Printing Office.

Smith, Stanley K. 1986. "A review and evaluation of the housing unit method of population estimation." *Journal of the American Statistical Association* 81 (394):287–296.

———. 1987. "Tests of forecast accuracy and bias for county population projections." *Journal of the American Statistical Association* 82 (400):991–1012.

Smith, Stanley K., and Mohammed Shahidullah. 1995. "An evaluation of population projection errors for census tracts." *Journal of the American Statistical Association* 90 (429):64–71.

Smith, Stanley K., and David A. Swanson. 1998. "In defense of the net migrant." *Journal of Economic & Social Measurement* 24 (3/4):249–264.

Smith, Stanley K., Jeff Tayman, and David A Swanson. 2001. *State and local population projections: Methodology and analysis.* New York: Kluwer Academic/Plenum.

Smith, Stanley K., Jeff Tayman, and David A. Swanson. 2013. *A practitioner's guide to state and local population projections.* New York: Springer.

Steinitz, Carl P., P. Parker, and L. Jordan. 1976. "Hand drawn overlays: Their history and prospective uses." *Landscape Architecture* 66:444–455.

Stevens, Benjamin H., and Craig L. Moore. 1980. "A critical review of the literature on shift-share as a forecasting technique." *Journal of Regional Science* 20:419–437.

Swanson, David A., and Jeff Tayman. 2013. The accuracy of the Hamilton-Perry method for forecasting state populations by age. Working Paper 13-01. Riverside, CA: Center for Sustainable Suburban Development, University of California.

Tomlin, C. Dana. 1990. *Geographic information system and cartographic modeling.* Englewood Cliffs, NJ: Prentice-Hall.

Tufte, Edward R. 1983. *The visual display of quantitative information.* Cheshire, CT: Graphics Press.

———. 1990. *Envisioning information.* Cheshire, CT: Graphics Press.

Tyrwhitt, Jaqueline. 1950. "Surveys for planning." In *Town and country planning handbook.* London: Architectural Press.

US Bureau of Labor Statistics. n.d.a. *Employment projections: Projections methodology.* Available from http://www.bls.gov/emp/ep_projections_methods.htm.

———. n.d.b. *BLS handbook of methods. Chapter 13: Employment projections.* Available from https://www.bls.gov/opub/hom/pdf/homch13.pdf.

US Bureau of the Census. 2003a. *Occupation by sex: 2000. Census 2000 Summary File 3 (SF 3)—Sample data.* Available from http://factfinder.census.gov/.

———. 2003b. *Profile of general population characteristics: 2000, Table DP-1.* Available from http://factfinder.census.gov/.

———. 2009. *2007 County Business Patterns: ZIP code business statistics: ZIP code. 2007 business patterns by employment size class: 2007. Table CB0700CZ2.* Available from http://factfinder.census.gov/.

———. 2010. *Standard hierarchy of census geographic entities.* Available from https://www2.census.gov/geo/pdfs/reference/geodiagram.pdf.

———. 2011. *2009 County Business Patterns: Geography area series: County Business Patterns, 2009 business patterns. Table CB0900A1.* Available from http://factfinder.census.gov/.

———. 2013a. *Total population: 2010 Census Summary File 1; Table P1.* Available from http://factfinder.census.gov/.

———. 2013b. *Age groups and sex: 2010, Table QT-P1.* Available from http://factfinder.census.gov/.

———. 2013c. *Profile of general population and housing characteristics: 2010, Table DP-1.* Available from http://factfinder.census.gov/.

———. 2013d. *Geographical mobility in the past year by age for current residence in the United States: 2008–2012 American Community Survey 5-year estimates; Report B07001.* Available from http://factfinder.census.gov/.

———. 2013e. *Geographical mobility in the past year by individual income in the past 12 months (in 2013 inflation-adjusted dollars) for current residence in the United States: 2008–2012 American Community Survey 5-year estimates; Report B07010.* Available from http://factfinder.census.gov/.

———. 2014a. Methodology, assumptions, and inputs for the 2014 national projections. Washington, DC: US Bureau of the Census.

———. 2014b. *ACS demographic and housing estimates: 2008–2012 American Community Survey 5-year estimates: Table DP05.* Available from http://factfinder.census.gov/.

———. 2015a. *Geography area series: County Business Patterns by employment size class, 2013 business patterns. Report CB1300A13.* Available from http://factfinder.census.gov/.

———. 2015b. *ZIP code business statistics: ZIP code business patterns by employment size class, 2013 business patterns. Report CB1300CZ21.* Available from http://factfinder.census.gov/.

———. 2015c. *Income in the past 12 months (in 2013 inflation adjusted dollars. 2010–2013 American Community Survey 3-year estimates. Table S1901.* Available from http://factfinder.census.gov/.

———. 2015d. *Occupation by class of worked for the civilian employed population 16 years and over: 2011–2013 American Community Survey 3-year estimates: Report B24060.* Available from http://factfinder.census.gov/.

———. 2016a. *ACS demographic and housing estimates: 2010–2014 American Community Survey 5-year estimates; Report DP05.* Available from http://factfinder.census.gov/.

———. 2016b. *Income in the past 12 months (in 2014 inflation adjusted dollars): 2010–2014 American Community Survey 5-year estimates. Report S1901.* Available from http://factfinder.census.gov/.

———. 2016c. *Selected housing statistics: 2010–2014 American Community Survey 5-year estimates; Report DP04.* Available from http://factfinder.census.gov/.

———. 2016d. *Total population: 2010–2014 American Community Survey 5-year estimates; Report B01003.* Available from http://factfinder.census.gov/.

———. 2016e. *Race: 2010–2014 American Community Survey 5-year estimates; Report*

B02001. Available from http://factfinder.census.gov/.

———. 2016f. *Hispanic or Latino origin: 2010–2014 American Community Survey 5-year estimates; Report B0300.* Available from http://factfinder.census.gov/.

———. 2016g. *Median household income in the past 12 months (in 2014 inflated-adjusted dollars): 2010–2014 American Community Survey 5-year estimates.* Available from http://factfinder.census.gov/.

———. 2016h. *Population projections.* Available from https://www.census.gov/population/projections/data/national/.

———. 2017. *Population and housing unit estimates.* Available from https://www.census.gov/programs-surveys/popest/about.html.

———. n.d.a. *Mean income in the past twelve months: American Community Survey 1-year estimates.* Available from http://factfinder.census.gov/.

———. n.d.b. *Geography area series: County Business Patterns by employment size class.* Available from http://factfinder.census.gov/.

———. n.d.c. *ZIP code business statistics: Total for zip code.* Available from http://factfinder.census.gov/.

———. n.d.d. *Geography atlas.* Available from https://www.census.gov/geo/reference/webatlas/.

Wachs, Martin. 1982. "Ethical dilemmas in forecasting for public policy." *Public Administration Review* 42 (6):562–567.

———. 1989. "When planners lie with numbers." *Journal of the American Planning Association* 55 (4):476–480.

———. 2001. "Forecasting versus envisioning: A new window on the future." *Journal of the American Planning Association* 67 (4):367–372.

Waddell, Paul. 2002. "UrbanSim: Modeling urban development for land use, transportation and environmental planning." *Journal of the American Planning Association* 68 (3):297–314.

Wang, Fahul. 2014. *Quantitative methods and socio-economic applications in GIS*, 2nd edition. New York: CRC Press.

Wang, Xinhao, and Ranier vom Hofe. 2007. *Research methods in urban and regional planning.* Beijing: Tsinghua University Press.

Wegener, Michael. 2014. "Land-use transport interaction models." In *Handbook of regional science*, edited by Manfred M. Fischer and Peter Nijkamp, 741–758. Heidelberg: Springer.

Whelpton, Pascal K., and Warren S. Thompson. 1933. *Population trends in the United States.* New York: McGraw-Hill.

Wilson, Tom, and Philip H. Rees. 2005. "Recent developments in population projection methodology: A review." *Population, Space and Place* 11:337–360.

Xiang, Wei-Ning, and Keith C. Clarke. 2003. "The use of scenarios in land-use planning." *Environment and Planning B: Planning and Design* 30 (6):855–909.

Index

Page numbers in italics refer to figures or tables.